INSECT–FUNGUS INTERACTIONS

INSECT–FUNGUS INTERACTIONS

Edited by

N. Wilding
Rothamsted Experimental Station, Harpenden, Hertfordshire, England

N. M. Collins
I.U.C.N. Conservation Monitoring Centre, Cambridge, England

P. M. Hammond
British Museum (Natural History), London, England

J. F. Webber
Forest Research Station, Alice Holt Lodge, Farnham, Surrey, England

14th Symposium of the
Royal Entomological Society of London
in collaboration with the
British Mycological Society
16–17 September 1987
at the
Department of Physics Lecture Theatre
Imperial College, London

1989

ACADEMIC PRESS

Harcourt Brace Jovanovich, Publishers
London San Diego New York Berkeley
Boston Sydney Toronto Tokyo

ACADEMIC PRESS LIMITED
24/28 Oval Road, London NW1 7DX

United States Edition published by
ACADEMIC PRESS, INC.
San Diego, CA 92101

British Library Cataloguing in Publication Data

Insect–fungus interactions.
1. Fungi. Interactions with insects
I. Wilding, N.
589. 2′045

ISBN 0–12–751800–2

Typeset by Mathematical Composition Setters, Salisbury, Wiltshire.
Printed and bound in Great Britain at The University Press, Cambridge

Contents

8. The Roles of the Bark Beetle *Ips cembrae*, the Woodwasp *Urocerus gigas* and Associated Fungi in Dieback and Death of Larches
D. B. REDFERN

9. Mycopathogens of Insects of Epigeal and Aerial Habitats
H. C. EVANS

10. Mycopathogens of Soil Insects
S. KELLER AND G. ZIMMERMANN

DR ROY WATLING

Appendix: Mycophagy in Insects: a Summary
P. M. HAMMOND AND J. F. LAWRENCE

Contributors

Numbers in parentheses indicate the page numbers on which the authors' contributions begin.

R. A. BEAVER (121), The University of the South Pacific, Laucala Bay, Suva, Fiji

A. A. BERRYMAN (145), Department of Entomology, Washington State University, Pullman, Washington 99164-6432, USA

J. M. CHERRETT (93), Department of Applied Zoology, University College of North Wales, Bangor, Gwynedd LL57 2UW, Wales

V. F. EASTOP (xv), Department of Entomology, British Museum (Natural History), Cromwell Road, London SW7 5BD, England

H. C. EVANS (205), C.A.B. International, Institute of Biological Control, Imperial College, Silwood Park, Ascot, Berkshire SL5 7PY, England

J. N. GIBBS (161), Forestry Commission, Forest Research Station, Alice Holt Lodge, Wrecclesham, Farnham, Surrey GU10 4LH, England

P. M. HAMMOND (275), Department of Entomology, British Museum (Natural History), Cromwell Road, London SW7 5BD, England

I. HANSKI (25), Department of Zoology, University of Helsinki, P. Rautatiekatu 13, SF-00100 Helsinki, Finland

S. KELLER (239), Swiss Federal Research Station for Agronomy, Reckenholzstr. 191/211, CH-8046 Zürich, Switzerland

J. F. LAWRENCE (1,275), C.S.I.R.O. Division of Entomology, Canberra, A.C.T. 2601, Australia

R. J. POWELL (93), Department of Biological Sciences, University of Exeter, Exeter EX4 4PS, England

D. B. REDFERN (195), Forestry Commission, Northern Research Station, Roslin, Midlothian EH25 9SY, Scotland

D. J. STRADLING (93), Department of Biological Sciences, University of Exeter, Exeter EX4 4PS, England

R. J. THOMAS (69), Overseas Development Natural Resources Institute, Central Avenue, Chatham Maritime, Chatham, Kent ME4 4TB, England

R. WATLING (271), Royal Botanic Garden, Edinburgh EH3 5LR, Scotland

J. F. WEBBER (161), Forestry Commission, Forest Research Station, Alice Holt Lodge, Wrecclesham, Farnham, Surrey GU10 4LH, England

T. G. WOOD (69), Overseas Development Natural Resources Institute, Central Avenue, Chatham Maritime, Chatham, Kent ME4 4TB, England

G. ZIMMERMANN (239), Federal Biological Research Centre for Agriculture and Forestry, Institute for Biological Pest Control, Heinrichstrasse 243, D-6100 Darmstadt, Federal Republic of Germany

Preface

The interactions between insects and fungi are many and varied, with total dependence of the insect on the fungus as food at one end of the spectrum to total dependence of the fungus on the insect at the other, and incorporating many shades of interdependence in between. As a consequence of this diversity, research workers investigating the relationships are from a wide range of different backgrounds (entomology, mycology, plant pathology, forestry, ecology, insect pathology) and tend to meet together seldom. The idea of a meeting to bring together these specialists had been under consideration by the Royal Entomological Society and the British Mycological Society independently, and it was appropriate that the societies should pool their resources and provide a common forum for their memberships. Although not the original intention of the organizers, it was decided that the meeting should form one of the biennial symposia of the Royal Entomological Society, the first to be jointly convened with another society, drawing on experts from all over the world.

This symposium volume, then, attempts to review the current state of our knowledge in four principal areas: mycophagy, mutualism, insect spread of plant fungal disease and insect mycopathology. Regrettably, for reasons beyond the control of the convenors, two aspects planned for discussion, mycophagy in the Diptera and mycopathogens of aquatic insects, were not considered. The reader is referred to the works of Chandler (1979), Hackman and Meinander (1979), P. M. Hammond and J. F. Lawrence (Appendix, this volume) and Lacey and Undeen (1986) which most closely and recently cover the missing aspects.

A feature of this volume is the use of code numbers to help readers 'place' the various insect and fungus names that occur in each chapter if they should wish to do so. The systematic position of any fungus (code numbers headed F) or insect (code numbers headed H) may be checked by consulting the outline classifications (Tables I (p. 283) and II (p. 287)) provided in the appendix. These tables also provide a summary of major mycophagous interactions (see the Introduction to the Appendix) and for this reason the insect classification is to the familial level where those families include mycophagous species. Some species and groups of insects, therefore, mentioned in the chapters are accompanied by code numbers that indicate the order of insects, others by numbers that indicate the family. The fungus classification is complete to the ordinal level only for the major groups.

NEIL WILDING
N. MARK COLLINS
PETER M. HAMMOND
JOAN F. WEBBER

REFERENCES

Chandler, P. (1979). Fungi. *In* "A Dipterist's Handbook" (A. Stubbs and P. Chandler, eds), pp. 199–211. Amateur Entomologists Society, Hanworth, Middlesex, UK.

Hackman, W. and Meinander, M. (1979). Diptera feeding as larvae on macro-fungi in Finland. *Ann. Zool. Fenn.* **16**, 50–83

Lacey, L. A. and Undeen, A. H. (1986). Microbial control of black flies and mosquitoes. *Annu. Rev. Entomol.* **31**, 265–296.

Acknowledgements

It is a pleasure to thank the speakers, the Presidents of the Royal Entomological Society and the British Mycological Society who respectively opened and closed the symposium, Mr P. M. Hammond, Dr W. A. Sands, Dr H. T. Tribe and Professor J. Webster who chaired the sessions, and the Symposium Committee of the Royal Entomological Society for their contributions which led to the scientific success of the Symposium.

The Societies gratefully acknowledge the financial support provided by Bayer U.K. Limited, British Crop Protection Council, Imperial Chemical Industries plc and Shell Research Limited, and warmly thank the Registrar of the Royal Entomological Society and his staff for the efficient organization of the Symposium.

Opening Remarks

Dr VICTOR EASTOP
President of the Royal Entomological Society of London

It is my pleasant duty to welcome you to the Society's 14th Symposium. This Symposium differs from the previous thirteen in being organized jointly with the British Mycological Society. Our symposia are international by tradition and by intent, and it is a great pleasure to welcome overseas visitors, including several of the speakers.

The consequences of one particular association between insects and fungi has been known for many years. Ruthven Todd (1967, Coleridge and Paracelsus; Honeydew and LSD. *London Magazine* **6**, 52–62) discusses the origin of "For he on honeydew hath fed, and drunk the milk of Paradise" with its reference back to Paracelsus's account of warriors before battle being fed honeydew contaminated with ergot spores by visiting sugar-seeking flies and beetles. This may also have been the reason for the instructions recorded in Exodus to collect fresh honeydew every day and for the order of the phrases in the plea "Give us this day our daily bread and forgive us our trespasses".

The increasing concern about our environment and awareness of ecology has increased interest in the interaction between disparate organisms and increased the need for information to be readily available. This symposium is our attempt to draw together some of the threads of a very diverse subject. Insects and fungi interact in a great variety of ways and the four sessions of this symposium reflect different aspects of this relationship. Firstly we consider mycophagous insects, then mutualistic relationships, thirdly insects as vectors of fungal diseases, and lastly, mycopathogens of insects.

Rotten wood has played an important part in the evolution of insects, and its availability has to a large extent been facilitated by fungi. This has resulted in a variety of mutualistic relationships, including the transmission of fungi that create a suitable habitat for immature beetles. The use of mycopathogens for the control of agricultural pests is an important concept in biological control.

We hope that by bringing together people from different disciplines and different parts of the world, we may fire enthusiasm and that the personal contacts made will facilitate further exchange of ideas and expertise.

1

Mycophagy in the Coleoptera: Feeding Strategies and Morphological Adaptations

J. F. LAWRENCE

I. INTRODUCTION

Mycophagy, the habit of feeding on mycelium, plasmodium, fruiting structures or spores of slime moulds (Myxomycetes—F1) or true fungi, is widespread within the order Coleoptera (H25). About half of the recognized beetle families are primarily mycophagous or feed on plant material which has been substantially altered by the action of fungal enzymes, although only about 25 of these families are mycophagous in the strict sense.

The distribution of feeding habits within the four suborders (Archostemata, Myxophaga, Adephaga, Polyphaga) of Coleoptera tends to support the hypothesis that general saprophagy or mycophagy was ancestral within the order. Although recent Archostemata have highly specialized, wood-boring larvae with enlarged, grinding mandibles, the larvae occur in old rotten wood, and one cupedid species, *Tenomerga mucida* (Chevrolat), has been associated with a wood-rotting hymenomycete (*Stromatoscypha* —F5.8) (Fukuda, 1941). Adephagan larvae are also highly specialized, in this case for liquid-feeding, and the great majority of them are predaceous, but the Rhysodidae (H25.2) breed in rotten wood and are known to be associated with Myxomycetes (see below). Recent Myxophaga are aquatic or semiaquatic and feed on algae, but their mouthparts are of the microphagous type (see below). The Mesozoic families Schizophoridae and Catiniidae, which are thought to be ancestral to Myxophaga, were quite diverse and may have had more varied feeding habits. Triassic Ademosynidae, which have plesiomorphic archostematan features but may represent ancestral Polyphaga, closely resemble modern Scirtidae and may have had similar saprophagous habits (Crowson, 1975; Lawrence and Newton, 1982; Ponomarenko, 1969). Although most scirtids are aquatic detritivores, some tropical and Southern Hemisphere forms occur in wet, terrestrial environments (Hudson, 1934; Lawrence and Britton, in press).

Within the suborder Polyphaga, saprophagy and mycophagy in the strict sense, occur widely among primitive Staphyliniformia (Hydrophilidae (H25.9), Ptiliidae (H25.3), Agyrtidae (H25.4a), Leiodidae (H25.4)), Eucinetiformia (Scirtidae (H25.11a), Eucinetidae (H25.10), Clambidae (H25.11)), and Bostrichiformia (Derodontidae (H25.20), Nosodendridae (H25.21), Endecatomidae (H25.22), Jacobsoniidae (H25.21b)), and fungal associations have been particularly important in the evolution of two large groups: the Staphylinoidea and Cucujiformia (see below). Within the Scarabaeoidea, some Geotrupidae (H25.12) are known to feed on hypogean fungi, while among the Elateriformia, many adult Ptilodactylidae (H25.16) have evolved spore-brushes for feeding on sooty moulds (F4.9) and hyphomycete (F6.2) conidia (Stribling and Seymour, in press), and larvae in several families feed in rotten wood (Callirhipidae, Ptilodactylidae, Cerophytidae) or leaf litter and flood debris (Dryopidae, Limnichidae, Eulichadidae, Ptilodactylidae). Even among the predominantly phytophagous Chrysomeloidea and Curculionoidea, many weevil and cerambycid larvae live in rotten wood, larvae of Clytrinae and Cryptocephalinae (Chrysomelidae) are litter feeders, and the most common and probably ancestral food sources of the primitive family Anthribidae (H25.62) are wood-rotting mycelia and ascocarps of Pyrenomycetes (F4.34, etc) (Crowson, 1984; Holloway, 1982).

Most of the published works on insect mycophagy have concentrated on the fauna inhabiting the larger fruiting bodies of Hymenomycetes (F5), either as general surveys or local censuses of the species occurring in the sporocarps of particular fungi (Benick, 1952; Graves, 1960; Klimaszewski and Peck, 1987; Matthewman and Pielou, 1971; Nuss, 1975; Paviour-Smith, 1961; Pielou and Verma, 1968; Rehfous, 1955; Roman, 1970; Scheerpeltz and Höfler, 1948). However, some more recent works have dealt with the beetle faunas associated with Ascomycetes (F4) (Crowson, 1984; Hingley, 1971; Lawrence, 1977a) and Myxomycetes (F1) (Lawrence and Newton, 1980), while others have dealt with particular groups of mycophagous beetles, such as Leiodidae (H25.4) (Wheeler, 1984), Staphylinoidea (Newton, 1984), Staphylinidae (H25.8): Gyrophaenina (Ashe, 1984a, 1984b, 1986), Phalacridae (H25.40) (Steiner, 1984), and Ciidae (H25.52) (Lawrence, 1973; Paviour-Smith, 1960). In general, those studies which have concentrated on smaller groups of beetles, in which both cladistic relationships and ecological roles are better defined (see Ashe's papers), have been the most useful in elucidating the nature of the beetle–fungus relationship and its role in evolution.

The subject of mycophagy may be approached by considering the physical nature of fungi as food sources. A single wood-rotting bracket fungus, for instance, may provide at least three different kinds of substrates or food sources, each of which supports a unique guild of mycophagous insects: the mycelium and chemically altered plant cells of the host tree, the spores produced by the hymenium, and the context tissue making up the body of the sporocarp. After some preliminary considerations, the various kinds of *substrates* produced by fungi will be surveyed, the *feeding strategies* adopted by different beetle groups for handling these substrates will then be considered, and finally specific *morphological adaptations* which have evolved to carry out these strategies will be examined. Throughout, the fungal class and order names of Ainsworth *et al.* (1973a, b) will be used. Larvae, rather than adults have been used in comparing different types of feeding adaptations because: (a) adult mouthparts tend to be more conservative than those of larvae; (b) larvae represent the major feeding stage in the life cycle and thus better reflect the adaptations of the species to a particular food type; and (c) adult mouthparts may have evolved specializations for activities other than feeding (emerging from site of eclosion, constructing egg cavities, courtship, defence, etc.). It hardly needs to be stressed that the ideas presented here on the functional aspects of larval feeding are based largely on circumstantial evidence and must be considered as preliminary hypotheses to be refuted or supported by those with the time, expertise and sophisticated equipment necessary for functional studies on minute organisms. Sources of data presented here include field observations on exact feeding sites, examinations of gut contents and comparisons with microscopic preparations of intact host tissue, examinations of larval mouthparts with light and scanning electron microscopes, and manipulations of cleared mouthparts.

II. FEEDING STRATEGIES AND THE LARVAL TROPHIC SYSTEM

A. Microphagy and Macrophagy

Microphagy is the habit of feeding on very small particles or loosely organized food masses, consisting of spores, hyphae or highly decomposed animal or plant tissue; it encompasses general detritus feeding or saprophagy, as well as filter feeding in a liquid medium. As mentioned above, this habit was probably ancestral in Coleoptera. The main problems facing a microphagous beetle are: (1) gathering the food items, which are often scattered over a surface, (2) concentrating them and moving them to a processing area near the mouth opening, and (3) feeding them into the gut, with or without further breakdown or trituration.

Macrophagy, or feeding on more compact or solid substrates, is a derived condition which has evolved a number of times in Coleoptera. The mechanisms for dealing with solid substrates vary from group to group. Phytophagous beetles are macrophagous, as are most predators, but the kinds of mechanisms employed by the two groups may differ greatly. Among the fungus feeders, macrophagy has developed to utilize the compact hyphal tissue making up the larger fruiting bodies of a few Ascomycetes (F4) and most Basidiomycetes (F5). Problems facing the macrophagous feeder are associated with the removal of manageable portions from a solid substrate. This removal often involves merely tearing or scraping of the substrate. Commonly, the freed material is swallowed whole, but in some cases food is triturated by enlarged and modified mandibular molae, while in others it is chemically altered. The removal process is accomplished by a pair of powerful mandibles constructed so that the maximum force is applied at or near the apices, which are adapted for cutting or scraping. If trituration is also involved, however, then the force must be distributed more evenly along the mesal edges of the mandibles.

B. Liquid Feeding

Liquid feeding involving extraoral digestion and a pharyngeal pump is most commonly encountered in predaceous insects, including many Adephaga, Staphyliniformia and Elateriformia. However, it also occurs in saprophagous Elateridae, Eucnemidae (H25.18) and Cerophytidae, and in probable fungivores such as Lycidae (H25.19) and Rhysodidae (H25.2). It may also be involved in the feeding mechanisms of a few Leiodidae (H25.4) and Eucinetidae (H25.10), many Cerylonidae (H25.41), and some Corylophidae (H25.42) which have evolved sucking tubes and styliform mandibles and maxillae (Besuchet, 1972; Lawrence and Stephan, 1975; Vit, 1977), but the feeding behaviour and the exact food sources of these minute beetles are still unknown.

C. Surface and Internal Feeding

Most microphagous larvae can be described as *surface feeders*, in that they are capable of moving from place to place and grazing on spores and hyphae. The legs are usually relatively long and the body is often somewhat flattened and more sclerotized dorsally than ventrally. Some species have protective devices, such as long hairs, setae, or spines, and the dorsal surfaces may be provided with defensive glands and aposematic coloration. Those insects occurring in litter or under bark are basically surface feeders, since they are usually small enough to move rapidly within interstitial spaces and feed on surface fungi. Some macrophagous forms living on pore surfaces of bracket fungi (F5.8) also have these protective devices. *Internal feeders*, on the other hand, are often highly modified for boring within a more or less solid substrate. The legs are shorter (or sometimes absent), the body is less flattened and more lightly sclerotized, sometimes with folds or ampullae, and the head is often enlarged and provided with endocarinae for the attachment of large mandibular muscles (Lawrence, in press).

D. The Larval Trophic System

What appears to be the basic, ancestral type of feeding mechanism in larval Coleoptera survives in Myxophaga and in several of the more primitive polyphagan families (e.g. Agyrtidae (H25.4a), Leiodidae (H25.4), Eucinetidae (H25.10), Derodontidae (H25.20)) (Lawrence and Hlavac,

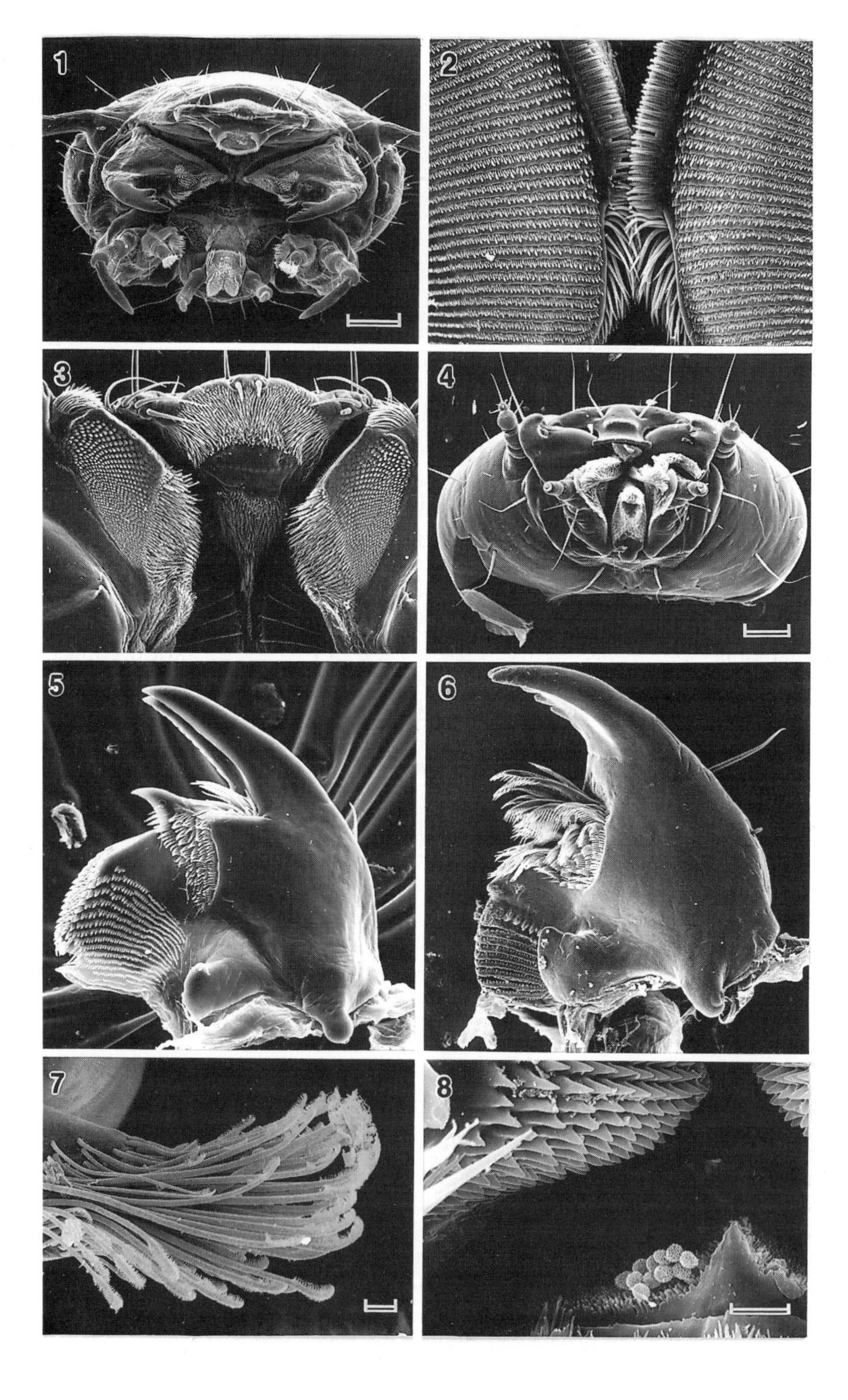
1
2
3
4
5
6
7
8

1979; Lawrence and Newton, 1980; Lawrence, 1988). This ancestral mechanism consists of the following (Figs 1–3, 5): (1) a pair of mandibles with relatively weak apices, no sharp incisor edges, small and simple prosthecae, and well-developed basal molae, which are ventrally and mesally asperate or tuberculate and dorsally armed with rows of microtrichia; (2) a pair of free and flexible, apically setose maxillae; (3) a setose and often complex hypopharynx, usually armed with a sclerome attached to the hypopharyngeal bracon; and (4) a cibarial region bearing a series of ridges lined with microtrichia. Food is gathered with the mandibular apices and the maxillae, moved towards the molar regions, and then fed into the gut by the intermeshing molar tubercles or asperities. Trituration may also take place, but it is not essential to the process. The maximum force is applied at the basomesal portion of the mandible and is probably relatively weak. The dorsal microtrichia of the mandible act in conjunction with the cibarial ridges to move small particles towards the mouth opening and away from the articular regions. The ventral tubercles or asperities of the mandibles act similarly in conjunction with the hypopharyngeal armature. The function of the prostheca is questionable, but it may act as a valve, preventing particles from escaping from the processing area. Although this type of mandible with a basal, tuberculate or asperate mola is characteristic of saprophagous and fungivorous beetles, it has been retained with little modification in certain predaceous forms, such as cucujine Cucujidae (H25.33b) (Figs 2, 3) and gempylodine Colydiidae (H25.27) (Lawrence, 1980). In most predators, however, the mandibular molae are reduced or absent.

From a trophic system of the kind described above all other more derived systems could have evolved to utilize different, more demanding substrates.

Fig. 1. *Necrophilus pettiti* Horn (Agyrtidae (H25.4a)). Larval head, anterior view, with mandibles spread. Line = 250 μm.

Fig. 2. *Cucujus clavipes* Fabricius (Cucujidae (H25.33b)). Larval mandibles, basomesal area, dorsal view. Mandible width = 0.6 mm.

Fig. 3. Same. Larval mandibles (spread), basomesal area, ventral view, showing cibarial ridges. Mandible width = 0.6 mm.

Fig. 4. *Omoglymmius hamatus* (LeConte) (Rhysodidae (H25.2)). Larval head, anterigo view, showing fringed membranes on labium and maxillae. Line = 100 μm.

Fig. 5. *Peltastica tuberculata* Mannerheim (Derodontidae (H25.20)). Larval mandible, ventral view. Mandible width = 0.25 mm.

Fig. 6. *Anchorius lineatus* Casey (Biphyllidae (H25.48)). Larval mandible, ventral view, showing prosthecal comb-hairs. Mandible width = 0.16 mm.

Fig. 7. *Nosodendron unicolor* Say (Nosodendridae (H25.21)). Larval maxillary apex, antero-lateral view, showing comb-hairs. Line = 10 μm.

Fig. 8. *Anisotoma* sp. (Leiodidae (H25.4)). Larval mandibular molae, hypopharynx, and spores of *Stemonitis* (F1.9). Line = 10 μm.

III. MARGINAL SUBSTRATES AND THE SAPROPHAGOUS HABIT

This term is used for substrates which consist of decomposing plant tissue which has been or is being penetrated by fungal mycelium, without any concentrations of hyphae or conidia being present. In most cases, the food available to an insect consists primarily of plant cells, with smaller components of hyphae, conidia or other fungal structures. Associated insects are usually considered saprophagous, although some may be able to subsist on fungal material alone.

A. Aquatic Detritus

This is formed by the deposition of decaying leaves or other plant parts, phytoplankton and zooplankton, in slow-moving streams, ponds, or other lentic environments. Following deposition it is attacked by various Hyphomycetes (F6.2) and water moulds (Saprolegniales (F2.3)). Among the Coleoptera utilizing this decomposing material are adult Hydrophilidae (H25.9), larval Scirtidae (H25.11a), Ptilodactylidae (H25.16) and Eulichadidae, and both larval and adult Elmidae.

B. Leaf Litter and Flood Debris

The terrestrial accumulation of decaying leaves and other plant material and its interface with the soil form an extremely rich environment in which fungi play a key role. As shown by Hudson (1968, 1972), leaves entering the litter already have a flora of phylloplane fungi, including Ascomycetes (F4) and Hyphomycetes (F6.2), many of which persist to become primary saprophytes. These are followed by secondary saprophytes, soil fungi, like *Mucor* (F3.1) and *Penicillium* (F6.2), and Hymenomycetes, like *Lactarius* (F5.12), *Collybia* (F5.10), and *Paxillus* (F5.11); slime mould (F1) plasmodia and fruiting bodies are also common. A large number of saprophagous Coleoptera inhabit this environment, but the exact feeding habits of many are unknown, and some may be true fungus feeders specializing in microsporocarps or spore fields on leaves.

C. Rotten Wood, Phloem and Cambium

Dead trees provide a rich habitat for all kinds of fungi, including slime moulds (F1), soft rots (yeasts (F4.11, etc) *Ceratocystis* (F4.24), *Chaetomium* (F4.32)), and the major groups of wood-rotting Pyrenomycetes (F4) and Hymenomycetes (F5), whose hyphae penetrate the plant cells and destroy cellulose (brown rots) or both lignin and cellulose (white rots). The subcortical region, consisting of phloem and cambium, is often considered separately, since a rich variety of both fungi and insects occur there, while a more restricted group are able to penetrate the sapwood and heartwood. Primary fungal invaders of the subcortical region live mainly on sugars and simple carbon compounds, but the subsequent succession of decay fungi is a complex one, which has been studied in some detail by Shigo (1967), Hudson (1968), Käärik (1975) and others. At present, there are few data on habitat preferences of subcortical or wood-inhabiting Coleoptera, but it is obvious to any collector that a single log will present an array of different microhabitats, each with a characteristic fauna. A particularly rich subcortical fauna, for instance, occurs with phloeophagous bark beetles (Curculionidae: Scolytinae (H25.63)) and their associated fungi. Also, white-rotten and brown-rotten wood may serve as food for different species of wood-boring beetle larvae, but this has not been well documented.

IV. LIQUID SUBSTRATES: FILTER-FEEDING ADAPTATIONS

A. Slime Mould Plasmodium

Although the plasmodia of Myxomycetes (F1) are common features of soil, litter and dead wood, they are usually difficult to observe. Exceptions are the large, yellow and fan-like phaneroplasmodia of *Fuligo* (F1.10) and related forms (Martin and Alexopoulos, 1969). Except for Wheeler's (1984) observations on species of *Agathidium* (H25.4), there is very little direct evidence that beetles feed on plasmodia. An adult rhysodid (H25.2) *Omoglymmius hamatus* (LeConte), however, was observed in the laboratory feeding on a yellow plasmodium (T. F. Hlavac and J. F. Lawrence, unpublished). The unusual larval mouthparts of rhysodids (Fig. 4) may well be adapted for plasmodiophagy—the labium and maxillae are lined with fringed membranes, which could act as a sponge similar to the dipteran labellum. Larvae of Lycidae (H25.19) and Endomychidae (H25.46) have also been seen in association with plasmodia, but there is no evidence that they utilize the food source.

B. Slime Fluxes and Fermenting Sap

In tree wounds, on cut stumps, and under certain conditions beneath bark, yeasts, such as *Dipodascus* and *Endomyces* (F4.11), and bacteria act on sap to form a thick fermenting liquid on which several specialized beetle larvae feed. Among the Coleoptera found in this habitat are Nosodendridae (H25.21) (*Nosodendron*), Derodontidae (H25.20) (*Peltastica*), Nitidulidae

(H25.30) (*Amphicrossus, Glischrochilus, Cryptarcha*), Helotidae (H25.33a) and Biphyllidae (H25.48) (*Anchorius*). In such a liquid medium, problems of food gathering are increased, since smaller food items are in suspension. Larvae occurring in such habitats all have more numerous hairs associated with the mandibles and maxillae (Fig. 5). In some groups, specialized comb-hairs are present on the mandibles (Fig. 6) or maxillae (Fig. 7), and these hairs act as filter-feeding devices. Nosodendrid larvae (Figs 7, 9, 10) have a complex feeding mechanism matched only in a few aquatic detritivores. In addition to the maxillary comb-hairs there are localized areas of specialized vestiture or armature on the mandible, one of which appears to be a food press (Fig. 10) of the type found in larval Scirtidae and in adult Hydrophilidae and Scarabaeidae that feed in fresh dung. Although Miller (1961) considered the fine ridges (which he called "tritors") in dung beetles to be part of the triturating mechanism, it is more likely that these serve to concentrate and compact solid particles, while allowing liquid to escape. An analogous type of filter-feeding mechanism has been described in detail by Beier (1952) for larval Scirtidae. Another adaptation commonly found in inhabitants of slime fluxes involves modifications in the respiratory system, which allow the larvae to gain access to the atmosphere while being immersed in fluid. In *Nosodendron* larvae the 8th abdominal spiracles are located at the end of a long process; in Nitidulidae, all the spiracles or the last pair only are placed at the ends of individual tubular projections; and in Biphyllidae, the apical spiracles may be displaced dorsally. *Nosodendron* is often considered to be a predator, and larvae of Diptera have been found in the gut (Costa *et al.*, 1986). Considering the complexity of the feeding system and its similarity to those of other known filter-feeders, however, it is likely that fly larvae are not the

---

Fig. 9. *Nosodendron unicolor* (H25.21). Larval mandible, ventral view. Line = 100 μm.

Fig. 10. Same. Larval mandible, basomesal area, ventral view, showing food press. Line = 10 μm.

Fig. 11. *Baeocera* sp. (Staphylinidae (H25.8): Scaphidiinae). Larval head, anterior view, showing subapical mandibular spore brushes. Line = 50 μm.

Fig. 12. *Dasycerus* sp. (Dasyceridae (H25.6)). Larval head, anterior view, showing apical mandibular spore brushes. Line = 50 μm.

Fig. 13. *Agaricomorpha* sp. (Staphylinidae (H25.8): Aleocharinae: Gyrophaenina). Larval head, anterior view, showing maxillary spore brushes. Head width = 0.3 mm.

Fig. 14. *Aphenolia monogama* (Crotch) (Nitidulidae (H25.30)). Larval maxillary apex, dorsal view, showing spatulate setae. Malar width = 0.12 mm.

Fig. 15. *Lycoperdina ferruginea* LeConte (Endomychidae (H25.46)). Larval mouthparts, anterior view. Line = 100 μm.

Fig. 16. Same, larval mandible, mesoventral view, showing molar spore mill. Mandible width = 0.25 mm.

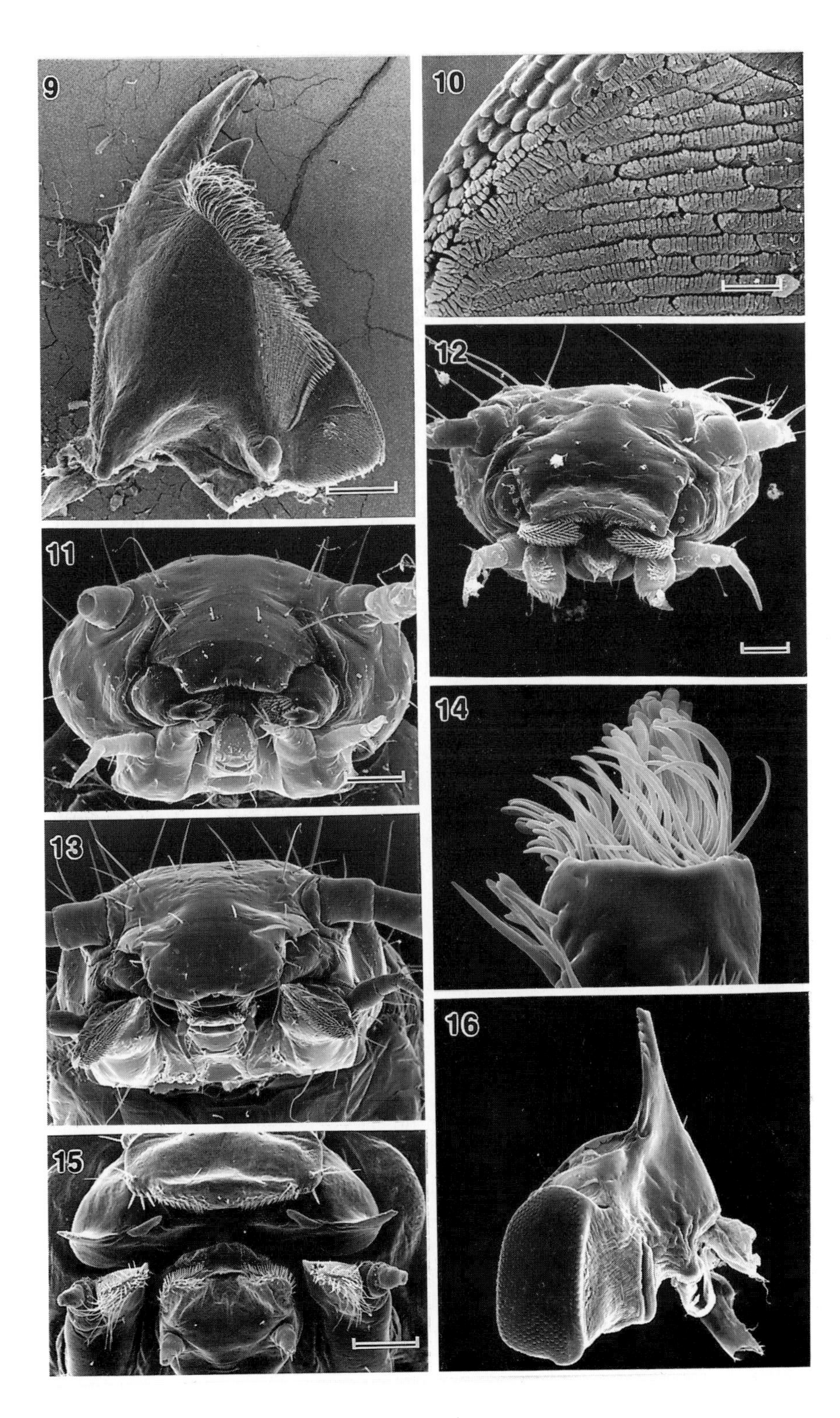
9
10
12
11
14
13
16
15

main food source. Guts of both adult and larval nosodendrids examined contained a large amount of fine-grained material, mixed with what might be insect cuticle or parts of plant cells, as well as hyphae and conidia (J. F. Lawrence, unpublished).

V. SMALL PARTICLES AND LOOSE SUBSTRATES: MICROPHAGOUS ADAPTATIONS

A. Surface Hyphae and Spores

Included in this group of similar substrates are the surface-inhabiting fungi, like moulds and mildews, the sporulating surfaces associated with various Pyrenomycetes (F4.34), the relatively loosely organized hyphal masses and associated conidia produced by sooty moulds (F4.9) (Loculoascomycetes), the sporulating surfaces associated with rusts (F5.22) and smuts (F5.23), the smaller fruiting bodies of Myxomycetes (F1), and finally the spore-producing surface or hymenium of basidiocarps (F5). The associated beetles tend to be exclusive at the species level to major groups of fungi, such as Myxomycetes, Pyrenomycetes, Hymenomycetes or Gasteromycetes, but specificity may occur at higher levels as well. For instance, myxomycete spore specialists include all members of the family Sphindidae (H25.29), the tribe Anisotomini and the Genus *Neopelatops* Jeannel (H25.4), the scaphidiine genera *Baeocera* and *Scaphobaeocera* (H25.8), the lathridiid genus *Revelieria* (H25.47), and some (but not all) species of the genera *Enicmus* (Lathridiidae), *Clambus* (H25.11) and *Eucinetus* (H25.10) (Lawrence and Newton, 1980; Newton, 1984). The faunas associated with Ascomycetes have been discussed in some detail by Crowson (1984). Among these are some of the smaller staphylinoids (such as Ptiliidae (H25.3) and Micropeplidae (H25.5)), some Clambidae (H25.11), many Cucujoidea (Rhizophagidae (H25.31), Laemophloeidae (H25.34), Phalacridae (H25.40), Silvanidae (H25.35), Cryptophagidae (H25.36), Corylophidae (H25.42), Lathridiidae (H25.47)), and a few tenebrionoids (Mycetophagidae (H25.50), Colydiidae (H25.57)). Mould feeders are common among clavicorn families, such as Rhizophagidae (*Monotoma*), Silvanidae, Cryptophagidae, Languriidae (H25.38) (*Cryptophilus*), Endomychidae (H25.46) (*Mycetaea*), Corylophidae (*Orthoperus, Sericoderus*), and Lathrididae (*Lathridius*). The fauna inhabiting sooty moulds in New Zealand is particularly rich, and includes *Nothoderodontus* (H25.20), *Agapytho* (H25.33) and *Cyclaxyra* (H25.40). The coccinellid tribe Psylloborini (H25.44) (Gordon, 1985) and at least some Corylophidae (*Corylophodes*) feed on powdery mildews (Erysiphales (F4.12)). Smut spores are consumed by the *Leucohimatium* (H25.38) and some Phalacridae (*Phalacrus*), but the latter family feed on a wide variety of other spore types, including those of rusts and Pyrenomycetes (Steiner, 1984). Hymenomycete spore grazers include the New Zealand leiodid *Zearagytodes* (H25.4) and Staphylinidae (H25.8) belonging to the subfamilies Scaphidiinae (*Scaphisoma* and some others), Tachyporinae (*Sepedophilus*) and Aleocharinae (subtribe Gyrophaenina), as well as various cucujoids (Lawrence, 1988; Newton, 1984).

Only minor changes in the generalized microphagous system are seen among most of these surface-feeding forms. In larvae of some Cryptophagidae and Rhizophagidae, the prostheca has become more complex and serrate and may be involved in cutting hyphae. In addition, some spore specialists tend to have more strongly developed molar teeth than occur in general detritivores (Fig. 8). In psylloborine Coccinellidae, however, and in some derived members of the Staphylinoidea, spore-feeding modifications are more extreme, because sporophagy has evolved in lineages which had already lost the basal mandibular mola. Thus in the Micropeplidae (H25.5) and Scaphidiinae (H25.8) (Fig. 11), a secondary, subapical pseudomola has evolved and serves as a spore brush. In Dasyceridae (H25.6) (Fig. 12) and in the Psylloborini (H25.44), the entire apical surface of the mandible has become modified to scrape or brush spores or hyphae from surfaces. In larvae of gyrophaenine staphylinids, the mandibles are of the falcate type normally found in predators, and the maxillary apices (Fig. 13) have taken over the role of spore brushes or scrapers (Ashe, 1986).

B. Epigean Spore Masses

A somewhat different type of substrate is formed when spores are produced in large numbers and contained within a compact fruiting structure. Examples are the epigean Gasteromycetes (Lycoperdales (F5.16) and Sclerodermatales (F5.13)) usually known as puffballs, the larger aethalia of Myxomycetes, and a peculiar species of polypore, *Cryptoporus volvatus* (Peck) Hubbard (F5.8), in which the pore surface is concealed by a mantle which allows spores to accumulate. Puffball spore feeders are found in the families Leiodidae (H25.4) (*Creagrophorus* and some *Nargomorphus*), Cryptophagidae (H25.36) (some *Cryptophagus*), Anobiidae (H25.33) (*Caenocara*), Nitidulidae (H25.30) (*Pocadius*), and Endomychidae (H25.46) (*Lycoperdina*), and larvae of the last three have special modifications probably associated with living enclosed within the spore mass. *Caenocara* larvae, for instance, have reduced legs, while in *Lycoperdina* (Figs 15, 16), the mandibles have reduced apices and greatly enlarged molae, which form a highly efficient mill for grinding spores (Lawrence,

1988; Newton, 1984; Pakaluk, 1984). Several spore feeders are associated with *Cryptoporus* in North America and Japan (Borden and McLaren, 1972; Hisamatsu, 1962); these include *Aphenolia* species (H25.30), in which the larval maxillae (Fig. 14) have spatulate setae for spore gathering.

C. Ambrosia Tunnels

The tunnels excavated into wood by ambrosia beetles (Lymexylidae, scolytine and platypodine Curculionidae) form a distinct substrate lined with yeasts of the genera *Ascoidea* and *Endomycopsis* (F4.11), and in some cases species of *Ceratocystis* (F4.24). These are cultivated by the beetles and serve as their food, but are also consumed by other tunnel-inhabiting fungivores such as teredine Bothrideridae (Pal and Lawrence, 1986) and probably some other non-predaceous, cylindrical beetles, e.g. Thioninae (H25.31) and some *Corticeus* (H25.60) (Schedl, 1962). Commensalism between fungi and ambrosia beetles has been discussed by Crowson (1981, 1984) and Francke-Grosmann (1967).

D. Hypogean Sporocarps

The unrelated orders Tuberales (F4.28) (Discomycetes) and Hymenogastrales (F5.20) (Gasteromycetes) produce fruiting bodies which remain below the surface of the ground. These fruiting bodies are fed upon by beetles capable (at least in the adult stage) of locating the concealed sporocarps and burrowing through soil to reach them. The food consists mainly of spores, embedded in a soft tissue (gleba) which undergoes autodigestion when the spores become mature. The major inhabitants of these subterranean fungi are members of the family Leiodidae (H25.4) (*Catopocerus, Hydnobius, Leiodes*, etc.), but some Geotrupidae (H25.12) (Bolboceratinae) and Nitidulidae (H25.30) (*Thalycra*) also utilize the food source (Arzone, 1970; Fogel and Peck, 1975).

E. Pore Tube Specialists

The hymenium formed by Polyporaceae (F5.8) is often enclosed within pores, which may be less than 0.3 mm in diameter. In general, spore feeders specializing on Polyporaceae feed on spores that gather on the pore surface, but at least two groups are able to gain access to the developing basidiospores within the pore tubes. In the subfamily Nanosellinae of Ptiliidae (H25.3), both adults and larvae are elongate, narrow and minute, so that they can enter the pores and feed directly on the hymenium. Dybas (1976) has recorded both *Nanosella* and *Throscoptilium* from *Phellinus* species. In the second group, larvae have evolved a different means of gaining access to the developing basidiospores. In some species of the New Zealand genus *Holopsis* (H25.42), larvae have developed a long rostrum (Figs 17, 18), which is curved at the apex, where there are a pair of rasping mandibles. Larvae occur on the undersides of *Ganoderma* sporocarps and are capable of inserting this rostrum into the pore tubes. It is interesting that some other *Holopsis* lack the rostrum, although the mouthparts are otherwise similar. Since *Ganoderma* spores have a reinforced spore coat, access to the tubes may make it possible to feed on spores which have not yet developed their full armature.

VI. COMPACT AND SOLID SUBSTRATES

This is the most demanding type of fungal substrate and is comparable to the woody tissue or foliage of higher plants. Included in this category is the context tissue making up the fruiting bodies of Hymenomycetes, the hyphae and pseudoparenchymatous tissue comprising the larger stromata or sclerotia of Pyrenomycetes, and the mutualistic associations of algae and fungi (mostly Ascomycetes) known as lichens. The insects utilizing these are usually macrophagous, but their general form depends on whether they are surface grazers or internal feeders. In some groups associated with the ascocarps of *Daldinia* (F4.34) and other Pyrenomycetes, the larvae may feed externally on spores or internally on the carbonaceous sporocarp tissue, but the mouthparts are of the microphagous type. Most species which feed on compact or solid substrates, however, have highly derived feeding systems in which the basal molae either form multiple shearing surfaces or are reduced or absent, and the mandibular apices and incisor regions have become strengthened and modified for cutting or scraping.

A. Ascomycete Stromata, Sclerotia and Lichens

The Ascomycetes provide some examples of compact substrates in the larger stromata of *Daldinia* and other Pyrenomycetes and in the sclerotia or ergots produced by *Claviceps* (F4.4). Steiner (1984) has shown that *Acylomus pugetanus* Casey (H25.40) has larvae which feed internally in ergots and differ from the spore grazing larvae of related species in general form and mandibular structure. Other phalacrids (*Litochropus*) and laemophloeids (*Placonotus*), which feed in *Daldinia* stromata, have reduced molae. In anthribid (H25.62) larvae, which are common pyrenomycete inhabitants, the mola is also highly reduced, the mandibles are of the macrophagous type, and the legs are reduced or absent. Lichens may be fed

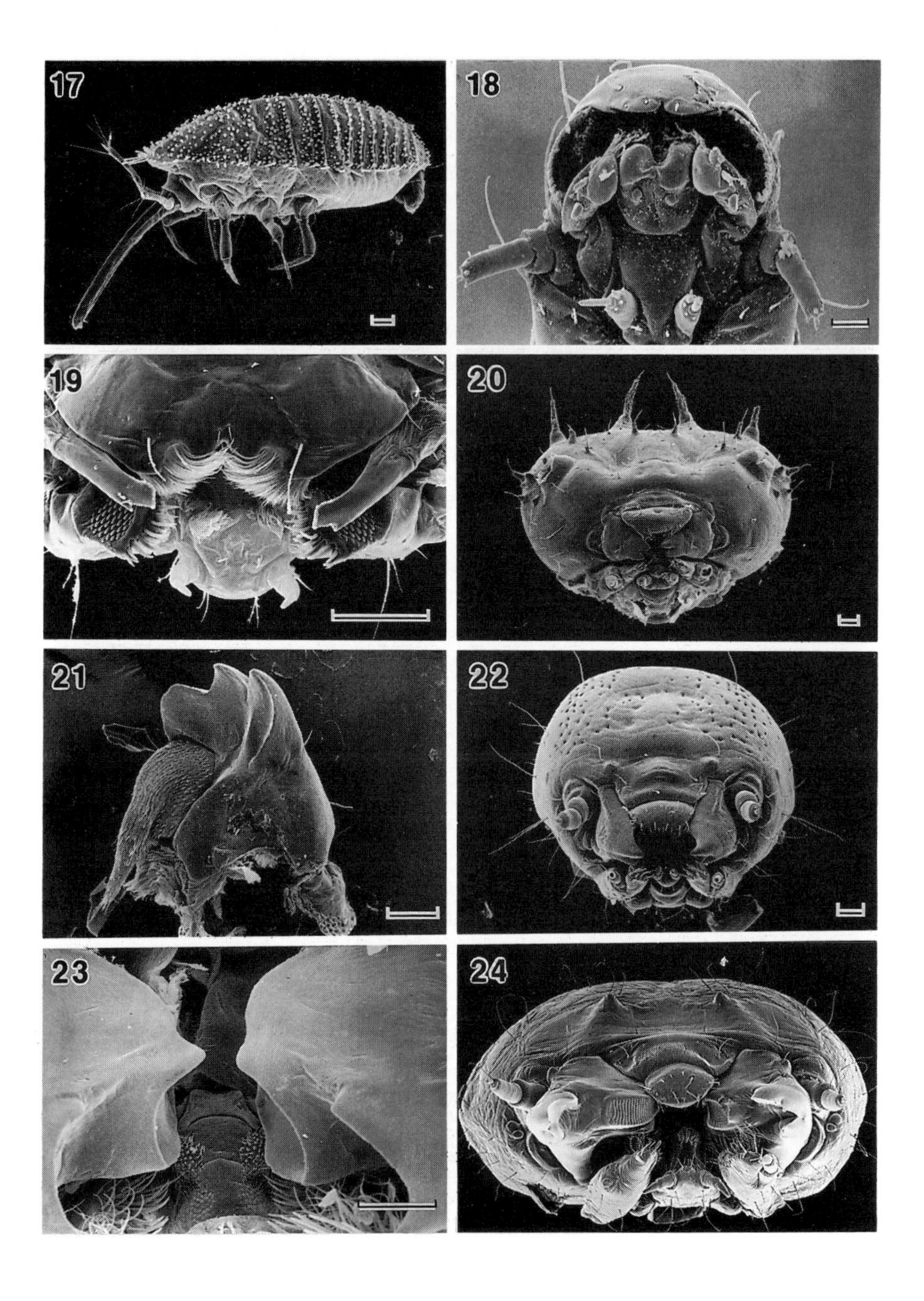
17
18
19
20
21
22
23
24

upon by a variety of adult beetles, but few are known to utilize them as larval food. *Orthocerus* (H25.57) and *Lichenobius* (H25.62) (Holloway, 1970) are well-known examples, but in Australia both larvae and adults of *Amarygmus* and *Titaena* (H25.60) have been seen feeding on lichens at night (J. F. Lawrence, unpublished).

B. Hymenial Grazing: Transition to Macrophagy

The basidiocarp hymenium was discussed above in connection with spore feeding, but it also serves as a food source for larvae which scrape surface tissue; this is particularly true for the pore surfaces of polypores, where most of the accessible tissue consists of sterile hyphae rather than spores. Notable examples of macrophagous feeders inhabiting sporocarp surfaces are in the families Erotylidae (Erotylini) and Endomychidae (*Eumorphus*), but Ashe (1986) has shown that some gyrophaenine Staphylinidae (H25.8), like *Agaricomorpha*, which inhabit pore surfaces, have more heavily sclerotized and densely packed maxillary spines (Fig. 13), and the unusual feeding mechanism in some *Sepedophilus* larvae (Staphylinidae: Tachyporinae) (Fig. 19) may also be used for hyphal scraping (Newton, 1984). In Erotylini (H25.39) and *Eumorphus* (H25.46), the larval head (Fig. 20) is usually strongly narrowed anteriorly and the large mandibles have multidentate, more or less perpendicularly oriented apices and a small gape; the molae have been lost and replaced with setose lobes (Fig. 21). Large pieces are removed from the substrate by these larvae and then swallowed whole. As mentioned above, these hymenial grazers often have protective devices, such as spines (Erotylini) or dehiscent lobes (*Eumorphus*), and they are usually brightly coloured.

Fig. 17. *Holopsis* sp. (Corylophidae (H25.42)). Larva, lateral view. Line = 100 μm.

Fig. 18. Same. Larval head, apex of rostrum, ventral view, with mandibles spread. Line = 10 μm.

Fig. 19. *Sepedophilus* sp. (Staphylinidae (H25.8): Tachyporinae). Larval mouthparts, anterior view, showing mandibles and maxillary rasping organ. Line = 100 μm.

Fig. 20. *Homoeotelus* sp. (Erotylidae (H25.39): Erotylini). Larval head, anterior view. Line = 100 μm.

Fig. 21. Same. Larval mandible, dorsal view. Line = 100 μm.

Fig. 22. *Platydema ellipticum* (Fabricius) (Tenebrionidae (H25.60): Diaperini). Larval head, anterior view. Line = 100 μm.

Fig. 23. Same. Larval mandibular molae and hypopharyngeal sclerome, anterodorsal view. Line = 100 μm.

Fig. 24. *Bolitotherus cornutus* (Panzer) (Tenebrionidae (H25.60): Bolitophagini). Larval head, anterior view, showing concave–convex, transversely ridged molae. Head width = 2.5 mm.

C. Internal Feeding in Basidiocarps

By far the largest number of macrophagous fungus feeders are of the internal type, which bore through the modified hyphae comprising the context of the basidiocarp. These hyphae may be thickened and solidly packed to form woody tissue (as in the dimitic system of Corner, 1953), while others have the thickened or skeletal hyphae bound together by anastomosing, binding hyphae producing a flexible, leathery tissue (as in the trimitic system of Corner). Examples of context feeders are common in the Tenebrionoidea, but also include the staphylinid genera *Cyparium* and *Oxyporus* (H25.8), the primitive bostrichoid genus *Endecatomus* (H25.22), various dorcatomine Anobiidae (H25.23), some Trogossitidae (H25.26) and many Erotylidae (H25.39).

Internal feeders have enlarged and strengthened mandibles, like those of the surface feeders, but the incisor edge is often well developed and the apical teeth fewer in number and more obliquely oriented. Some of these larvae have retained the primitive basal, symmetrical, tuberculate molae; examples may be found in Nitidulidae (H25.30) (*Phenolia, Cychramus, Pallodes*), Hobartiidae (H25.23), Erotylidae (Dacnini), and Endomychidae (H25.46) (Leiestinae). However, there are three distinct, derived conditions, two of which involve modifications of the basal mola, and one its loss. In the tenebrionid (H25.60) *Platydema ellipticum* (Fabricius) (Figs 22, 23), both molae are concave and unarmed, and the hypopharyngeal sclerome is well developed; these larvae are found in the woody tissue of the polypore *Phellinus gilvus* (Schweinitz) Patouillard (F5.8) and the molae appear to act with the hypopharyngeal sclerome to compact the moderately large pieces of hyphal tissue into a food bolus. Gut contents of *P. ellipticum* larvae include intact hyphal masses. More commonly the molae are asymmetrical and the tubercles or asperities have joined to form transverse ridges; the resulting structure acts as a multiple shearing device, which serves as an effective triturator, allowing only small fragments into the gut (Lawrence, 1977b). This is best seen in the tenebrionid *Bolitotherus cornutus* (Panzer) (Fig. 24), which feeds on fruiting bodies of *Ganoderma* and *Fomes* (F5.8), but it occurs in other tenebrionids (including some *Platydema*) and in the Pterogeniidae (H25.56). An intermediate condition between the symmetrical, tuberculate mola and this derived type can be seen in some Tenebrionoidea (Mycetophagidae (H25.50), Archeocrypticidae (H25.55), Colydiidae (H25.57), pisenine Tetratomidae (H25.51), and hallomenine Melandryidae (H25.53)). Internally feeding larvae which have lost the molae entirely may be found in a number of families, including Anobiidae (H25.23) (*Dorcatoma, Byrrhodes*, etc.), Trogossitidae (H25.26) (*Ostoma, Thymalus, Calitys*), Erotylidae (H25.39) (Megalodacnini and Triplacini), Ciidae

(H25.52), Tetratomidae (Tetratominae), Melandryidae (Eustrophinae and Orchesiini), Mordellidae (H25.54) (*Curtimorda*), and Anthribidae (H25.62) (Eupariini).

VII. CONCLUSION

The above discussion attempts to summarize an enormously complex set of relationships between fungi and beetles, on the basis of the patchy data available. Generalizations on the co-evolution of the Coleoptera and their fungal hosts would be premature, given our present poor knowledge of phylogenetic relationships within both groups. Although significant advances have been made in our understanding of the Cucujoidea by Crowson and his students and a major revision of the superfamily is nearing publication, there are still doubts about the limits of this group and its relationships to Cleroidea and Tenebrionoidea. Furthermore, the basal lineages of Cucujiformia and Bostrichiformia are far from being resolved. Also within the Staphylinoidea, the second most important group of fungivores, relationships at the family level are still somewhat controversial (compare Lawrence and Newton, 1982, with Naomi, 1985). The author is not in a position to speak for the mycologists, but it is generally agreed that many higher taxa of fungi are artificial assemblages and generic limits in important host groups, such as the Polyporaceae and their relatives, are still disputed (Peterson, 1971). Clearly there is a need for further rigorous cladistic analyses of both beetles and fungi. An attempt has been made here to emphasize the importance of defining the exact nature of the food substance taken by mycophagous insects and establishing exactly how the insects obtain and process this food, as a necessary prerequisite to understanding the nature of mycophagy and its role in co-evolution.

Acknowledgements

I am grateful to many entomologists and mycologists who have helped me through the years by collecting critical material, identifying insects and fungi, and discussing topics of mutual interest. Included are the following: R. A. Crowson, V. De Moulin, M. Farr. O. Fidalgo, R. Fogel, R. L. Gilbertson, T. F. Hlavac, S. J. Hughes, J. Kukalova-Peck, M. Martin, A. F. Newton, Jr., M. Nobles, I. Pascoe, S. B. Peck, D. Reid, J. D. Rogers, J. A. Stevenson, M. K. Thayer, J. Walker, N. Walters, and T. A. Weir. Thanks are given to E. S Nielsen and I. D. Naumann for critically reading the manuscript, J. Green for photography, and W. Dressler for technical assistance. Scanning electron micrographs in Figs 1–6, 8, 11–14, 16 and 24 were produced by E. Seling, Museum of Comparative Zoology, Harvard University, and those in Figs 7, 9, 10, 15 and 17–23 by K. Pickerd and E. Hines, Division of Entomology, C.S.I.R.O.

Facilities were provided by C.S.I.R.O., and early portions of this work were aided by National Science Foundation grants (BMS 7502606 and BMS 7412494), and supported by Harvard University.

REFERENCES

Ainsworth, G. C., Sparrow, F. K., and Sussman, A. S. (eds.) (1973a). "The Fungi: an Advanced Treatise. Volume IVA. A Taxonomic Review with Keys: Ascomycetes and Fungi Imperfecti". Academic Press, New York & London.

Ainsworth, G. C., Sparrow, F. K., and Sussman, A. S. (eds.) (1973b). "The Fungi: an Advanced Treatise. Volume IVB. A Taxonomic Review with Keys: Basidiomycetes and Lower Fungi". Academic Press, New York & London.

Arzone, A. (1970). Reperti ecologici ed etologici di *Liodes cinnamomea* Panzer vivente su *Tuber melanosporum* Vittadini (Coleoptera Staphylinoidea). *Annali Fac. Sci. agr. Univ. Studi Torino* **5**, 317–357.

Ashe, J. S. (1984a). Generic revision of the subtribe Gyrophaenina (Coleoptera: Staphylinidae: Aleocharinae) with review of the described subgenera and major features of evolution. *Quaest. Ent.* **20**, 129–349.

Ashe, J. S. (1948b). Major features of the evolution of relationships between gyrophaenine staphylinid beetles (Coleoptera: Staphylinidae: Aleocharinae) and fresh mushrooms. *In* "Fungus–Insect Relationships: Perspectives in Ecology and Evolution". (Q. D. Wheeler and M. Blackwell, eds), pp. 227–255. Columbia Univ. Press, New York.

Ashe, J. S. (1986). Structural features and phylogenetic relationships among larvae of genera of gyrophaenine staphylinids (Coleoptera: Staphylinidae: Aleocharinae). *Fieldiana, Zool.* (N.S.) **30**, i–iv, 1–60.

Beier, M. (1952). Bau und Funktion der Mundwerkzeuge bei den Helodiden-Larven (Col.) *Proc. Int. Congr. Entomol., 9th*, Vol. 1, 135–138.

Benick, L. (1952). Pilzkäfer und Käferpilze: Ökologische und statistische Untersuchungen. *Acta zool. fenn.* **70**, 1–250.

Besuchet, C. (1972). Les Coléoptères Aculagnathides. *Rev. Suisse Zool.* **79**, 99–145.

Borden, J. H., and McClaren, M. (1972). Biology of *Cryptoporus volvatus* (Peck) Shear (Agaricales, Polyporaceae) in southwestern British Columbia: life history, development and arthropod infestation. *Syesis* **5**, 66–72.

Corner, E. J. H. (1953). The construction of polypores—1. Introduction: *Polyporus sulphureus, P. squamosus, P. betulinus* and *Polystictus microcyclus. Phytomorphology* **3**, 152–167.

Costa, C., Casari-Chen, S. A., and Teixeira, E. P. (1986). Larvae of Neotropical Coleoptera. XVI. Nosodendridae. *Revta bras. Ent.* **30**, 291–297.

Crowson, R. A. (1975). The evolutionary history of Coleoptera, as documented by fossil and comparative evidence. *Atti Congresso naz. Ital. Ent.* **10**, 47–90.

Crowson, R. A. (1981). "The Biology of Coleoptera". Academic Press, London.

Crowson, R. A. (1984). The associations of Coleoptera with Ascomycetes. *In* "Fungus–Insect Relationships: Perspectives in Ecology and Evolution". (Q. D. Wheeler and M. Blackwell, eds), pp. 256–285. Columbia Univ. Press, New York.

Dybas, H. S. (1976). The larval characters of featherwing and limulodid beetles and their family relationships in the Staphylinoidea (Coleoptera: Ptiliidae and Limulodidae). *Fieldiana, Zool.* **70**, 29–78.

Fogel, R., and Peck, S. B. (1975). Ecological studies of hypogeous fungi. I. Coleoptera associated with sporocarps. *Mycologia* **67**, 741–747.
Francke-Grosmann, H. (1967). Ectosymbiosis in wood-inhabiting insects. *In* "Symbiosis. Volume II. Associations of Invertebrates, Birds, Ruminants, and other Biota". (S. M. Henry, Ed.), pp. 141–205. New York.
Fukuda, A. (1941). Some ecological studies on *Cupes clathratus. Trans. nat. Hist. Soc. Formosa* **31**, 394–399.
Gordon, R. D. (1985). The Coccinellidae (Coleoptera) of America north of Mexico. *J. N.Y. Entomol. Soc.* **93**, 1–912.
Graves, R. C. (1960). Ecological observations on the insects and other inhabitants of woody shelf fungi (Basidiomycetes: Polyporaceae) in the Chicago area. *Ann. Entomol. Soc. Amer.* **53**, 61–78.
Hingley, M. R. (1971). The ascomycete fungus, *Daldinia concentrica*, as a habitat for animals. *J. Anim. Ecol.* **40**, 17–32.
Hisamatsu, S. (1962). On some beetles of the pouch fungus. *Ageha* **10**, 8–9.
Holloway, B. A. (1970). A new genus of New Zealand Anthribidae associated with lichens (Insecta: Coleoptera). *N.Z. J. Sci.* **13**, 435–446.
Holloway, B. A. (1982). Anthribidae (Insecta: Coleoptera). *Fauna of New Zealand* **3**, 1–269.
Hudson, G. V. (1934). "New Zealand Beetles and their Larvae". Ferguson and Osborne, Wellington.
Hudson, H. J. (1968). The ecology of fungi in plant remains above the soil. *New Phytol.* **67**, 837–874.
Hudson, H. J. (1972). "Fungal Saprophytism". E. Arnold, London.
Käärik, A. (1975). Succession of microorganisms during wood decay. *In* "Biological Transformation of Wood by Microorganisms". (W. Liese, ed.), pp. 39–51. Springer, Berlin, Heidelberg, New York.
Klimaszewski, J., and Peck, S. B. (1987). Succession and phenology of beetle faunas (Coleoptera) in the fungus *Polyporellus squamosus* (Huds.: Fr.) Karst. (Polyporaceae). *Can. J. Zool.* **65**, 542–550.
Lawrence, J. F. (1973). Host preference in ciid beetles (Coleoptera: Ciidae) inhabiting the fruiting bodies of Basidiomycetes in North America. *Bull. Mus. Comp. Zool.* **145**, 163–212.
Lawrence, J. F. (1977a). Coleoptera associated with an *Hypoxylon* species (Ascomycetes: Xylariaceae) on oak. *Coleopt. Bull.* **31**, 309–312.
Lawrence, J. F. (1977b). The family Pterogeniidae, with notes on the plylogeny of the Heteromera. *Coleopt. Bull.* **31**, 25–56.
Lawrence, J. F. (1980). A new genus of Indo-Australian Gempylodini with notes on the constitution of the Colydiidae (Coleoptera). *J. Entomol. Soc. Aust.* **19**, 293–310.
Lawrence, J. F. (1988). Coleoptera. *In* "Immature Insects". Vol. 2. (F. W. Stehr, ed.). Kendall-Hunt, Dubuque, Iowa.
Lawrence, J. F., and Britton, E. B. (in press). Coleoptera. *In* "Insects of Australia". Revised Edition. (C.S.I.R.O., ed.) Melbourne.
Lawrence, J. F., and Hlavac, T. F. (1979). Review of the Derodontidae (Coleoptera: Polyphaga) with new species from North America and Chile. *Coleopt. Bull.* **33**, 369–414.
Lawrence, J. F., and Newton, A. F., Jr. (1980). Coleoptera associated with fruiting bodies of slime molds (Myxomycetes). *Coleopt. Bull.* **34**, 129–143.
Lawrence, J. F., and Newton, A. F., Jr. (1982). Evolution and classification of beetles. *Annu. Rev. Ecol. Syst.* **13**, 261–290.
Lawrence, J.F., and Stephan, K. (1975). The North American Cerylonidae (Coleoptera: Clavicornia). *Psyche* **82**, 131–166.

Martin, G. W., and Alexopoulos, C. J. (1969). "The Myxomycetes". University of Iowa Press, Iowa City, Iowa.
Matthewman, W. G., and Pielou, D. P. (1971). Arthropods inhabiting the sporophores of *Fomes fomentarius* (Polyporaceae) in Gatineau Park, Quebec. *Can. Entomol.* **103**, 775–847.
Miller, A. (1961). The mouth parts and digestive tract of adult dung beetles (Coleoptera: Scarabaeinae), with reference to the ingestion of helminth eggs. *J. Parasitol.* **47**, 735–744.
Naomi, S.-I. (1985). The phylogeny and higher classification of the Staphylinidae and their allied groups (Coleoptera, Staphylinoidea). *Esakia* **23**, 1–27.
Newton, A. F., Jr. (1984). Mycophagy in Staphylinoidea (Coleoptera). *In* "Fungus–Insect Relationships: Perspectives in Ecology and Evolution". (Q. D. Wheeler and M. Blackwell, eds.), pp. 302–353. Columbia Univ. Press, New York.
Nuss, I. (1975). "Zur Ökologie der Porlinge: Untersuchungen über die Sporulation einiger Porlinge und die an ihnen gefundenen Käferarten". (Bibliotheca Mycologica. Band 45. Vaduz.
Pakaluk, J. (1984). Natural history and evolution of *Lycoperdina ferruginea* (Coleoptera: Endomychidae) with descriptions of immature stages. *Proc. Entomol. Soc. Wash.* **86**, 312–325.
Pal, T. K., and Lawrence, J. F. (1986). A new genus and subfamily of mycophagous Bothrideridae (Coleoptera: Cucujoidea) from the Indo-Australian region, with notes on related families. *J. Entomol. Soc. Aust.* **25**, 185–210.
Paviour-Smith, K. (1960). The fruiting-bodies of macrofungi as habitats for beetles of the family Ciidae (Coleoptera). *Oikos* **11**, 1–71.
Paviour-Smith, K. (1961). Insect succession in the "birch-bracket fungus" *Polyporus betulinus. Proc. Int. Congr. Entomol., 11th, 1960* Vol. 1, 792–796.
Peterson, R. H., ed. (1971). "Evolution in the Higher Basidiomycetes: an International Symposium". University of Tennessee Press, Knoxville.
Pielou, D. P., and Verma, A. N. (1968). The arthropod fauna associated with the birch bracket fungus, *Polyporus betulinus*, in eastern Canada. *Can. Entomol.* **100**, 1179–1199.
Ponomarenko, A. G. (1969). Historical development of the Coleoptera-Archostemata. *Tr. paleont. Inst.* **125**, 1–240.
Rehfous, M. (1955). Contribution à l'élude des insectes des champignons. *Mitt. Schweiz. Entomol. Ges.* **28**, 1–106.
Roman, E. (1970). Observations sur divers Coléoptères évoluant dans les polypores (champignons Basidiomycetes). *Bull. Mens. Soc. Linn. Lyon* **39**, 300–307.
Schedl, K. E. (1962). Forstentomologische Beiträge aus dem Kongo. Ráuber und Kommensalen. *Ent. Abh. Mus. Tierk. Dresden* **28**, 37–84.
Scheerpeltz, O., and Höfler, K. (1948). "Käfer und Pilze". Verlag für Jugend und Volk, Vienna.
Shigo, A. L. (1967). Succession of organisms in discoloration and decay of wood. *In* "International Review of Forestry Research II". (J. A. Romberger and P. Mikola, eds), pp. 237–299. New York.
Steiner, W. E., Jr. (1984). A review of the biology of phalacrid beetles (Coleoptera). *In* "Fungus–Insect Relationships: Perspectives in Ecology and Evolution". (Q. D. Wheeler and M. Blackwell, eds), pp. 424–445. Columbia Univ. Press, New York.
Stribling, J. B., and Seymour, R. I. (in press). Evidence of mycophagy in Ptilodactylidae (Coleoptera: Dryopoidea) with notes on phylogenetic implications *Coleopt. Bull.*
Vit, S. (1977). Contribution à la connaissance des Eucinetidae (Coleoptera). *Rev. Suisse. Zool.* **84**, 917–935.

Watt, J. C. (1974). Chalcodryidae: a new family of heteromerous beetles (Coleoptera: Tenebrionoidea). *J. R. Soc. N.Z.* **4**, 19–38.

Wheeler, Q. D. (1984). Evolution of slime mold feeding in leiodid beetles. *In* "Fungus–Insect Relationships: Perspectives in Ecology and Evolution". (Q. D. Wheeler and M. Blackwell, eds.), pp. 446–477. Columbia Univ. Press, New York.

2

Fungivory: Fungi, Insects and Ecology

I. HANSKI

I. INTRODUCTION

Fungal fruiting bodies present a mixture of resource characteristics for fungivorous insects. Sporophores resemble higher plants, from an insect's viewpoint, in their immobility and lack of physical defences. Yet the chemical composition of fungal cell walls, and hence their sporophores, is so different from the chemical composition of vascular plants that herbivorous and fungivorous insects require quite different digestive capabilities (Kukor and Martin, 1987), the fungivores being in this respect more closely related to detritivorous than herbivorous insects (Martin *et al.*, 1981). Cellulose, lignin and pectin are unknown in the higher macrofungi (Aronson, 1965); instead, the fungivorous insect needs the capacity to digest chitin and non-cellulosic β-(1,3)- and β-(1,6)-glucans, and it is well advised to learn to use urea as a source of nitrogen (Martin, 1979). Provided that the right enzymes are available, fungal tissue has a high nutritional value (Martin, 1979), though substantial variation exists between species, e.g. the nitrogen concentration of whole sporophores varies from less than 1% to more than 10% (Vogt and Edmonds, 1980). Another difference with higher plants is the markedly unpredictable occurrence of sporophores in space and time, even if much variation occurs in this respect in fungi as well as in vascular plants.

Fungivorous insects provide good examples and pose problems for two current themes in ecology, around which this chapter will revolve. The first one is the supposedly evolved defensive function of plant secondary compounds (Feeny, 1975; Cates and Rhoades, 1977; Rhoades, 1979; Haukioja, 1980; and many others), and the "predictability" (Rhoades and Cates, 1976) and "apparency" (Feeny, 1976) of vascular plants to herbivores. These ideas at first sight apply equally well to fungi and fungivores as to higher plants and herbivores. Nonetheless, the concept of a sporophore trying to "escape from" and defending itself against fungivorous insects may be misleading or unnecessary—this question will be touched on in Section III. Sidestepping issues concerning the evolutionary past of the biochemical and structural differences between fungal sporophores, the large numbers of macrofungi often growing together provide alternative hosts to fungivores and pose challenges to ecologists and geneticists working on host selection, insect life histories, genetic diversity in insect populations versus host diversity, sympatric speciation, etc. The comparison between vascular plants and herbivores on the one hand and fungi and fungivores on the other is an attractive prospect for research in this perspective, and an approach that has remained little studied, perhaps because of the greater practical difficulties of identifying fungi and fungivores than vascular plants and herbivores (e.g. Graves and Graves, 1985). It should also be noted that while the rearing of some fungivorous insects is straightforward (e.g. many Ciidae (H25.52); Klopfenstein, 1971) some others pose practical problems (e.g. many Mycetophilidae (H28.15); Väisänen, 1981).

The second major topic dealt with in this chapter is of interest to population ecologists: the dynamics of populations and communities in patchy environments (Horn and MacArthur, 1972; Slatkin, 1974; Levin, 1974, 1978; Hanski, 1981, 1987a; Atkinson and Shorrocks, 1981; Shorrocks and Rosewell, 1988; Chesson, 1986; and others). Section IV comprises a brief review, applied to fungivores, of some general life history and population dynamic questions about survival on a markedly ephemeral and patchy resource, while Section V is devoted to the ecology of guilds of co-occurring species. Meanwhile, before turning to the insects themselves, let us consider the characteristics of their habitat. Fungal fruiting bodies encompass an unrivalled range of heterogeneity amongst patchy and ephemeral microhabitats at several spatial and, especially, temporal scales. This is the perspective from which this chapter has been written, by an ecologist rather than an entomologist, in the hope of introducing some general ecological thinking to entomological observations and vice versa.

II. HABITAT VARIABILITY

Fungivorous insects are faced with an exceptionally variable resource. Habitat variability is manifested in the quality of the resource: in different fungal species and different stages of sporophore development; and in the temporal and spatial occurrence of the sporophores.

This chapter is limited to the "macrofungi", and principally those two groups of Hymenomycetes with by far the largest numbers of insect associates: the gilled mushrooms—Agaricales, *sensu lato* (F5.10–5.12)—and the pored mushrooms, Polyporaceae (Aphyllophorales (F5.8)). Luckily for ecologists, the two groups provide a contrasting pair of microhabitats for insects: while Agaricales sporophores are ephemeral to very ephemeral, Polyporaceae sporophores last from months (annual species) to several years (perennial species) (Table I). Lawrence and Newton (1980), Blackwell (1984) and Wheeler (1984) have summarized the meagre data available for insects associated with Myxomycetes (F1), and Crowson (1984) presents a review of beetles associated with Ascomycetes (F4) (see also Lawrence, 1977).

Temperate forests have typically several hundred species of Agaricales (e.g. Ulvinen, 1976) and tens of species of Polyporaceae (e.g. Kotiranta and Niemelä, 1981), though many of them have such small (especially Agaricales) or such dry and hard (Polyporaceae) sporophores, that only small numbers of insects, if any, are able to colonize them (Buxton, 1960). The small sporophores of *Omphalina* (Tricholomataceae (F5.10)) have at most a few quickly developing polyphagous species, e.g. *Exechia fusca*

TABLE I. Characteristics of gilled and pored mushrooms as insect microhabitats. Modified from Ashe (1984).

	Pored mushrooms: Polyporaceae (F5.8)	Gilled mushrooms: Agaricales (F5.10)
Durational stability	Months to years	Days to weeks
Spore production	Inside tubes, over a long period	On surface of gills, large numbers over a short period
Major problems for fungivores	Often dry and hard	Often small and very ephemeral

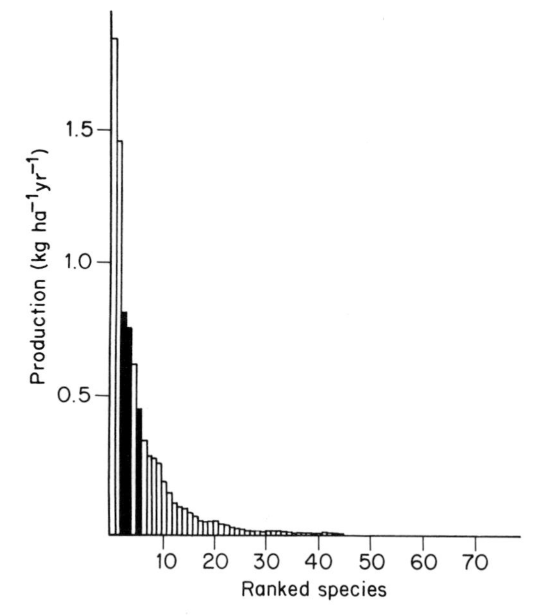

Fig. 1. Distribution of annual sporophore production in 79 species of Agaricales *sensu lato* in a spruce forest in central Finland. The shaded columns are *Leccinum versipelle, L. scabrum*, and *Suillus variegatus* (all Boletaceae (F5.11)). Data from Ohenoja (1974).

(H28.15) or *Drosophila* (H28.36), often just a single larva per sporophore (Hackman and Meinander, 1979). *Fomes everhartii* (F5.8) has extremely hard conks and is little or not at all used by insects in nature. Klopfenstein (1971), however, managed to feed *Hadraule blaisdelli* (H25.52) on the powdered tissue of this species, indicating that hardness is a major reason for lack of infestation in nature. It should also be remembered that although the number of macrofungi at one site may be very large, most of the suitable sporophores and most of the resource is accounted for by a

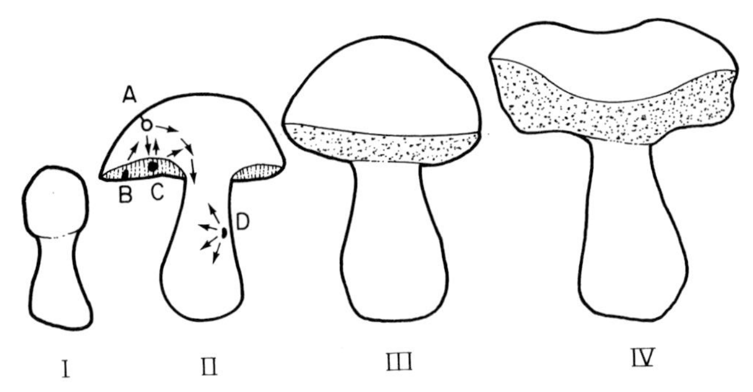

Fig. 2. The four stages in the development of *Boletus* (F5.11) sporophores according to Hackman and Meinander (1979). The drawing of the stage II sporophore has information on the sites of oviposition by *Pegomya* flies (H28.39) (from Hackman, 1976): (A) Eggs deposited in a cluster through a hole in the cap surface (*P. notabilis*); (B) Eggs deposited singly in the spore layer (*P. winthemi*); (C) Larger clutches in the spore layer (several species); and (D) Eggs deposited in the stipe (*P. incisiva*).

small number of species, e.g. in the example in Fig. 1 four species accounted for more than 50% of the sporophore production of the gilled mushrooms in a spruce forest in Finland.

The development of single sporophores from their first appearance to final decay has been classified into a number of stages, e.g. into four stages by Hackman and Meinander (1979) (Fig. 2; see also Bruns, 1984). Different species have different growth rates, with an increase of 1 cm per day in cap diameter being typical for large and medium-sized gilled mushrooms (Hiukko, 1978). The rate of sporophore development is primarily dependent on soil and litter temperatures (Hiukko, 1978). Different species of fungivorous insects colonize sporophores at characteristic times during their development, undoubtedly depending on exactly what resource in the sporophore the species is exploiting (Section III). Large sporophores may provide several microhabitats for fungivores; e.g. in the anthomyiid genus *Pegomya* (H28.39) breeding in Boletaceae (F5.11) one species (*P. incisiva*) has specialized in the stipe, while most species use various parts of the large cap (Fig. 2).

A. Spatial Variation

The spatial distribution of fruiting bodies in many species of fungus is markedly aggregated, the sporophores growing in tight clusters, while in others the sporophores occur singly and more or less haphazardly. A further aspect of spatial variation is the predictability of occurrence at certain sites. Such variation is small in the mycorrhizal species associated with long-living

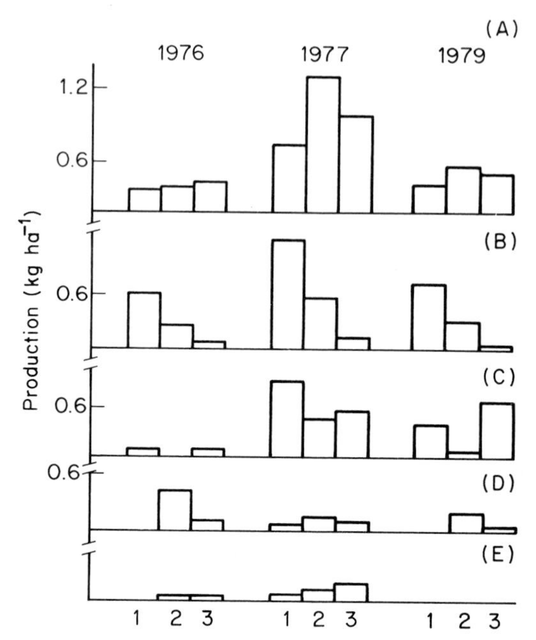

Fig. 3. Annual sporophore production in five species of edible mushrooms in three forest types in 3 years. The forest types are: (1) barren pine heaths; (2) dry pine forests and (3) moist coniferous and mixed forests. The species are (A) *Suillus variegatus* (F5.11); (B) *Lactarius rufus* (F5.12); (C) *Leccinum versipelle* and *L. vulpinum* (F5.11); (D) *Russula paludosa* (F5.12); and (E) *Lactarius trivialis* (F5.12). Data from Ohenoja and Koistinen (1984).

trees. Differences in sporophore microdistribution is one level of resource variation typical for fungi, but practically neglected by ecologists and entomologists working with fungivorous insects. Bruns (1984) has suggested that gregarious fruiting in *Suillus* (F5.11) may be the reason why only a few individuals of *Mycetophila fisherae* (H28.15) usually occur in *Suillus* sporophores while tens emerge from other host species.

Many fungal species have restricted macrohabitat distribution (Fig. 3). The mycorrhizal species belong to this category as they cannot develop where their vascular host plants do not occur. Many Polyporaceae (F5.8) are more or less restricted to old forests (Kotiranta and Niemelä, 1981).

B. Temporal Variation

Three scales of temporal variation are of obvious significance to fungiv-

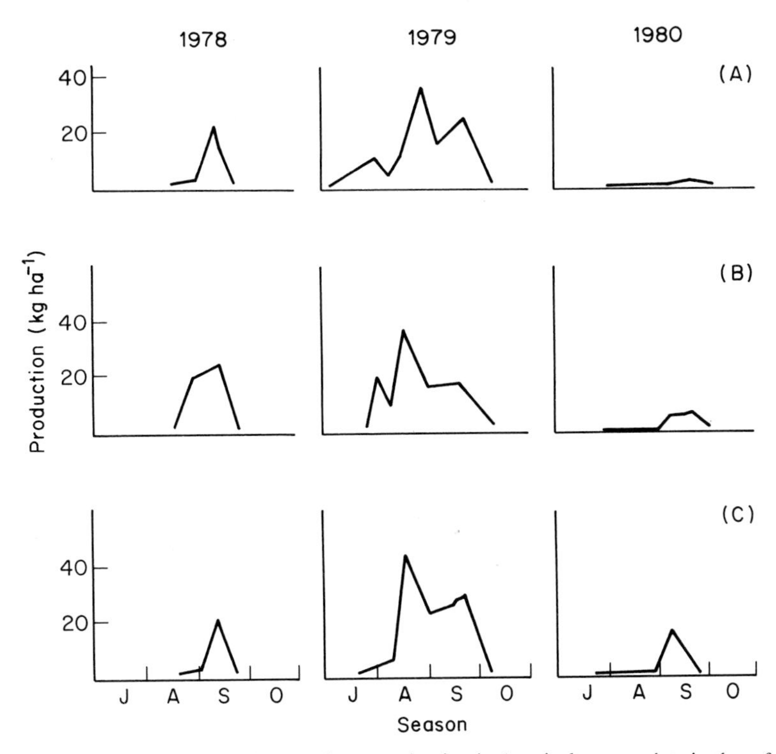

Fig. 4. Seasonal distribution of sporophore production in Agaricales *sensu lato* in three forest stands (A–C) in central Finland in 3 years: J, July; A, August; S, September; O, October. Data from Ohenoja (1983).

orous insects: durational stability of single sporophores, their seasonal occurrence (phenology), and year-to-year variability in sporophore production (Fig. 4). Species differ markedly in all these respects. It has already been observed that the gilled mushrooms (F5.10, etc.) have generally shorter durational stability than the pored ones (Table I). Amongst the former, Richardson (1970) reported variation from 4 to 19 days, while Polyporaceae (F5.8) sporophores last from several weeks to many years.

One could expect that species with small sporophores would tend to be more abundant and have shorter durational stability than species with large sporophores. The author explored these relationships using the data of Richardson (1970) for 10 to 15 species from a pine plantation in Scotland. No significant correlations were found, but this may be because of the small numbers of species included. It would be interesting to repeat such analyses for larger sets of data.

III. HOST SELECTION

A. Monophagy Versus Polyphagy

Fungivorous insects are generally very polyphagous. A survey of Diptera breeding in Agaricales *sensu lato* (F5.10–5.12) shows that the number of host species increases rapidly with sample size (Fig. 5), suggesting that many records of specialist fungivores may be artefacts of insufficient sampling. Oligophagous and even monophagous fungivores do exist (e.g. *Pegomya* (H28.39), Fig. 5), but they are uncommon and tend to be restricted to single genera rather than species of host fungi (Buxton, 1960; Hackman, 1976, 1979; Russell-Smith, 1979).

Lack of specialization characterizes not only species of fungivores, but individuals as well. Jaenike (1978a) has experimentally shown that individuals of *Drosophila falleni, D. putrida,* and *D. testacea* (H28.36) use sporophores of the different host species more or less indiscriminately. Electrophoretic studies on *D. falleni* breeding in seven morphologically, chemically, and ecologically distinct species of *Amanita* (F5.10) revealed that only 2–5% of the genetic diversity at four structural loci could be

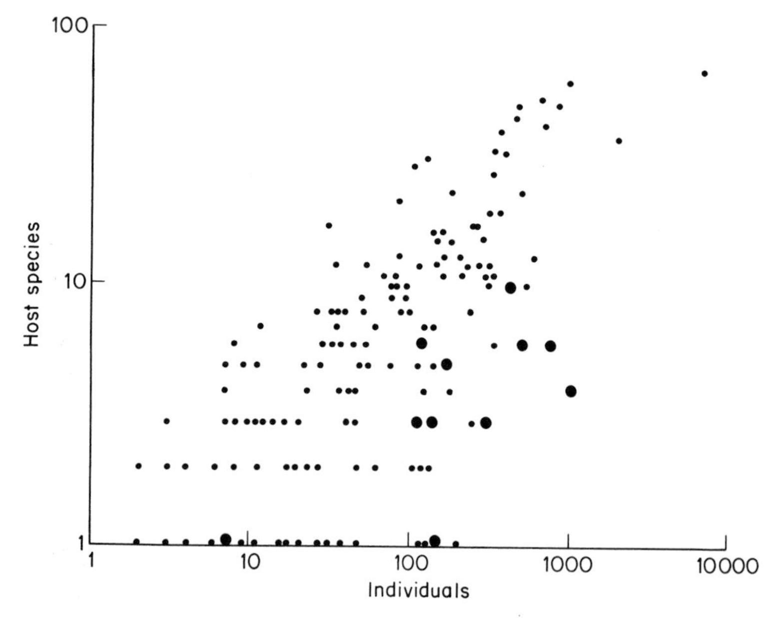

Fig. 5. Numbers of host fungi recorded for Diptera as a function of the total number of individuals reared. The large dots are *Pegomya* species (Anthomyiidae (H28.39)). Data from Hackman and Meinander (1979).

accounted for by differences between flies reared from different host species (Jaenike and Selander, 1979; see also Lacy, 1982, 1983), suggesting that host race formation is unlikely in polyphagous fungivores. In spite of these results, Lacy (1982) found the more polyphagous species to have more electrophoretically detectable variation than the less polyphagous species, in support of the niche-variation hypothesis of Van Valen (1965). Jaenike and Grimaldi (1983) report on genetic variation for host preference in a population of *D. tripunctata* (H28.36), but it is noteworthy that this species is not strictly fungivorous, as it commonly uses, for example, decomposing fruits.

Why is the vast majority of fungivores polyphagous? To clarify the possible explanations, let us look at a simple model (for a somewhat different formulation see Levins and MacArthur, 1969). Let N be the total number of sporophores available, and denote by p_1 the proportion of the fungal species currently used by the fungivore. The expected number of offspring F by one female may be modelled using Holling's (1959) "disc equation",

$$F = Bp_1N\ (p_1N + C), \tag{1}$$

where B is maximal fecundity and C is a parameter related to the dispersal ability and longevity of the fungivore. Let us now consider another female (genotype) with a wider host selection, using the fraction $p_1 + p_2 < 1$ of the fungal species. Widening of host selection increases the number of acceptable host sporophores, but with the cost that some unsuitable ones may be oviposited on in error (Levins and MacArthur, 1969). Let q denote the probability that a species in the extended diet (fraction p_2 of the host species) is suitable. All the "old" host species (fraction p_1 of the species) are assumed to be suitable. The expected number of offspring by the second female is given by

$$F = B(p_1 + qp_2)N/((p_1 + p_2)N + C). \tag{2}$$

Selection favours the genotype with a wider host selection if

$$1/p_1 > (1 - q)N/qC. \tag{3}$$

Figure 6 illustrates this condition for three values of q. Selection towards increasing polyphagy occurs: (1) when a large fraction of the currently unused host species is in fact suitable (q large); (2) when the density of sporophores is low (N small); and (3) when the dispersal ability of the insect is poor and/or its expected life time is short (C large). Note that selection also depends on the current degree of polyphagy, p_1 (Fig. 6). This model makes a number of simplifications: single eggs are laid on sporophores, all host species are equally abundant, and all the suitable hosts are equally

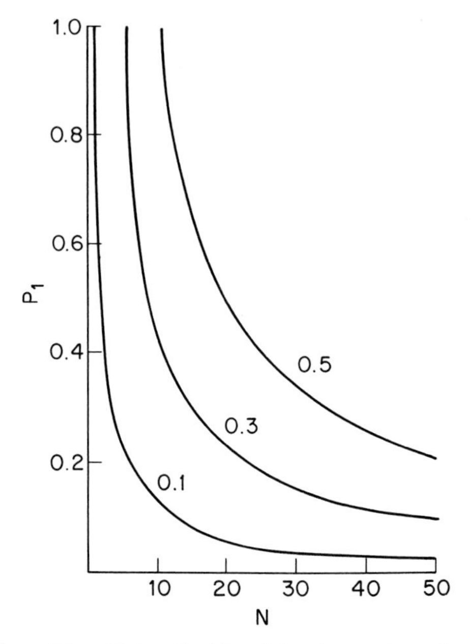

Fig. 6. Illustration of condition (3) for $C = 10$ and for three values of q. The horizontal axis gives the number of sporophores (N) and the vertical axis the current degree of polyphagy (p_1). Selection favours increasing polyphagy on the left side and decreasing polyphagy on the right side of the lines.

suitable. There appear to exist no data for polyphagous fungivores on the important third point.

Within this framework, there appear two major explanations for polyphagy in fungivores. The quantity hypothesis states that the number of sporophores (N) is often small and selects for polyphagy, while the quality hypothesis postulates that there are only minor differences in the quality of different host species (q large) making widening of host selection easy.

B. The Quantity Hypothesis

This hypothesis has often been stated in the form that the ephemeral and unpredictable occurrence of sporophores selects for polyphagy in fungivores (e.g. Jaenike, 1978b). In support of this hypothesis, the degree of polyphagy in North American fungivorous Drosophilidae (H28.36) decreases with increasing durational stability of the host sporophores (Fig. 7; see also Lacy, 1984a). In Europe, the relative frequency of polyphagous Diptera (H28) decreases with the size of the sporophore (Table V), which may be related to durational stability and unpredictable occurrence.

Oligophagous fungivores are more frequent in species colonizing Polyporaceae (F5.8) than in species colonizing Agaricales *sensu lato* (F5.10–5.12) with more ephemeral sporophores (Table I). Examples of the former include several ciid beetles (H25.52) (Paviour-Smith, 1960; Lawrence, 1973), the beetles *Bolitotherus cornutus* (H25.60) (Heatwole and Heatwole, 1968), and *Episcaphula australis* (H25.39) (Hawkeswood, 1986), the moth *Morophaga cryptophori* (H32.4) (Lawrence and Powell, 1969), and the fly *Mycodrosophila claytonae* (H28.36) (Lacy, 1984a). Possible examples of the latter are *Bolitophila hybrida* (H28.15), apparently specializing on *Paxillus involutus* (F5.11) in Finland (Hackman and Meinander, 1979), and the beetle *Atomaria fimetarii* (H25.36) specializing on *Coprinus comatus* (Benick, 1952). The species of Myxomycetes (F1) that are most often used by beetles and other arthropods are not closely related taxonomically but share the following characteristics: common occurrence, long fruiting season, persistent sporangia, and numerous spores in large aethalia (Blackwell, 1984)—all factors making the host less ephemeral from the insects' viewpoint.

Two further comparisons are illuminating. In *Drosophila* with both fungivores and species breeding in other microhabitats, specialists are rare or unknown in the fungivores but common in species exploiting predictable resources (Jaenike, 1987b) such as fermenting material in abundant tree species in Hawaii (Hee, 1968). And in general, herbivorous insects, most of which use more predictable resources than fungivores, stand in sharp contrast to the latter. The incidence of monophagy is often high in entire faunas of herbivorous insects. For example, of the some 250 species of herbivorous Agromyzidae (H28) in Britain not less than 60% are thought to be strictly monophagous (Spencer, 1972; Lawton and Price, 1979).

In conclusion, the quantity hypothesis is supported by a range of

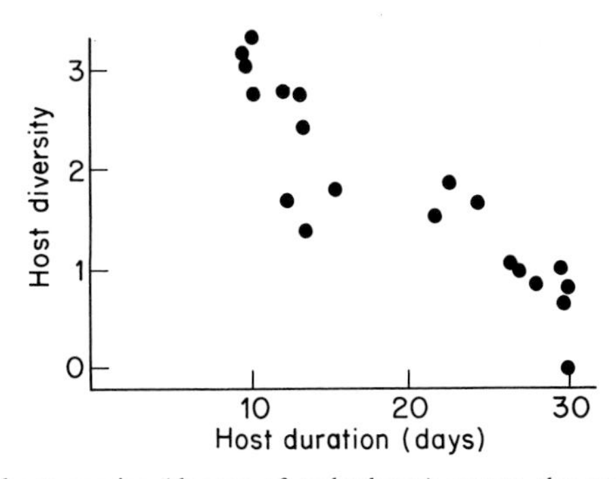

Fig. 7. Diversity of host species (degree of polyphagy) versus the average duration of the sporophores used by fungivorous Drosophilidae (H28.36) in North America. Data from Lacy (1984a).

observations, and there is little doubt that the unpredictable occurrence of host fungi, most of which are in any case uncommon (Fig. 1), contributes to polyphagy in fungivores.

C. The Quality Hypothesis

Assessing the merits of the quality hypothesis is complicated by variation in the actual food material used by insects and their larvae in the sporophores. Bruns (1984) distinguishes four major trophic levels in fungal-breeding insects: primary fungivores, secondary fungivores, detritivores, and predators. The first three groups represent a more or less arbitrary division of a continuum from strictly fungivorous species that use only fresh host tissue (primary fungivores) to species that increasingly prefer decaying to fresh fruiting bodies (secondary fungivores) to species that breed in a wide variety of decomposing organic matter, including mushrooms (detritivores). Secondary fungivores and detritivores may largely feed on the microorganisms and various products of fermentation present in the decomposing sporophores (e.g. Kearney and Shorrocks, 1981). Even the predators do not form a distinct trophic group, as many species (e.g. *Mydaea* and *Muscina assimilis* (H28.40)) are predatory only in their third larval instars. Incidentally, it is probable that Bruns' (1984) classification is related to the evolution of fungivory in insects, detritivores representing the primitive and primary fungivores the most evolved feeding habit (Hackman and Meinander, 1979).

If the quality hypothesis is not correct, primary fungivores may be expected to be more specialized than secondary fungivores and saprophagous species, because any chemical differences between fungal species in the food material actually used by the insects should be greater for primary than for secondary fungivores. This prediction is not easy to test with large numbers of species as the feeding habits of most "fungivores" are not sufficiently well known. An extreme example of monophagy in a primary fungivore is *Agathomyia wankowiczi* (H28.21), a fly whose larvae develop in galls on *Ganoderma applanatum* (F5.8) (Eisfelder and Herschel, 1966; Kotiranta and Niemelä, 1981). Turning to more orthodox species, Kimura (1980) found that in *Drosophila* (H28.36) host specialization was greater in three species of primary fungivores (*Hirtodrosophila* spp.) than in the fungus-breeding species of the *D. immigrans* radiation, in which the species retain their original saprophagous feeding habits. Nonetheless, larvae of even the *Hirtodrosophila* species preferred decaying to fresh mushrooms, which Kimura (1980) considers as an adaptation to the decomposition of sporophores that inevitably occurs during larval development (7–9 days in four North American species; Grimaldi, 1985). Another unusually oligophagous group of primary fungivores is the dozen *Pegomya* species (H28.39) breeding in a few species of Boletaceae (F5.11) (Fig. 5; Section V).

Although the degree of polyphagy is generally high in fungivores, it is not uniformly high in all species assemblages associated with different host fungi. This is a reflection of the fact that host associations in fungivores are complex but not random (Ashe, 1984). Table II presents an analysis based on Hackman and Meinander's (1979) data on Diptera breeding in Agaricales *sensu lato* (F5.10–5.12) in Europe (most data come from northern Europe). Polyphagous species are here defined as species reared from 20 or more host species. As the percentage of polyphagous species thus defined decreases with increasing total number of species recorded from the host, probably because of incomplete records for uncommon species, the degree of polyphagy in the different host assemblages was calculated as the deviation from the regression of polyphagous against all species. Differences between the genera are highly significant (Table II), with *Amanita* (F5.10) having the most and *Leccinum* (F5.11) the least polyphagous assemblage of Diptera. If toxic chemicals in the sporophores were an important barrier to polyphagous species, the result in Table II is diametrically the opposite to one's intuitive expectation. Genera with the highest proportion of polyphagous species include *Amanita* and *Cortinarius* (F5.10), well known to include species toxic to mammals, while Russulaceae (F5.12) and Boletaceae (F5.11) with mostly edible (and often delicious) mushrooms have the smallest proportions of polyphagous species. Small genera with perhaps biochemically relatively distinct species

TABLE II. The degree of polyphagy in Diptera breeding in nine genera of Agaricales *sensu lato* (F5.10–5.12). Fungal species with at least five Diptera recorded are included. For calculations see the text. Differences between the genera are highly significant (ANOVA, $P < 0.008$). Data from Hackman and Meinander (1979).

Genus	Family	Number of species	Polyphagy	
			Mean	SE
Amanita	Amanitaceae	6	1.96	0.59
Hygrophorus	Hygrophoraceae	4	1.15	0.73
Cortinarius	Cortinariaceae	5	0.79	0.65
Russula	Russulaceae	17	0.21	0.35
Boletus	Boletaceae	5	0.12	0.65
Tricholoma	Tricholomataceae	6	0.11	0.59
Lactarius	Russulaceae	14	− 0.21	0.39
Suillus	Boletaceae	5	− 0.31	0.65
Leccinum	Boletaceae	5	− 2.12	0.73
Small genera	Mixed	37	− 0.31	0.24

have fewer than the average number of polyphagous species, but the deviation is not significant (Table II). It may be added that the abundance of the host species does not appear to explain well the degree of polyphagy; Cortinariaceae (F5.10) and Boletaceae (F5.11) include several of the most abundant species of Agaricales *sensu lato* in temperate forests, yet they lie at the opposite ends of the continuum in Table II.

Assuming that the chemical compounds present in the sporophores have an important effect on host selection by fungivores, we would expect that the insect assemblages colonizing different families of fungi with more or less different chemical composition would be different. Such differentiation has been observed in some herbivorous insects (Gilbert, 1977; Gilbert and Smiley, 1978; Berenbaum, 1981). The results of Hackman and Meinander (1979) on Diptera breeding in Agaricales were again used to test this prediction. The result shows little if any separation between the different host families, with the exception of Boletaceae (Fig. 8), which is considered

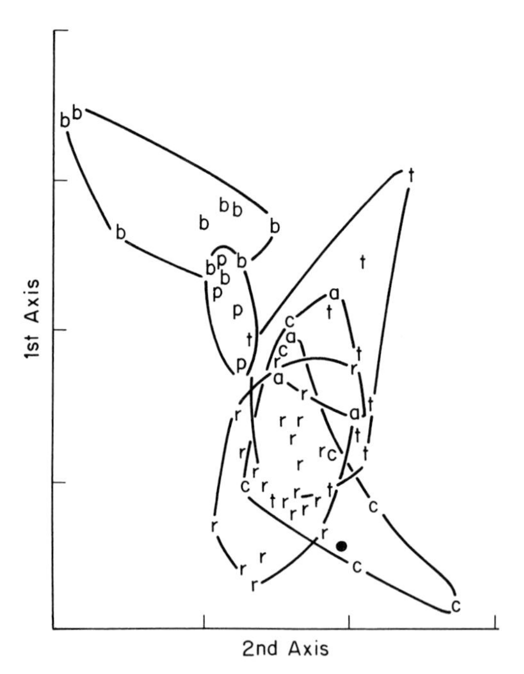

Fig. 8. The first two axes of DECORANA ordination of host fungi (Agaricales *sensu lato*) on the basis of their Dipteran assemblages. Each character represents one fungal species. The symbols are: (B) Boletaceae (F5.11); (T) Tricholomataceae (F5.10); (C) Cortinariaceae (F5.10); (R) Russulaceae (F5.12); (A) Amanitaceae (F5.10); and (P) Paxillaceae (F5.11) and Gomphidiaceae (F5.11) (the point is *Paxillus atrotomentosus*). Data from Hackman and Meinander (1979).

by many mycologists to form the separate order Boletales (F5.11). Paxillaceae and Gomphidiaceae are often included in Boletales, and these families lie between Boletales and Agaricales in the present ordination (Fig. 8), with the exception of *Paxillus atrotomentosus*, which for some reason has a quite different fungivore assemblage (Eisfelder, 1961; Hackman and Meinander, 1979).

Boletaceae have a unique chemistry (Bruns, 1984), and their sporophores are colonized by several species of *Pegomya* (H28.39), a guild of primary fungivores that are often so abundant that they may competitively exclude other, more polyphagous species. However, biochemistry is not the only possible explanation of the relatively distinct fauna of Boletaceae. Their sporophores have relatively predictable occurrence as most species are mycorrhizal; the sporophores are long lasting (slowly developing), large, and have a unique pored structure in Agaricales *sensu lato*, all of which are factors that may facilitate specialization. Paviour-Smith (1960) has observed that in the primary fungivorous beetles of the family Ciidae (H25.52) living in Polyporaceae (F5.8) host selection is largely based on a structural characteristic, the complexity of hyphal structure and weaving. Perhaps a fair comparison here is the relatively distinct macrolepidopteran fauna of conifers amongst trees and shrubs (Neuvonen and Niemelä, 1983), possibly due to the specialized structure of conifer leaves.

In summary, there is little indication of great importance of fungal chemistry in host selection in this broad survey of host associations. Ashe (1984) reached a similar conclusion from a careful analysis of host preferences in the gyrophaenine staphylinids (H25.8). These results support the quality hypothesis: chemical differences between host species are not a severe barrier to widening of host selection; structural differences if any may be more important. This conclusion however begs the question of why the chemical differences that presumably exist should be so unimportant.

D. Coping with Fungal Chemistry

It is clear that many of the secondary metabolites present in fungi are harmful or toxic to insects in general, though little specific information is available (Kukor and Martin, 1987). Some fungi may be free of fungivores because of specific chemicals, e.g. L-dopa in *Strobilomyces floccopus* (F5.11) (Bruns, 1984) and (possibly) anthraquinone pigments in *Dermocybe* (F5.10) (Besl and Blumreisinger, 1983). In this context, the high degree of polyphagy in species breeding in *Amanita* (F5.10) (Table II), for example, raises an apparent difficulty, exemplified by Jaenike's *et al.* (1983) study on the α-amanitin tolerance in six species of *Drosophila* (H28.36). Three non-fungivorous species did not survive even a low concentration of the chemical, while three fungivores were little if at all affected. The puzzling question is, as Jaenike *et al.* (1983) point out, why the highly polyphagous species should be so little affected even though only a small fraction of their host species contains α-amanitin at all?

Let us make a digression to insect resistance to pesticides. The pesticide industry is becoming increasingly concerned about the extent of resistance, and about the growing number of species that are simultaneously resistant to two to five classes of chemicals (DDT, cyclodienes, organophosphates, carbamates, and pyrethroids; Georghiou, 1986). Negative cross-resistance, the dream of the pesticide user, appears uncommon in comparison with the reverse, accumulation of several mechanisms of resistance in the same population. A case in point is resistance to pyrethroids that has evolved rapidly on the foundation of DDT resistance. A semirecessive gene, *kdr*, which has been detected as one of the components of DDT resistance, is selected for and provides protection against pyrethroids (Georghiou, 1986; Georghiou and Taylor, 1986; and references therein). Such rapidly evolving pesticide resistance may be often based on existing, general detoxification mechanisms. Such mechanisms include microsomal (mixed-function) oxidases (Terriere, 1984) that have been found to have higher activity in polyphagous than in monophagous Lepidoptera (H32) (Krieger *et al.* 1971), and *trans*-epoxide hydrolase, with a higher activity in a polyphagous herbivorous mite than in a predatory mite (Mullin *et al.*, 1982). Versatility of the microsomal oxidase system is increased by the possibility of induction, enhanced activity of the detoxification system in the presence of a chemical stimulus (Terriere, 1984).

I suggest that fungivorous insects can be regarded as a somewhat analogous group to "pest" insects in agricultural habitats. In the fungivores, spatial and temporal unpredictability of the resource is the force that primarily selects for polyphagy, and exposes the populations to a wide range of chemicals. Detoxification mechanisms evolve against several kinds of chemicals and, although selection for further increase in polyphagy should decrease with increasing polyphagy (Eq. (3) and Fig. 6), the possibility of accumulating detoxification mechanisms on the foundation of existing mechanisms may make it increasingly easy to widen host selection.

This run-away process of increasing cross-resistance may be modelled by letting q in Eq. (1) to be a function of p_1 instead of a constant; p_1 is now assumed to be correlated with the range of existing detoxification mechanisms. Amongst a large number of possible relationships between q and p_1 the following one is used here,

$$q = \alpha + \beta p_1^{\gamma}, \tag{4}$$

where $\alpha + \beta < 1$. We may further simplify the analysis by assuming that

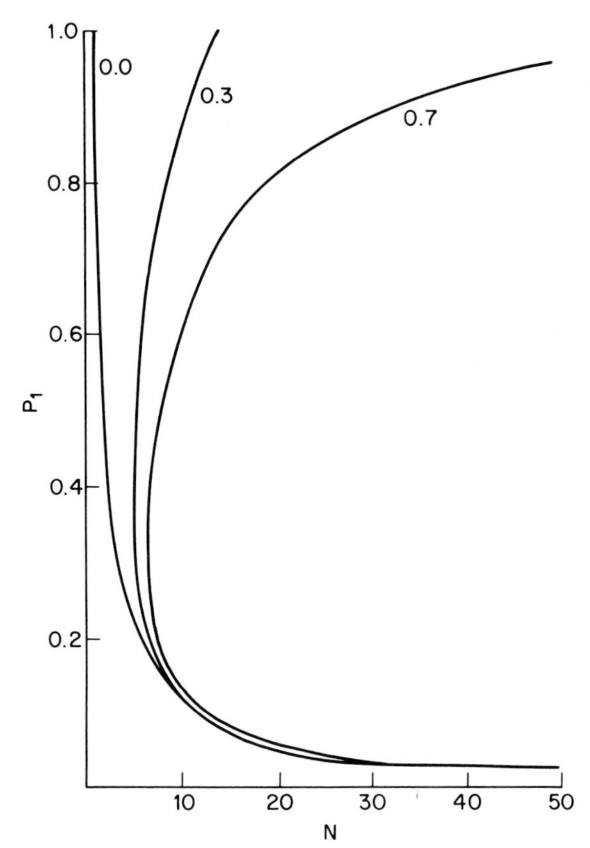

Fig. 9. Illustration of condition (4) for three values of β, and for $\alpha = 0.1$ and $C = 10$. The horizontal axis gives the number of sporophores (N) and the vertical axis the current degree of polyphagy (p_1). Selection favours increasing polyphagy on the left side and decreasing polyphagy on the right side of the lines. Compare with Fig. 6.

$\gamma = 2$, which retains qualitatively the intended relationship between q and p_1. Figure 9 illustrates how this modification alters the predictions of the original model. With assumption (4), alternative stable states in the degree of polyphagy become possible; to which one of them the population has evolved depends on its evolutionary past. Unlike the original model, where the degree of polyphagy represents an evolutionary compromise between conflicting selection pressures, the population may now evolve towards extreme polyphagy. Species whose N/C ratio is smaller, or is occasionally smaller, than a critical value are likely to evolve towards ever-increasing polyphagy. Incidentally, the well-known contrast between the generally polyphagous lepidopteran larvae (H32) and the generally monophagous or oligophagous sawfly larvae (H33) (e.g. Neuvonen and Niemelä, 1983) may be due to differences in the types of detoxification mechanisms in these taxa.

Other factors contributing to extreme polyphagy in fungivores should not be overlooked. The main volatile compound in the seven unrelated mushrooms studied by Pyysalo (1976) was 1-octen-3-ol. If fungivorous insects are attracted to sporophores by this or other common compounds (see Kukor and Martin, 1987), widespread polyphagy could be expected. Second, it is questionable whether the chemical compounds that are potentially toxic to insects in sporophores have evolved for that purpose (below). If not, there has been no counter-evolution in the fungi to evolution of resistance in the fungivores, unlike the situation in at least some vascular plant–herbivore systems (e.g. Strong *et al.* 1984 and references therein). And third, the secondary fungivores breeding in decomposing sporophores may not consume the fungal tissue at all, so that the problem of detoxification may be non-existent in the first place. For instance, the degree of specialization in many fungivorous *Drosophila* (H28.36) may reflect specialization in the use of different types of yeasts (Kearney and Shorrocks, 1981).

E. Fungivory and Host Fitness

Secondary fungivores and detritivores breeding in decaying sporophores are unlikely to affect host fitness at all. In contrast, the primary fungivores that colonize sporophores before or during spore production have the potential to decrease the fitness of their host. Damage that would affect spore production is more likely the earlier the insects colonize the sporophore. Hackman and Meinander (1979) list four fungal adaptations that make early colonization by fungivores more difficult:

1. The sporophore appears from the soil at stage II (Fig. 2). *Lactarius necator* (F5.12) is an example. It may be noted here that underground fungi (truffles) are nonetheless colonized by a number of beetle species and other insects (Fogel and Peck, 1975).
2. The unripe basidia are protected by a volva or cortina, or by the fringes of an inwardly bent cap edge, preventing oviposition on the basidial layer at stage I. This is a common situation, but does not prevent oviposition in the stipe by flies: *Cordyla* species (H28.15) and *Pegomya incisiva* (H28.39), which may badly damage young sporophores of *Gomphidius glutinosus* (F5.11) and *Leccinum versipelle* (F5.11) respectively.
3. Rich milky sap in young sporophores. Most *Lactarius* (F5.12) species.

4. Toxic or repellent chemicals, which may not be restricted to young sporophores. Hackman and Meinander (1979) give *Tylopilus felleus* (F5.11) and *Paxillus atrotomentosus* (F5.11) as examples.

Two cases in which the fungivore clearly interferes with host reproduction are the erotylid beetle *Cypherotylus californicus* (H25.39) consuming entire sporophores of *Polyporus adustus* (F5.8) before spore production has occurred (Graves, 1965), and the platypezid fly *Agathomyia wankowiczi* (H28.21) inducing massive gall formation on the underside of its host, the perennial *Ganoderma applanatum* (F5.8) (see photograph in Kotiranta and Niemelä, 1981). Both species are monophagous. It should be added that many mammals, including *Homo sapiens*, are probably much more serious enemies of fungi than insects. If the chemical compounds in fungi have an evolved defensive role it may be against vertebrates rather than insects: the former consume entire sporophores and are much more likely than insects to affect spore production and successful dispersal. Many of the toxic chemicals in the sporophores are indeed extremely poisonous to mammals.

In some cases fungal-visiting insects play a role in the dispersal of host spores. Thus the stinkhorn fungi (Phallales (F5.18)) use odour to attract spore dispersing insects; spores develop after being discharged in the insect faeces (Ramsbottom, 1953). Another example is *Hylobius abietis* (H25.63), which disperses *Fomes annosus* (F5.8) spores (Nuorteva and Laine, 1972). Sivinski (1981) has discussed the role of luminescence in fungi, but cannot decide between the contrasting hypotheses of attraction of spore dispersers and warning signals (other hypotheses are also discussed). Generally, however, it is assumed that fungivorous insects play little or no role in spore dispersal (Hackman and Meinander, 1979).

IV. POPULATION ECOLOGY

A. Life Histories: Diapause

Ecologists often contrast two means of avoiding unfavourable conditions "here and now"—dispersal and diapause. Which one is the better strategy largely depends on habitat and resource characteristics (e.g. Southwood, 1977). In fungivorous insects, dispersal is closely related to the monophagy versus polyphagy continuum: the greater the number of host species used, the shorter the average distance that needs to be travelled between two acceptable sporophores. It has been suggested above that fungivores have largely opted for polyphagy to overcome the problems of habitat variability in time and space. Is there any role left for diapause?

An oligophagous fungivore could solve the problem of year-to-year variability in resource availability (Fig. 4) by delaying the emergence of some of its offspring by one or more years. Prolonged diapause is not uncommon in insects that use temporarily unpredictable resources (Danks, 1987 lists tens of examples; see Hanski, 1988 for a review). A well-known example is provided by the insects developing in conifer seeds and cones that often display dramatic year-to-year variation in availability (Annila, 1981, 1982).

Few observations of prolonged diapause, or lack of it, have been published for fungivorous insects. The *Pegomya* (H28.39) flies breeding in *Leccinum* (F5.11) provide an interesting example; it may be significant that *Pegomya* spp. are exceptionally oligophagous fungivores (Fig. 5). In southern Finland, prolonged diapause has been recorded, rarely, for only one species, *P. furva* (Hackman, 1976). In striking contrast to this, a recent rearing experiment by Ståhls (1987) revealed that prolonged diapause occurs in all seven species present in Finnish Lapland, though the frequency of prolonged diapause varied widely and significantly between the species, from 4–68% (25 *Leccinum* fruiting bodies, 1984 individuals of *Pegomya*). It is probable that the reason for the difference between southern and northern Finland is the greater year-to-year variability of fruiting body production in the north than in the south. Ohenoja and Metsänheimo (1982) give data on fruiting body production of macrofungi during six years in Lapland, and they found the maximum and minimum figures in birch forests to be 363 and 0.3 kg ha^{-1}, respectively. Similar extreme low values have not been reported for southern Finland (Ohenoja and Koistinen, 1984).

Central European *Drosophila phalerata* and *D. testacea* (H28.36) are polymorphic with respect to diapause at all daylengths, and unlike most insects studied in this respect, there is no clear north–south cline in the photoperiodic response of these species (Muona and Lumme, 1981). In *Pegomya winthemi*, some first generation pupae emerge in late summer to produce a second generation while the rest enter into an early diapause and emerge the following year (Hackman and Meinander, 1979). Muona and Lumme (1981) suggest that such polymorphisms are an adaptation to spatially and temporally varying resources.

Apart from year-to-year variation in resource availability, fungivorous insects need to cope with variation at shorter time-scales, e.g. the phenology of sporophore production depends on weather conditions during summer. Fungivorous *Drosophila* have a flexible reproductive development, maturation of females being immediate or delayed depending upon the presence or absence of suitable host fungi (Charlesworth and Shorrocks, 1980). Many *Pegomya* species, e.g. *P. calyptrata*, emerge many weeks earlier than their usual breeding season (Hackman, 1979), which may be an adaptation to take advantage of summers when the sporophores appear exceptionally early. It is also possible that the females need to feed for a prolonged period of time to reach maturity (Tiensuu, 1935; Hackman and Meinander, 1979). Larval development may be geared towards fast development in the several co-existing *Pegomya* species on *Leccinum* because of the often severe competition, and this may leave little time for the building up, during larval development, of metabolic reserves for adult life. A similar ecological situation occurs in the highly competitive carrion and dung insect communities (Hanski, 1987b).

Several polyphagous fungus gnats, e.g. *Exechia fusca* and *Mycetophila fungorum* (H28.15), have short generation times of 2–3 weeks, and several generations may develop in one year even in northern temperate regions (Hackman and Meinander, 1979). *Bolitotherus cornutus* (H25.60), a temperate tenebrionid breeding in the perennial sporophores of *Ganoderma applanatum* (F5.8), represents the opposite extreme: adult beetles may live for at least three years, and females may oviposit in at least two summers (Pace, 1967). Fecundity is low (Liles, 1956). Some observations indicate that although the beetle can fly it seldom moves from one tree to another (Heatwole and Heatwole, 1968), unless no resources are left on the host tree. In a laboratory experiment, large males with large horns had greater access to sporophores and mated more frequently than small males (Brown, 1980), in support of Pace's (1967) earlier suggestion that males are territorial and defend single sporophores. Such behaviour is not unexpected for a species using a patchy resource (Emlen and Oring, 1977; Krebs and Davies, 1981), and has been documented, for example, for some carrion (Otronen, 1984) and dung flies (Borgia, 1980, 1981). *Bolitotherus cornutus* is exceptional amongst fungivores in possessing the capacity to defend resource patches, *G. applanatum* sporophores. The Australian *Drosophila polypori* establishes courting territories (leks) on the underside of the same host species (Parsons, 1977).

B. Clutch Size

Another life history trait that deserves a comment here is clutch size. Recent theoretical work (e.g. Charnov and Skinner, 1984, 1985; Iwasa *et al.*, 1984; Parker and Courtney, 1984) has treated clutch size in insects in the way Lack (1947) originally analysed clutch size in birds. These models attempt to solve the clutch size that maximizes the fitness of egg-laying females. Assuming that the fitness of individual larvae decreases with clutch size, e.g.

because of larval competition in resource patches, an upper limit to the optimal clutch size, appropriately called the "Lack solution", is reached when the fitness returns from a single clutch are maximized. Several other factors, such as the average distance between resource patches and possible egg limitation, will affect the optimal clutch size, as reviewed by Godfray (1987). Furthermore, the variance of fitness is likely to increase with clutch size, and selection to decrease variance (Gillespie, 1974, 1975; Real, 1980; Bulmer, 1984) may lead to a decrease in clutch size.

Fungivorous insects provide an interesting testing ground for clutch size theories, though very little information is currently available. Hackman (1979) found that while some species of *Pegomya* (H28.39) lay their eggs singly, others deposit larger clutches, even when breeding on the same host species. As the sporophores (*Leccinum* (F5.11)) have enough resources for several hundred larvae, the reason for depositing eggs singly is unlikely to be larval competition within clutches; a more likely explanation is "spreading of risk" by depositing eggs in many sporophores. Bruns (1984) has observed that while only a few individuals of *Mycetophila fisherae* (H28.15) occur per sporophore in the *Suillus* (F5.11) species that fruit gregariously, tens of individuals may develop in other, non-gregarious host species. This observation is consistent with the theory: smaller clutches on the gregarious *Suillus* are beneficial because, for minimal cost of movement, the female may decrease competition amongst the offspring and may enhance "risk-spreading".

C. Localized Interactions

Fungivorous insects are often found in such high densities that resource competition must occur. Hundreds of first instar larvae of Mycetophilidae (H28.15) may occur in individual sporophores of *Russula* (F5.12) and *Amanita* (F5.10) species (Russell-Smith, 1979); hundreds of *Pegomya* larvae have been observed in single *Leccinum* (F5.11) sporophores (W. Hackman, pers. comm.); hundreds of ciid larvae and adults may crowd together in the conks of Polyporaceae (Lawrence, 1973); and Ashe (1984) collected over 700 individuals of gyrophaenine staphylinids (H25.8) representing 13 species from a single sporophore of *Amanita verna*. Increasing competition is likely, first, to decrease pupal and adult size and hence, for example, female fecundity (Atkinson, 1979; Grimaldi and Jaenike, 1984); more severe competition increases mortality. Grimaldi and Jaenike (1984) have demonstrated both effects in fungivorous *Drosophila* (H28.36) by experimentally manipulating the amount of resource available to larvae. However, they also observed "substantial variation among individual mushrooms in the degree of food depletion by larvae; while some mushrooms were completely devoured, others appeared to provide more than enough for the larvae." Such variation in density between individual sporophores is indeed commonplace. As larvae can only compete with individuals in the same sporophore, variation in density between sporophores may be expected to have important population ecological consequences.

Localized interactions introduce two new terms to models of competition: the between-sporophore variation in the numbers of individuals of the same species, and the degree of covariation between pairs of species. If local (within-sporophore) interactions were density-independent, the spatial variance–covariance structure of populations would not matter, but as just indicated, local densities often rise to such high levels that density-dependent effects will operate. Even if high local densities occur infrequently, the effects on populations can be substantial because of the large numbers of individuals present in the high-density sporophores.

The spatial variance–covariance structure of populations can most simply be incorporated into a competition model by calculating per capita growth rates of populations as weighted averages of the growth rates in individual habitat patches (Hanski, 1981). Adding the variance–covariance structure into the Lotka–Volterra competition model for two species thus gives the equations:

$$dn_1/dtN_1 = r_1(1 - (1/k_1)(n_1 + \mathrm{var}\ (n_1)/n_1 + (n_2 + \mathrm{cov}\ (n_1,\ n_2)/n_1))) \quad (5a)$$

$$dn_2/dtN_2 = r_2(1 - (1/k_2)(n_2 + \mathrm{var}\ (n_2)/n_2 + (n_1 + \mathrm{cov}\ (n_1,\ n_2)/n_2))). \quad (5b)$$

The lower case ns denote the average local population sizes of species 1 and 2, while the upper case Ns give the respective total (regional) population sizes. The local carrying capacities k are assumed to be equal in all patches. As will be apparent in a comparison with the standard Lotka–Volterra model, the new elements in Eqs (5) are the variance and the covariance terms, both of which will affect the chances of species 1 and 2 co-existing regionally in systems of habitat patches. Increasing variance amplifies intraspecific competition and makes, other things being equal, co-existence easier. Increasing covariance increases interspecific competition, and in the extreme case where two species are aggregated with complete spatial correlation, intraspecific variances make no difference to co-existence. However, complete spatial correlation hardly occurs in natural populations (e.g. Hanski, 1987a), and co-existence is facilitated by large variances even with substantial positive covariances between species (Ives, 1988).

Shorrocks and co-workers have developed a simulation model that is more detailed and mechanistic than Eqs (5) but focuses on the same factor, spatial aggregation of competitors (Atkinson and Shorrocks, 1981, 1984;

see also the analytical model of Ives and May, 1985). Shorrocks and co-workers have applied their model to fungivorous *Drosophila*. Several species typically co-exist, as will be described in the next section, often without obvious niche differences that could "explain" their co-existence. Atkinson and Shorrocks (1984) have demonstrated that *Drosophila* larvae are more or less independently aggregated among individual sporophores, in other words that the variance terms in Eqs (5) are large and increase with mean abundance (Shorrocks and Rosewell, 1988; see also Hanski, 1986, 1987a) but that the covariance term is small and often does not differ significantly from zero (Atkinson and Shorrocks, 1984). This kind of spatial variance–covariance structure facilitates co-existence by intensifying intra-specific relative to interspecific competition.

Predation and parasitism are likely to be often important processes in insect communities inhabiting sporophores, though little information is available beyond lists of species recorded in general surveys. Welch (1959), Montague and Jaenike (1985) and Jaenike (1985) report that 10–20% of females of fungivorous *Drosophila*, in their respective studies, had burdens of abdominal nematode parasites that decreased fecundity significantly (Montague and Jaenike, 1985). The exception was flies reared from toxic *Amanita* species (high concentrations of α-amanitin), and it has been suggested by Jaenike (1985) that breeding in toxic host species is beneficial to fungivorous *Drosophila* in decreasing parasitism.

Most predatory insects in sporophores are beetles, and amongst them Staphylinidae (H25.8) are generally the most diverse. In Benick's (1952) survey of fungivorous beetles in northern temperate regions, 455 out of 1116 species were Staphylinidae, and in Rehfous' (1955) study 277 out of 585 beetles collected from 130 fungi were Staphylinidae. Rehfous (1955) estimated that one-third of beetle species in fungi are predators.

How do predation and parasitism fit into the model of Eqs (5)? It is relatively simple to incorporate a predator population or populations into this theoretical framework (Hanski, 1981). Apart from the parameters describing the predators' functional and numerical responses (Hassell, 1982), we now have terms covering the spatial covariance between the predator and its prey populations. The most interesting result to emerge from this model is that even generalist predators may inflict frequency-dependent mortality upon their prey species, and hence maintain prey species diversity, provided that the prey species are aggregated with little covariance, and that predators congregate in the habitat patches with the highest *total* prey density. The latter situation is predicted by optimal foraging theories and is often observed in predators and parasitoids (Stephens and Krebs, 1986). Vet (1983) has however shown that *Leptopilina clavipes* (H33.2), a parasitoid of fungivorous *Drosophila*, is attracted by odours to decaying sporophores, and that the parasitoids could not distinguish between sporophores with and without host larvae. However, it is possible that parasitoids would stay and oviposit longer in sporophores with more host larvae. Vet (1983) also demonstrated associative learning in the parasitoids during oviposition, and she suggests that such learning facilitates utilization of ephemeral resources.

V. GUILDS AND COMMUNITIES

Theoretical studies of community structure predict that many communities should consist of loosely connected sets (guilds) of possibly strongly interacting species, because such community structure increases stability (May, 1973; Pimm, 1982). The community of fungivorous insects, however, does not represent a highly structured community of this type.

It is difficult to delineate guilds of fungivorous insects because of their broadly overlapping host selection. Papers have been written on the ecology of fungivorous *Drosophila* (H28.36) (Burla and Bächli, 1968; Shorrocks and Charlesworth, 1980, 1982; Shorrocks and Rosewell, 1986; Grimaldi, 1985), Mycetophilidae (Russell-Smith, 1979) and *Pegomya* (H28.39) (Anthomyiidae; Hackman, 1976, 1979), but an examination of host selection amongst 182 mycophagous species of flies from Agaricales indicates that all but one fail to form clusters that could be identified as guilds (Fig. 10). The exception is a monophyletic species group in *Pegomya*.

Although taxonomic guilds associated with particular host species cannot be generally identified, it is instructive to examine more closely the ecology of some of the better studied taxa of fungivores.

A. Drosophila (*H28.36*)

In Europe, about 15 species breed in fungi, mostly in Agaricales. The most regular and abundant fungivorous species is *D. phalerata*. *Drosophila transversa* (Finland: Hackman and Meinander, 1979; Hungary: Dely-Draskovits and Papp, 1973), *D. testacea* (Hungary; Switzerland: Burla and Bächli, 1968), *D. busckii* (Switzerland), and *D. cameraria* (England: Shorrocks and Charlesworth, 1980) have been reared abundantly in some regions, while the remaining species are either rare fungivores (e.g. *D. confusa*; Dely-Draskovits and Papp, 1973) or generalist detritivores. In the eastern United States there are about a dozen fungivorous drosophilids (Lacy, 1984b), with a guild of four abundant species occurring in north-eastern USA (Grimaldi, 1985). Kimura (1976) reports more species from Japan, with 14 Drosophilidae strongly attracted to fungi, and another 19 species collected in small numbers. The regional densities of fungivorous

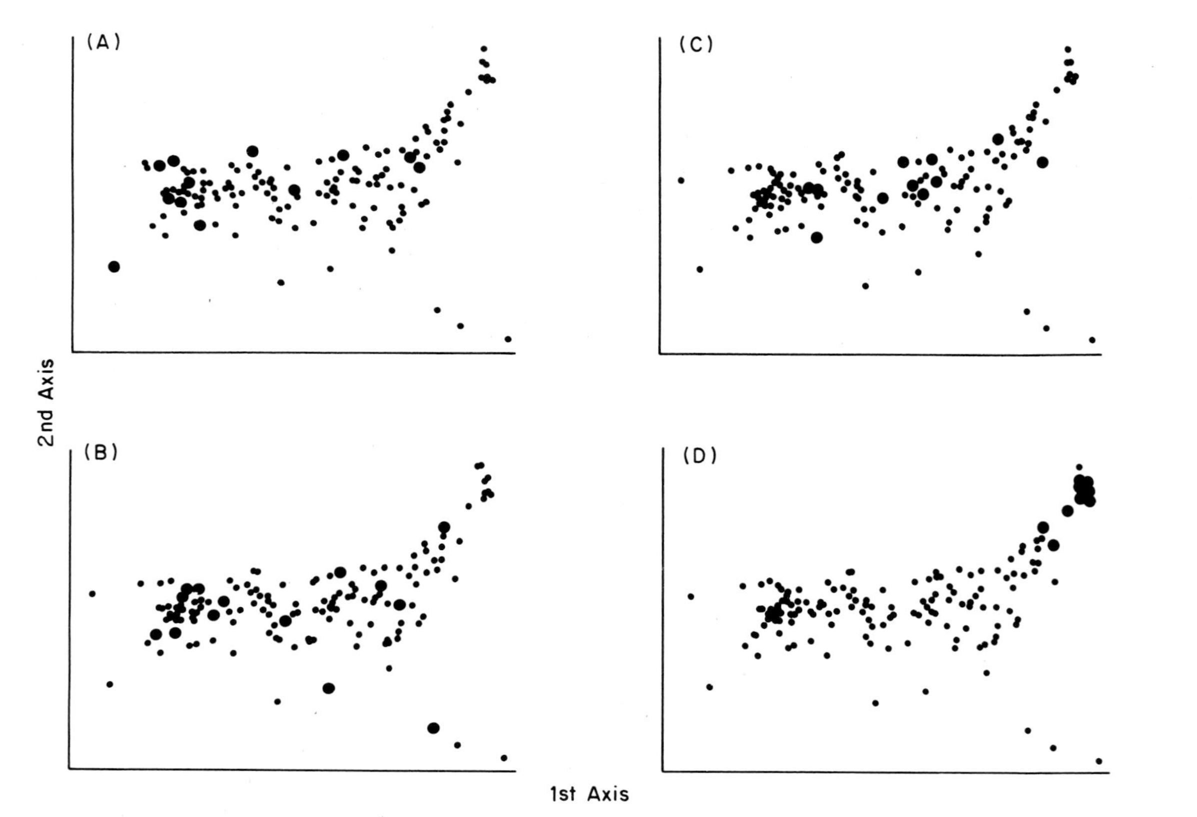

Fig. 10. DECORANA ordination of Diptera breeding in Agaricales *sensu lato* (the same data as in Fig. 8). The species in four "guilds" are identified by larger points amongst the rest of the species. (A) *Exechia* (Mycetophilidae (H28.15)); (B) *Mycetophila*; (C) *Drosophila* (H28.36); (D) *P. tabida* subgroup of *Pegomya* (Anthomyiidae (H28.39)). Note that only species in the *P. tabida* subgroup form a distinct ''guild'' in terms of host selection.

Drosophila are of the order of thousands per hectare, e.g. 2000–3000 ha^{-1} of *D. phalerata* and *D. subobscura* breeding mainly in *Phallus impudicus* (F5.18) in England (Shorrocks and Charlesworth, 1982).

Different species often occur in different seasons and most frequently in different fungal species, which can indeed be called the two major "niche axes" in fungivores (see e.g. Russell-Smith, 1979 for Mycetophilidae). Furthermore, as Dely-Draskovits and Babos (1976) point out and demonstrate with numerous examples, fungivore seasonality and host selection are related niche dimensions, one often determining the other, though it may be difficult to tell which one is the cause and which one the effect. Shorrocks and Charlesworth (1980) describe a seasonal separation between the two most abundant species in their study, *D. phalerata* and *D. cameraria*. The dominant species generally emerge most abundantly from different species of fungi; in Burla and Bächli's (1968) study *D. phalerata* from *Russula linnaei* (F5.12) and *Oudemansiella platyphylla* (F5.10), while *D. busckii* came from *R. linnaei* and *R. alutacea*; in the study of Shorrocks and Charlesworth (1980) *D. phalerata* from *Phallus impudicus* and *Polyporus squamosus* (F5.8), *D. cameraria* from *Lactarius quietus* (F5.12), *Amanita rubescens* (F5.10) and *Phallus impudicus*, while *D. confusa* specialized on *Polyporus squamosus*.

Despite these differences in host selection, several species of *Drosophila* often occur together in the same sporophores and may compete for the same resources (Grimaldi and Jaenike, 1984). Grimaldi (1985) has discussed the various factors that facilitate co-existence, but one may go a step further and ask whether the number of co-existing species is limited by interspecific competition. Jaenike (1985) suggests that interspecific competition rather than lack of appropriate physiological adaptations prevents three species of *Drosophila* from utilizing mushrooms.

Shorrocks and Rosewell (1986) have studied this question of community saturation with a simulation model based on the two-species competition model of Atkinson and Shorrocks (1981)—see Section IV, C. Their model predicts a modal number of five to six species in guilds in which competition sets the limit to guild size. This compares very favourably with the observed guild size in natural drosophilid assemblages with a mode of seven species (including species assemblages from various breeding substrates). The excellent quantitative correspondence between the prediction and the observations may be fortuitous, however, for two reasons. First, as described above, quantitatively important differences in host selection usually occur in fungivorous *Drosophila*, and the number of species that interact strongly is small, often one to three. Second, although fungivorous drosophilids undoubtedly do compete every now and then amongst themselves (Jaenike, 1978b; Grimaldi and Jaenike, 1984), it remains to be shown that resource competition with aggregated distributions is the main reason for the very small numbers that usually emerge from sporophores (Shorrocks and Charlesworth, 1980), and that drosophilids compete primarily amongst themselves, not, for example, with the multitude of other flies with which they co-occur in fungi (Hackman and Meinander, 1979). Burla and Bächli (1968), for instance, suggest that drosophilids may compete with mycetophilids (H28.15, etc.), with at least 90 fungal-breeding species in Europe (Hackman and Meinander, 1979), while in Grimaldi's (1985) study four *Drosophila* species co-occurred with one abundant tipulid (H28.2) and an abundant anisopid (H28.7). There may also occur important geographical differences: the drosophilids seem to co-occur with many more abundant fly species in northern Europe (Hackman and Meinander, 1979) and in north-eastern USA (Grimaldi, 1985).

While many *Drosophila* are saprophagous (e.g. Kearney and Shorrocks, 1981), primary fungivorous species also exist, e.g. the *Hirtodrosophila* species in Japan (Kimura, 1980; see also Grimaldi, 1985). Kimura (1980) suggests that the primary fungivorous habit may have evolved because of competition: as the duration of most sporophores is relatively short, and the amount of resources is limited, the first species to start development has a competitive advantage (the "priority effect"; Gilpin, 1974; Slatkin, 1974; Hanski and Kuusela, 1977; Hanski, 1987b).

B. Pegomya (*H*28.39)

Pegomya is a large anthomyiid genus with herbivorous and primary fungivorous species. The fungivorous species are more diverse in northern than in central or western Europe (Hackman, 1979), perhaps because their main host species in the Boletaceae (F5.11) are especially abundant in northern coniferous forests (Fig. 1; Ohenoja, 1974; Ohenoja and Koistinen, 1984). All the European fungivorous *Pegomya* form a monophyletic group in the *P. geniculata* species group (Hennig 1966–1976; though see Griffiths 1982–1984, p. 14). *Pegomya geniculata* itself is an abundant generalist but does not use Boletaceae. Five other species breed in various Agaricales *sensu lato*, e.g. *P. deprimata* specializes on *Suillus* (F5.11), while the remaining 14 species all belong to the *P. tabida* subgroup and breed exclusively in Boletaceae, especially in *Leccinum versipelle* and *L. scabrum* (Table III; Hackman and Meinander, 1979; Ståhls, 1987). All or most of these species also occur in North America where they have similar ecologies to those in Europe (Bruns, 1984).

The group of 14 species breeding in *Leccinum* is unusual in several respects. No other such group of congeners with similar and narrow host selection can be found amongst the flies breeding in Agaricales in Europe.

TABLE III. Ecology and biology of *Pegomya* (H28.39) in Finland. The abundance columns give the numbers of individuals reared from various Boletales (F5.11) by Hackman and Meinander (1979): 80 sporophores of *Boletus edulis, Leccinum versipelle* and *L. scabrum* (South Finland), and by Ståhls (1987): 133 sporophores of *L. versipelle* and *L. scabrum* (North: Finnish Lapland). Life history data are from Hackman (1979) and Ståhls (1987).

Species	Abundance		Hosts						Life history	Site of oviposition
	South	North	V	S	C	B	P	S[c]		
P. circumpolaris Ackland	–	40	*	*	–	–	–	–	PD[d]: 4%	?
P. flavoscutellata Zett.	114	459	*	*	*	–	–	–	PD: 68%	Spore layer
P. fulgens (Meig.)	139	–	*	*	*	–	–	–		Spore layer
P. furva Ringd.	297	–	*	*	–	–	–	–	PD: rare	Spore layer
P. incisiva Stein	1013	80	*	*	–	–	–	*	Early; PD: 13%	Stipe
P. notabilis Zett.[b]	481	144	*	*	*	*	–	*	Early; PD: 7%	Cap tissue
P. pulchripes Loew	–	–	–	*	–	–	–	–		?
P. scapularis Zett.	139	2649	*	*	*	*	–	*	PD: 16%	Spore layer
P. tabida (Meig.)	152	–	–	*	–	–	–	–		?
P. tenera	–	–	–	*	–	–	–	–		?
P. transgressa Zett.	–	654	*	*	–	–	–	–	PD: 29%	Spore layer
P. vittigera Zett.	7	+	–	*	–	–	–	–		?
P. winthemi (Meig.)	385	–	*	*	*	*	*	–	Multivoltine	Spore layer
P. zonata Zett.[a]	875	636	*	*	*	*	–	–	PD: 20%	Spore layer

[a] *tenera* Zett.
[b] *zonata* Zett. (*tenera* auctt., non Zett.).
[c] The host fungi are: V, *Leccinum versipelle*; S, *L. scabrum*; C, *L. carpini*; B, *Boletus edulis*; P, *B. piperatus*; and S, *Suillus* spp.
[d] Prolonged diapause; all figures except for *P. furva* refer to Lapland and are based on Ståhls' (1987) results; percentage of prolonged diapause and abundance are significantly correlated at 5% level (Olmstead-Tukey's corner test)

Unlike other genera of fungivorous flies, Pegomya has no species strongly dominating its guild and only a few species seem to be rare, suggesting that the species have similar ecologies. Certain differentiation has, however, occurred in the site of oviposition (Fig. 2, Table III; Hackman, 1976, 1979). Two species, *P. incisiva* and *P. notabilis*, generally colonize sporophores earlier than the others (Ståhls, 1987), and may thus gain a competitive advantage. Differences in oviposition behaviour may reflect competitive pressures as, for example, are found in blowflies (Hanski, 1987b).

The abundance distribution of *Pegomya* appears to be less even in northern than in southern Finland (Table III). Ståhls' (1987) results on prolonged diapause in *Pegomya* in northern Finland suggest an interesting explanation. The frequency of prolonged diapause amongst the seven species for which data are available varies widely, from 4% (*P. circumpolaris*) to 68% (*P. flavoscutellata*; this figure is based on a small sample of 44 flies and may be an overestimate), and differences between the abundant species are highly significant. Prolonged diapause was significantly correlated with species' abundance (Table III), suggesting that without prolonged diapause *Pegomya* do poorly in Lapland, where year-to-year variation in fruiting body production is very pronounced and greater than for example, southern Finland (see Section IV).

Individual sporophores may have several hundred *Pegomya* larvae, and competition is likely to occur within and between species. Ståhls' (1987) study from Lapland indicates that the size of the emerging adults decreases with increasing density of larvae (Table IV), supporting the competition hypothesis. More unexpectedly, the size of the emerging adults also decreased with the size of the sporophore (Table IV). Sporophore diameter is, however, not only an indication of the amount of resource, but also of the age of the sporophore. Large and old sporophores decompose faster than small ones, and Ståhls' (1987) result suggests that the primary fungivores such as *Pegomya* develop relatively poorly in old sporophores.

Larval interactions are localized within sporophores, and the theory outlined in Section IV is relevant when considering the regional outcome of competition. Figure 11 shows that, in an interspecific comparison, variance in the numbers of emerging flies from sporophores increases rapidly with increasing average abundance, which should facilitate co-existence (Eqs. (5) in Section IV). The slope of log variance against log mean regression is significantly less than 2, the null hypothesis for density-independent dynamics (Hanski, 1982), perhaps because of mortality due to intraspecific competition in the most abundant species. Interspecific correlations tend to increase with species' average abundance (Fig. 11); such correlations,

TABLE IV. Results of ANCOVA (individual flies) and multiple regression (means for sporophores) of the thorax length in *Pegomya scapularis* (H28.39), the dominant species in Lapland, in 17 sporophores of *Leccinum versipelle* and *L. scabrum* (F5.11). The pooled numbers of *Pegomya* vary from less than 10 to 300 per sporophore. Data from Ståhls (1987).

	Individual flies ($n = 299$)			Sporophores ($n = 17$)		
	MS	F	P	Coefficient	t-value	P
Main effect: sex	1496	92.5	0.000	(Sex ratio: no effect)		
Covariates:						
Numbers of *Pegomya*	99	6.1	0.014	− 1.57	− 3.73	0.002
Cap diameter	304	18.8	0.000	− 0.87	− 3.96	0.001

though not preventing co-existence mediated by intraspecific aggregation, nonetheless make this effect a less powerful one (cf. the covariance terms in Eqs (5) in Section IV). Unfortunately, without detailed knowledge of the distribution of the eggs of different species in sporophores, it is not possible to draw more quantitative conclusions.

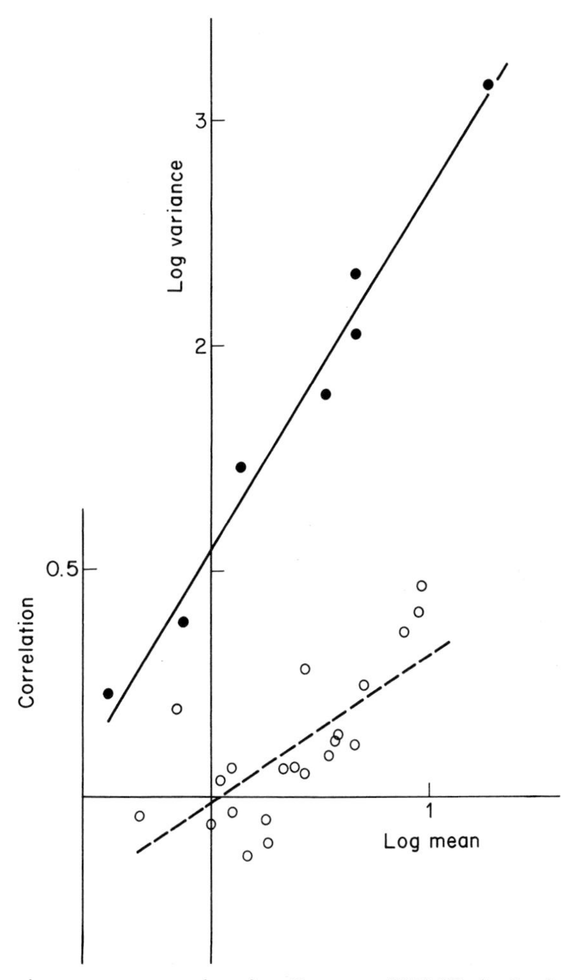

Fig. 11. The variance-mean regression for *Pegomya* (H28.39) in Lapland (logarithmic transformation of both axes; each point represents one species). The regression line is log $V = 1.13 + 1.62$ log mean. The open circles give pair-wise correlations of the numbers of emerging flies against the average log density of the two species (correlations calculated from log-transformed data). Correlation increases with species' average abundance (t of the slope $= 4.54$, $P < 0.0002$). Correlations greater than 0.32 are significant at 1% level. Data from Ståhls (1987).

Pegomya fit better than fungivorous *Drosophila* the assumptions of the competition model of Shorrocks and Rosewell (1986), which predicted a modal guild size of five species, substantially less than the observed 12 in *Pegomya* in southern Finland. The model of Shorrocks and Rosewell (1986) assumes no spatial correlations between the species, which makes the discrepancy even greater.

C. Ciidae (*H*25.52)

Although little ecological information, apart from host preferences, has been published for this beetle family, it clearly has considerable potential interest for guild and community studies.

Unlike fungivorous *Pegomya* and most Drosophilidae, ciids specialize on the relatively long-lasting sporophores of Polyporaceae (F5.8) rather than those of the ephemeral Agaricales *sensu lato* (F5.10–5.12). The generation time is relatively long, over 2 months in *Hadraule blaisdelli* (H25.52) at room temperature, with adult beetles surviving up to 7 months (Kopfenstein, 1971). Most species have relatively narrow host ranges for fungivores (Lawrence, 1973).

Paviour-Smith (1960) divided the 10 species that occurred in her study area in Wytham Wood, southern England, into two groups breeding primarily in two sets of host fungi, differing in the complexity of hyphal structure and weaving (as described by e.g. Corner, 1932 and Cunningham, 1947, 1954). Lawrence (1973), working with the 74 North American species, divided them into four host preference groups. Paviour-Smith's (1960) and Lawrence's (1973) results indicate that the structure of the sporophore is the decisive factor in host selection. However, the nutritive value may also be of some importance, as *H. blaisdelli* showed preferences between fungal species even if the fungal tissue had been powdered (Klopfenstein, 1971).

Many of the sporophores used by Ciidae are perennial and could be used for several years and generations. Klopfenstein (1971) reports on a colony of *Hadraule blaisdelli* that survived in a sealed bag for 13 years, on the same piece of fungus without added food or moisture! It would be interesting to know how frequently beetles breed in the same sporophores for several generations in nature (Graves, 1960; Klopfenstein, 1971). Are populations monomorphic with respect to dispersal behaviour, or polymorphic as found for many insects breeding in dead trees (Hamilton, 1978)? Do females mate before or after dispersal? The population structure of Ciidae may fit the assumptions of the Local Mate Competition (Hamilton, 1967) or the Haystack (Bulmer and Taylor, 1980) models, and one could expect surprises, for example, in sex ratios (Charnov, 1982). It may be significant that at least one species, *Cis fuscipes* (H25.52), is known to be parthenogenetic (Lawrence, 1967). Incidentally, the tenebrionid beetle *Bolitotherus cornutus* (H25.60) appears to have highly structured populations with possibly substantial inbreeding in subpopulations living in the sporophores of individual trees (Heatwole and Heatwole, 1968).

Further questions arise about the co-occurrence of several species in the same sporophores. Do different species differ in terms of dispersal and competitive abilities, which could facilitate their co-existence (cf. the models of Horn and MacArthur, 1972; Slatkin, 1974; Hanski, 1983; Hanski and Ranta, 1983)? Paviour-Smith (1960) found that *Cis bilamellatus* was always the first species to colonize *Polyporus betulinus* (F5.8) conks, while *Cis bidentatus* and *Ennearthron cornutum* (H25.52) appeared several months later. Competition may be severe, as up to 1000 individuals representing several species have been collected from single conks (Graves, 1960). These and other questions await further study.

D. Questions about Species Richness

No comprehensive survey of insect species associated with macrofungal sporophores exists, but some figures for restricted taxa can be cited. The review of Hackman and Meinander (1979) on Diptera (H28) breeding in Agaricales *sensu lato* (F5.10–5.12) in northern Europe includes some 230 species. Benick's (1952) work covering the entire Holarctic region and including literature data lists 1116 species of Coleoptera (H25) of which 202 were considered to be "truly" fungivorous, 767 entirely accidental and the remaining 147 species intermediate. Paviour-Smith (1960) estimated that some 600 species of insects had been recorded from fungi in the United Kingdom. The order of magnitude of insect species associated with fungi in Europe is thus one thousand; a much more accurate estimate would require drawing an arbitrary line between "true" fungivores and the more or less accidental species.

It may not be a coincidence that the total number of insect species associated with dung and carrion, two other ephemeral and patchy microhabitats, is of the same order of magnitude (Hanski, 1987b). As the chemical composition of these microhabitats is radically different, both from each other and that of fungal sporophores, but the size and occurrence of the resource units is roughly similar, one is tempted to suggest that the latter characteristics are the important ones in the determination of the sizes of regional and geographical species pools.

The total number of species occurring in macrofungi is not increased by host specialization nearly as dramatically as is the number of herbivorous insects, but the two main groups of macrofungi that are important for insects, Agaricales and Polyporaceae (F5.8), have relatively distinct faunas, probably because of the marked differences in the structure and durational stability of their sporophores. As a rule, the ephemeral Agaricales *sensu lato* are dominated by Diptera, while the longer-lasting Polyporaceae are dominated by Coleoptera. The same difference can be seen in other patchy microhabitats, for example carcasses, dung and decaying wood, the more ephemeral microhabitats having relatively more flies and the more persistent ones being dominated by the generally more slowly developing beetles (Hanski, 1987b).

Polyporaceae tend to have more species of insects associated with them than Agaricales. A comprehensive survey of insects and mites occurring in, on, or emerging from the bracket fungus *Polyporus betulinus* in Canada reached 198 species of Hexapoda and 59 species of mites, most of which were however extremely scarce and many accidental (Pielou and Verma, 1968). Benick's (1952) work includes an even higher figure, 246 beetles recorded from *Polyporus squamosus*. No species of Agaricales has an equivalent number of associated insect species, though it must be stressed that there are often fewer individual insects in Polyporaceae than in Agaricales (Rehfous, 1955), and the majority of the insect species recorded from Polyporaceae are more or less accidental. A comparison can again be made with dung and carrion communities. Small carcasses are dominated by a relatively small number of specialist species, in temperate regions especially by *Nicrophorus* (Silphidae (H25.7)) and calliphorid flies (H28.42). The small carcass community is intensely competitive (Hanski, 1987c). Large carcasses, if not consumed by vertebrate scavengers, and relatively large dung sources, cow pats for example, have a mixed community of several hundred species, with a large number of generalist predators that frequent many different microhabitats. Predation is believed to be an important process in these relatively long-lasting species assemblages (Hanski, 1987b).

The number of flies breeding in Agaricales *sensu lato* increases with the size of the sporophore (Table V). This is expected, as host species with large sporophores provide generally more resources for fungivores than species with small sporophores, and because large sporophores probably tend to last longer than small ones. Note that the frequency of polyphagous species decreases with increasing sporophore size (Table V). The most unpredictable resources, making host specialization difficult, are species of Agaricales with small sporophores.

In temperate regions the numbers of fungivores peak, not surprisingly, in late summer and autumn, at the time when the availability of sporophores is greatest (Fig. 4; Dely-Draskovits and Babos, 1976; Hackman and Meinander, 1979). Dely-Draskovits and Babos (1976) have reported on the phenology of 120 species of fungivorous flies in Hungary. There is a

TABLE V. Numbers of Diptera (H28) breeding in Agaricales *sensu lato* (F5.10–5.12) versus the logarithm of average sporophore size. Polyphagous species have been defined as in Table II, and their frequency was calculated as the residual from the regression of polyphagous against all species (see text). Data for flies from Hackman and Meinander (1979) and for sporophores from Ohenoja (unpublished).

	Coefficient	*t*-value	Significance level	R^2
All species	10.14	9.81	0	0.35
Polyphagous species	2.63	7.77	0	0.25
Frequency of polyphagous species	− 0.53	− 3.19	0.002	0.05

tendency for the peak occurrences of the 23 most abundant species to be relatively evenly spread from July to October, while the clear majority of the rarer species peak in September–October (comparing the numbers peaking before and after the beginning of September gives $P < 0.1$). In the pooled numbers of individuals there is a highly significant ($P < 0.001$) difference between the abundant and rare species, the latter occurring mostly in the autumn. A possible though nearly untestable explanation of these results is (past) competition that has led to a divergence in the phenologies of the abundant species.

VI. ECONOMIC IMPORTANCE OF FUNGIVOROUS INSECTS

Edible wild mushrooms are an important though often little-used human resource. It has been estimated that the annual crop of edible mushrooms in Finland is 1.5–5.0 million tonnes (Rautavaara, 1947), of which only 2–3% are used (Ohenoja and Koistinen, 1984). One difficulty in using mushrooms on a large scale is the marked year-to-year variation in their availability, e.g. in the birch forests in Finnish Lapland from 0.3 to 363 kg ha^{-1} (Ohenoja and Metsänheimo, 1982).

Fungivorous insects pose another problem for those using wild mushrooms. In Estonia, one third of the crop is infested by insects (Kalamees, 1979). The degree of infestation naturally increases with the developmental stage of the sporophore, and in many commercial species practically all old sporophores contain insect larvae (Hackman and Meinander, 1979). There is not much that could (or should) be done with this "problem". People may concentrate on the species that are less used by insects (see Hackman and Meinander, 1979), and they may and do collect relatively young sporophores.

Another set of fungivorous insects are economically important in causing damage to cultivated mushrooms. The most important pests, apart from mites and nematodes, are sciarid, phorid and cecid flies (H28.16, 10 and 22) (Snetsinger, 1972), though of the flies, only sciarids with chewing larvae are capable of directly damaging the crop (Binns, 1980). Sciarids are also important vectors of various mites (Clift, 1978) and the bacterium *Pseudomonas tolaasi* (Wong and Preece, 1980). An interesting point about these flies is that their larvae may feed on fungal mycelia in the mushroom beds, suggesting that the same may happen also in nature. Mycelium feeding has been reported in phorids (Robinson, 1977; Snetsinger, 1972), cecids (Wyatt, 1959), and sciarids (Clift, 1978).

The problem of cecid flies is aggravated by paedogenesis, or larval reproduction (Wyatt, 1959), which greatly increases population growth rate. A normal life cycle with pupae and adult flies occurs only under exceptional circumstances, for example, crowding. Paedogenesis should give an advantage in competition for resources, if any occurs. Competition may partly explain the changing phorid composition that occurs in mushroom cultivations in North America, one species generally strongly dominating, although several may be present (Robinson, 1977). Clift and Larsson (1984) report similar observations for sciarids.

VII. SUMMARY

The high degree of polyphagy observed in fungivorous insects can be explained by strong selection pressure towards polyphagy, by low cost of polyphagy, or by both. In this connection, it is particularly important to know whether the breeding success of polyphagous fungivores varies much between different host fungi. If it does, counterselection due for example to the unpredictable occurrence of sporophores, is needed to maintain the observed level of polyphagy. If it does not, the implication is that the cost of widening host selection is low, perhaps because of general detoxification mechanisms as suggested earlier in this chapter. Comparative studies on the detoxification capacities of herbivorous, fungivorous and detritivorous insects would be illuminating.

Ephemeral and patchy habitat structure has important life history and population and community ecological consequences. An example is presented in which the frequency of prolonged diapause is positively correlated with species' abundance, suggesting that prolonged diapause as well as polyphagy is an important risk-spreading strategy in fungivores. This chapter has emphasized the role of independently aggregated spatial distributions in facilitating co-existence in competing species. More empirical studies are needed; it is fortunate that the assumptions of the theory outlined here are easily tested in the field (for an experiment on blowflies see Hanski, 1987c). Further work on food web structure (e.g. Graves, 1960; Ackerman and Shenefelt, 1973) and community composition in the fungivores should pay particular attention to localized interactions—the use of regional food webs, pooling and implicitly averaging data from numerous microhabitats, gives an incomplete picture of the dynamics in the community.

Acknowledgements

Several colleagues helped in the preparation of the manuscript. Special thanks are due to Pekka Niemelä for useful discussions on polyphagy and for comments on the manuscript, to Esteri Ohenoja for unpublished data on Agaricales, to Gunilla Ståhls for unpublished data on *Pegomya*, and to Erkki Haukioja for comments on the manuscript.

REFERENCES

Ackerman, J. K., and Shenefelt, R. D. (1973). Organisms, especially insects, associated with wood rotting higher fungi (Basidiomycetes) in Wisconsin forests. *Trans. Wis. Acad. Sci. Arts Lett.* **61**, 185–206.

Annila, E. (1981). Fluctuations in cone and seed insect populations in Norway spruce. (In Finnish with English summary.) *Commun. Inst. For. Fenn.* **101**, 1–32.

Annila, E. (1982). Diapause and population fluctuations in *Megastigmus specularis* Walley and *Megastigmus spermotrophus* Wachtl. (Hymenoptera, Torymidae). *Ann. Entomol. Fenn.* **48**, 33–36.

Aronson, J. M. (1965). The cell wall. *In* "The Fungi" (G. C. Ainsworth and A. S. D. Sussman, eds), Vol. 1, pp. 49–76. Academic Press, New York.

Ashe, J. S. (1984). Major features of the evolution of relationships between gyrophaenine staphylinid beetles (Coleoptera: Staphylinidae: Aleocharinae) and fresh mushrooms. *In* "Fungus–Insect Relationships, Perspectives in Ecology and Evolution" (Q. D. Wheeler and M. Blackwell, eds), pp. 227–255. Columbia Univ. Press, New York.

Atkinson, W. D. (1979). A comparison of the reproductive strategies of domestic species of *Drosophila. J. Anim. Ecol.* **48**, 53–64.

Atkinson, W. D., and Shorrocks, B. (1981). Competition on a divided and ephemeral resource: a simulation model. *J. Anim. Ecol.* **50**, 461–471.

Atkinson, W. D., and Shorrocks, B. (1984). Aggregation of larval Diptera over discrete and ephemeral breeding sites: the implications for coexistence. *Am. Nat.* **124**, 336–351.

Benick, L. (1952). Pilzkäfer und Käferpilze. *Acta zool. fenn.* **70**, 1–250.

Berenbaum, M. (1981). Patterns of furanocoumarin production and insect herbivory in the Umbelliferae: plant chemistry and community structure. *Ecology* **62**, 1254–1266.

Besl, H., and Blumreisinger, M. (1983). *Drosophila melanogaster* suitable for testing the susceptibility of higher fungi to larval feeding. *Z. Mykol.* **49**, 165–170.

Binns, E. S. (1980). Field and laboratory observations on the substrates of the mushroom fungus gnat *Lycoriella auripila* (Diptera: Sciaridae). *Ann. Appl. Biol.* **96**, 143–152.

Blackwell, M. (1984). Myxomycetes and their arthropod associates. *In* "Fungus–Insect Relationships, Perspectives in Ecology and Evolution" (Q. D. Wheeler and M. Blackwell, eds), pp. 67–90. Columbia Univ. Press, New York.

Borgia, G. (1980). Size and density-related changes in male behaviour in the fly *Scatophaga stercoraria*. *Behaviour* **75**, 185–206.

Borgia, G. (1981). Mate selection in the fly *Scatophaga stercoraria*: female choice in a male-controlled system. *Anim. Behav.* **29**, 71–80.

Brown, L. (1980). Aggression and mating success in males of the forked fungus beetle *Bolitotherus cornutus* (Coleoptera: Tenebrionidae). *Proc. Entomol. Soc. Wash.* **82**, 430–434.

Bruns, T. D. (1984). Insect mycophagy in the Boletales: fungivore diversity and the mushroom habitat. *In* "Fungus–Insect Relationships, Perspectives in Ecology and Evolution" (Q. D. Wheeler and M. Blackwell, eds), pp. 91–129. Columbia Univ. Press, New York.

Bulmer, M. G. (1984). Risk avoidance and nesting strategies. *J. Theor. Biol.* **110**, 529–535.

Bulmer, M. G., and Taylor P. D. (1980). Sex ratio under the haystack model. *J. Theor. Biol.* **86**, 83–89.

Burla, H., and Bächli, G. (1968). Beiträg zur Kenntnis der schweizerischen Diptera, inbesondere *Drosophila*-arten, die sich in Fruchtkörpen von Hutpilzen entwickeln. *Vierteljahrsschr. Naturforsch. Ges. Zuerich* **113**, 311–336.

Buxton, P. A. (1960). British Diptera associated with fungi. III. Flies of all families reared from about 150 species of fungi. *Entomol. Mon. Mag.* **96**, 61–94.

Cates, R. G., and Rhoades, D. F. (1977). Patterns in the production of antiherbivore chemical defenses in plant communities. *Biochem. Syst. Ecol.* **5**, 185–193.

Charlesworth, P., and Shorrocks, B. (1980). The reproductive biology, and diapause of the British fungal breeding *Drosophila. Ecol. Entomol.* **5**, 315–326.

Charnov, E. L. (1982). "The Theory of Sex Allocation". Princeton Univ. Press, Princeton.

Charnov, E. L., and Skinner, S. W. (1984). Evolution of host selection and clutch size in parasitoid wasps. *Fla. Entomol.* **67**, 5–21.

Charnov, E. L., and Skinner, S. W. (1985). Complementary approaches to the understanding of parasitoid oviposition decisions. *Ecol. Entomol.* **14**, 383–391.

Chesson, P. L. (1986). Environmental variation and the coexistence of species. *In* "Community Ecology" (J. Diamond and T. J. Case, eds), pp. 240–256. Harper & Row, New York.

Clift, A. D. (1978). The identity, economic importance and control of insect pests of mushrooms in New South Wales, Australia. *Mushroom Sci.* **10**, 367–383.

Clift, A. D., and Larsson, S. F. (1984). The incidence and ecology of *Lycoriella mali* (Diptera: Sciaridae) in the commercial culture of 2 species of mushroom in New South Wales. *Gen. Appl. Entomol.* **16**, 49–56.

Corner, E. J. H. (1932). A *Fomes* with two systems of hyphae. *Trans. Br. Mycol. Soc.* **17**, 51–81.

Crowson, R. A. (1984). The associations of Coleoptera with Ascomycetes. *In* "Fungus–Insect Relationships, Perspectives in Ecology and Evolution" (Q. D. Wheeler and M. Blackwell, eds), pp. 256–285. Columbia Univ. Press, New York.

Cunningham, G. H. (1947). Notes on classification of the Polyporaceae. *N.Z. J. Sci. Technol.* **28**, 238–251.

Cunningham, G. H. (1954). Hyphal systems as aids in identification of species and genera of the Polyporaceae. *Trans. Br. Mycol. Soc.* **37**, 44–50.

Danks, H. V. (1987). "Insect Dormancy: An Ecological Perspective". Biological Survey of Canada No. 1, Ottawa.

Dely-Draskovits, A., and Babos, M. (1976). Fenologische Zusammenhänge zwischen Fliegen und Hutpilze. Part I. *Folia Entomol. Hung.* **19**, 23–38.

Dely-Draskovits, A., and Papp, L. (1973). Systematical and ecological investigations of fly pests of mushrooms in Hungary. V. Drosophilidae (Diptera). *Folia Entomol. Hung.* **26**, 21–29.

Eisfelder, I. (1961). Käferpi'ze und Pilzkäfer. *Z. Pilzkd.* **27**, 44–54.

Eisfelder, I., and Herschel, K. (1966). *Agathomyia wankowiczi* Schnabl, die "Zitzengallenfliege" ans *Ganoderma applanatuum,. Westfäl. Pilzbriefe 6*, 5–10.

Emlen, S. T., and Oring, L. W. (1977). Ecology, sexual selection and the evolution of mating systems. *Science* **197**, 215–223.

Feeny, P. P. (1975). Biochemical coevolution between plants and their insect herbivores. *In* "Coevolution of Animals and Plants" (L. E. Gilbert and P. H. Raven, eds), pp. 3–19. Univ. of Texas Press, Austin, Texas.

Feeny, P. P. (1976). Plant apparency and chemical defence. *Recent Adv. Phytochem.* **10**, 1–40.

Fogel, R., and Peck S. B. (1975). Ecological studies on hypogeous fungi. *Mycologia* **67**, 741–747.

Georghiou, G. P. (1986). The magnitude of the resistance problem. *In* "Pesticide Resistance. Strategies and Tactics for Management", pp. 14–44. National Academy Press, Washington.

Georghiou, G. P., and Taylor, C. E. (1986). Factors influencing the evolution of resistance. *In* "Pesticide Resistance. Strategies and Tactics for Management", pp. 157–169. National Academy Press, Washington.

Gilbert, L. E. (1977). The role of insect–plant coevolution in the organization of ecosystems. *Colloq. Int. C.N.R.S.* **265**, 399–413.

Gilbert, L. E., and Smiley, J. T. (1978). Determinants of local diversity in phytophagous insects: host specialists in tropical environments. *Symp. R. Entomol. Soc. London* **9**, 89–104.

Gillespie, J. H. (1974). Natural selection for within-generation variance in offspring number. *Genetics* **76**, 601–608.

Gillespie, J. H. (1975). Natural selection for within-generation variance in offspring number. II. Discrete haploid models. *Genetics* **81**, 403–413.

Gilpin, M. E. (1974). Stability of feasible predator–prey systems. *Nature* (*London*) **254**, 137–138.

Godfray, H. C. J. (1978). The evolution of clutch size in invertebrates. *In* "Oxford Surveys in Evolutionary Biology", Vol. 4 (P. H. Harvey and L. Partridge, eds). Oxford Univ. Press, Oxford.

Graves, R. C. (1960). Ecological observations on the insects and other inhabitants of woody shelf fungi (Basidiomycetes: Polyporaceae) in the Chicago area. *Ann. Entomol. Soc. Am.* **53**, 61–68.

Graves, R. C. (1965). Observations on the ecology, behavior and life cycle of the fungus-feeding beetle, *Cypherotylus californicus*, with a description of the pupa (Coleoptera: Erotylidae). *Coleopt. Bull.* **19**, 117–122.

Graves, R. C., and Graves, A. C. F. (1985). Diptera associated with shelf fungi and certain other micro-habitats in the highlands area of western North Carolina. *Entomol. News* **96**, 87–92.

Griffiths, G. C. D. (1982–1984). Anthomyiidae. Cyclorrhapha II (Schizophora: Calyptratae). *In* "Flies of the Nearctic Region" (G. C. D. Griffiths, ed.) Vol. 8 (2) pp. 1–351. E. Schweizerbart'sche Verlagsbuchhandlung, Stuttgart.

Grimaldi, D. (1985). Niche separation and competitive coexistence in mycophagous *Drosophila* (Diptera: Drosophilidae). *Proc. Entomol. Soc. Wash.* **87**, 498–511.

Grimaldi, D., and Jaenike, J. (1984). Competition in natural populations of mycophagous *Drosophila. Ecology* **65**, 1113–1120.

Hackman, W. (1976). The biology of anthomyiid flies feeding as larvae on fungi (Diptera). *Notulae. Entomol.* **56**, 129–134.

Hackman, W. (1979). Reproductive and developmental strategies of fungivorous *Pegomya* species (Diptera, Anthomyiidae). *Aquilo Ser. Zool.* **20**, 62–64.

Hackman, W., and Meinander, M. (1979). Diptera feeding as larvae on macrofungi in Finland. *Ann. Zool. Fenn.* **16**, 50–83.

Hamilton, W. D. (1967). Extraordinary sex ratios. *Science* **156**, 477–488.

Hamilton, W. D. (1978). Evolution and diversity under bark. *Symp. R. Entomol. Soc. London* **9**, 154–175.

Hanski, I. (1981). Coexistence of competitors in patchy environment with and without predation. *Oikos* **37**, 306–312.

Hanski, I. (1982). On patterns of temporal and spatial variation in animal populations. *Ann. Zool. Fenn.* **19**, 21–37.

Hanski, I. (1983). Coexistence of competitors in patchy environment. *Ecology* **64**, 493–500.

Hanski, I. (1986). Individual behaviour, population dynamics and community structure of *Aphodius* (Scarabaeidae) in Europe. *Acta Oecologica.* **7**, 171–187.

Hanski, I. (1987a). Colonization of ephemeral habitats. *In* "Colonization, Succession and Stability" (A. J. Gray, M. J. Crawley and P. J. Edwards, eds), pp. 155–185. Blackwell, Oxford.

Hanski, I. (1987b) Nutritional ecology of dung- and carrion-feeding insects. *In* "Nutritional Ecology of Insects, Mites, Spiders and Related Invertebrates" (F. Slansky Jr. and J. G. Rodriguez, eds), pp. 837–884. John Wiley & Sons, New York.

Hanski, I. (1987c). Carrion fly community dynamics: patchiness, seasonality and coexistence. *Ecol. Entomol.* **12**, 257–266.

Hanski, I. (1988). Four kinds of extra long diapause in insects: A review of theory and observations. *Ann. Zool. Fenn.* **25**, 37–53.

Hanski, I., and Kuusela, S. (1977). An experiment on competition and diversity in the carrion fly community. *Ann. Entomol. Fenn.* **43**, 108–115.

Hanski, I., and Ranta, E. (1983). Coexistence in a patchy environment: three species of *Daphnia* in rock pools. *J. Anim. Ecol.* **52**, 263–279.

Hassell, M. P. (1982). Patterns of parasitism by insect parasitoids in patchy environments. *Ecol. Entomol.* **7**, 365–377.

Haukioja, E. (1980). On the role of plant defences in the fluctuation of herbivore populations. *Oikos* **35**, 202–289.

Hawkeswood, T. J. (1986). Notes on two species of Australian fungus beetles (Coleoptera: Erotylidae). *Coleopt. Bull.* **40**, 27–28.

Heatwole, H., and Heatwole, A. (1968). Movements, host–fungus preferences, and longevity of *Bolitotherus cornutus* (Coleoptera: Tenebrionidae). *Ann. Entomol. Soc. Am.* **61**, 18–23.

Heed, W. B. (1968). Ecology of the Hawaiian Drosophiliidae. *Studies in Genetics IV., Univ. Texas. Publ.* **6818**, 387–420.

Hennig, W. (1966–1976). Anthomyiidae. *In* "Die Fliegen der Palaearktischen Region" (E. Lindner, ed.) Vol. 7 (1) pp. 1–974. E. Schweizerbart'sche Verlagsbachhandlung, Stuttgart.

Hiukko, E. (1978). Helttasienten itiöemien kasvunopeudesta. *Sienilehti* **30**, 40–42.

Holling, C. S. (1959). Some characteristics of simple types of predation and parasitism. *Can. Entomol.* **91**, 385–398.

Horn, H. S., and MacArthur, R. H. (1972). Competition among fugitive species in a harlequin environment. *Ecology* **53**, 749–752.

Ives, A. R. (1988). Covariance, coexistence and the population dynamics of two competitors using a patchy resource. *J. Theor. Biol.* (in press).

Ives, A. R., and May, R. M. (1985). Competition within and between species in a patchy environment: relations between microscopic and macroscopic models. *J. Theor. Biol.* **115**, 65–92.

Iwasa, Y., Suzuki, Y., and Matsuda, H. (1984). Theory of oviposition strategy of parasitoids. I. Effect of mortality and limited egg number. *Theor. Populat. Biol.* **26**, 205–227.

Jaenike, J. (1978a). Host selection by mycophagous *Drosophila. Ecology* **59**, 1286–1288.

Jaenike, J. (1978b). Resource predictability and niche breadth in the *Drosophila quinaria* species group. *Evolution* **32**, 676–678.

Jaenike, J. (1985). Parasite pressure and the evolution of amanitin tolerance in *Drosophila. Evolution* **39**, 1295–1301.

Jaenike, J., and Grimaldi, D. (1983). Genetic variation for host preference within and among populations of *Drosophila tripunctata. Evolution* **37**, 1023–1033.

Jaenike, J., and Selander, R. K. (1979). Ecological generalism in *Drosophila falleni*: genetic evidence. *Evolution* **33**, 741–748.

Jaenike, J., Grimaldi, D., Sluder, A. E., and Greenleaf, A. L. (1983). Alpha-amanitin tolerance in mycophagous *Drosophila. Science* **221**, 165–167.

Kalamees, K. (1979). Kas seeni jätkub? *Eesti Loodus* **12**, 551–558.

Kearney, J. N., and Shorrocks, B. (1981). The utilization of naturally occurring yeasts by *Drosophila* species using chemically defined substrates. *Biol. J. Linn. Soc.* **15**, 39–56.

Kimura, M. T. (1976). *Drosophila* survey of Hokkaido. XXXII. A field survey of fungus preferences of drosophilid flies in Sapporo. *J. Fac. Sci., Hokkaido Univ.* **20**, 288–298.

Kimura, M. T. (1980). Evolution of food preference in fungus feeding *Drosophila*: an ecological study. *Evolution* **34**, 1009–1018.

Klopfenstein, P. C. (1971). "The Ecology, Behavior, and Life Cycle of the Mycetophilous Beetle, *Hadraule blaisdelli* (Casey) (Insecta: Coleoptera: Ciidae)". Thesis, Bowling Green State Univ., Ohio.

Kotiranta, H., and Niemelä, T. (1981). Composition of the polypore communities of four forest areas in southern Central Finland. *Karstenia* **21**, 31–48.

Krebs, J. R., and Davies, N. B. (1981). "An Introduction to Behavioural Ecology". Blackwell, Oxford.

Krieger, R. I., Feeny, P. P., and Wilkinson, C. F. (1971). Detoxication enzymes in the guts of caterpillars: an evolutionary answer to plant defenses? *Science* **172**, 579–580.

Kukor, J. J., and Martin, M. M. (1987). Nutritional ecology of fungus-feeding arthropods. *In* "Nutritional Ecology of Insects, Mites, Spiders, and Related Invertebrates" (F. Slansky Jr. and J. G. Rodriguez, eds), pp. 791–814. John Wiley & Sons, New York.

Lack, D. (1947). The significance of clutch size. *Ibis* **89**, 309–352.

Lacy, R. C. (1982). Niche breadth and abundance as determinants of genetic variation in populations of mycophagous drosophilid flies (Diptera: Drosophilidae). *Evolution* **36**, 1265–1275.

Lacy, R. C. (1983). Structure of genetic variation within and between populations of mycophagous *Drosophila. Genetics* **104**, 81–94.

Lacy, R. C. (1984a). Ecological and genetic responses to mycophagy in Drosophilidae (Diptera). *In* "Fungus–Insect Relationships, Perspectives in Ecology and Evolution" (Q. D. Wheeler and M. Blackwell, eds), pp. 286–301. Columbia Univ. Press, New York.

Lacy, R. C. (1984b). Predictability, toxicity and trophic niche breadth in fungus feeding Drosophilidae (Diptera). *Ecol. Entomol.* **9**, 43–54.

Lawrence J. F. (1967). Biology of the parthenogenetic fungus beetle *Cis fuscipes* Mellié (Coleoptera: Ciidae). *Breviora* **258**, 1–14.

Lawrence, J. F. (1973). Host preference in ciid beetles (Coleoptera: Ciidae) inhabiting the fruiting bodies of Basidiomycetes in North America. *Bull. Mus. Comp. Zool.* **145**, 163–212.

Lawrence, J. F. (1977). Coleoptera associated with an *Hypoxylon* species (Ascomycetes: Xylariaceae) on oak. *Coleopt. Bull.* **31**, 309–312.

Lawrence, J. F., and Newton, A. F. Jr. (1980). Coleoptera associated with the fruiting bodies of slime molds (Myxomycetes). *Coleopt. Bull.* **34**, 129–143.

Lawrence, J. F., and Powell, J. (1969). Host relationships in North American fungus-feeding moths (Oecophoridae, Oinophilidae, Tineidae). *Bull. Mus. Comp. Zool.* **138**, 29–51.

Lawton, J. H., and Price, P. W. (1979). Species richness of parasites on hosts: agromyzid flies on the British Umbelliferae. *J. Anim. Ecol.* **48**, 619–637.

Levin, S. A. (1974). Dispersion and population interactions. *Am. Nat.* **108**, 207–228.

Levin, S. A. (1978). Population models and community structure in heterogeneous environments. *In* "Populations and Communities", (S. A. Levin, ed.), Vol. II. Math. Ass. Am., Washington.

Levins, R., and MacArthur, R. H. (1969). An hypothesis to explain the incidence of monophagy. *Ecology* **50**, 910–911.

Liles, M. P. (1956). A study of the life history of the forked fungus beetle, *Bolitotherus cornutus* (Panzer) (Coleoptera: Tenebrionidae). *Ohio J. Sci.* **56**, 329–337.

Martin, M. M. (1979). Biochemical implications of insect mycophagy. *Biol. Rev. Cambridge Philos. Soc.* **54**, 1–21.

Martin, M. M., Kukor, J. J., Martin, J. S., O'Toole, T. E., and Johnson, M. W. (1981). Digestive enzymes of fungus-feeding beetles. *Physiol. Zool.* **54**, 137–145.

May, R. M. (1973). "Complexity and Stability in Model Ecosystems". Princeton Univ. Press, Princeton.

Montague, J. R., and Jaenike, J. (1985). Nematode parasitism in natural populations of mycophagous Drosophilids. *Ecology* **66**, 624–626.

Mullin, C. A., Croft, B. A., Strickler, K., Matsumura, F., and Miller, J. R. (1982). Detoxification enzyme differences between a herbivorous and predatory mite. *Science* **217**, 1270–1271.

Muona, O., and Lumme, J. (1981). Geographical variation in the reproductive cycle and photoperiodic diapause of *Drosophila phalerata* and *Drosophila transversa* (Drosophilidae, Diptera). *Evolution* **35**, 158–167.

Neuvonen, S., and Niemelä, P. (1983). Species richness and faunal similarity of arboreal insect herbivores. *Oikos* **40**, 452–459.

Nuorteva, M., and Laine, L. (1972). Lebensfahige Diasporen des Wurzelschwamms (*Fomes annosus* (Fr.) Cooke) in den Exkrementen von *Hylobius abeietis* L. (Col., Curculionidae). *Ann. Entomol. Fenn.* **38**, 119–121.

Ohenoja, E. (1974). Metsäsienten määrän mittaamisesta. *Karstenia* **14**, 46–53.

Ohenoja, E. (1983). Documentation on the research into the fruiting body production of forest fungi. *Memo. Soc. Fauna Flora fenn.* **59**, 112–115.

Ohenoja, E., and Koistinen, R. (1984). Fruiting body production of larger fungi in Finland. 2. Edible fungi in northern Finland 1976–1978. *Ann. Bot. Fenn.* **21**, 357–366.

Ohenoja, E., and Metsänheimo, K. (1982). Phenology and fruiting body production of macrofungi in subarctic Finnish Lapland. *In* "Arctic and Alpine Mycology" (G. A. Laursem and J. F. Ammirati, eds), pp. 390–409. Univ. Washington Press, Seattle.

Otronen, M. (1984). Male contests for territories and females in the fly *Dryomyza anilis*. *Anim. Behav.* **32**, 891–898.

Pace, A. E. (1967). Life history and behavior of a fungus beetle, *Bolitotherus cornutus* (Tenebrionidae). *Occ. Pap. Mus. Zool. Univ. Mich.* **653**, 1–15.

Parker, G. A., and Courtney, S. P. (1984). Models of clutch size in insect oviposition. *Theor. Populat. Biol.* **26**, 27–48.

Parson, P. A. (1977). Lek behavior in *Drosophila polypori*, an Australian rain forest species. *Evolution* **31**, 223–225.

Paviour-Smith, K. (1960). The fruiting-bodies of macrofungi as habitats for beetles of the family Ciidae (Coleoptera). *Oikos* **11**, 43–71.

Pielou, D. P., and Verma, A. N. (1968). The arthropod fauna associated with the birch bracket fungus, *Polyporus betulinus*, in eastern Canada. *Can. Entomol.* **100**, 1179–1199.

Pimm, S. L. (1982). "Food Webs". Chapman & Hall, London.

Pyysalo, H. (1976). Identification of volatile compounds in seven edible fresh mushrooms. *Acta Chem. Scand.* **30**, 235–244.

Ramsbottom, J. (1953). "Mushrooms and Toadstools". Collins, London.

Rautavaara, T. (1947). "Suomen Sienisato". WSOY, Porvoo.

Real, L. A. (1980). Fitness, uncertainty, and the role of diversification in evolution and behavior. *Am. Nat.* **115**, 623–638.

Rehfous, M. (1955). Contribution à l'étude des insectes des champignons. *Mitt. Schweiz. Entomol. Ges.* **28**, 1–106.

Rhoades, D. F. (1979). Evolution of plant chemical defense against herbivores. *In* "Herbivores. Their Interaction with Secondary Plant Metabolites" (G. A. Rosenthal and D. H. Janzen, eds), pp. 3–54. Academic Press, New York.

Rhoades, D. F., and Cates, R. G. (1976). Toward a general theory of plant antiherbivore chemistry. *Recent Adv. Phytochem.* **10**, 168–213.

Richardson, M. J. (1970). Studies on *Russula emetica* and other agarics in a Scots pine plantation. *Trans. Br. Mycol. Soc.* **55**, 217–229.

Robinson, W. (1977). Phorids associated with cultivated mushrooms in the eastern United States. *Proc. Entomol. Soc. Wash.* **79**, 452–462.

Russell-Smith, A. (1979). A study of fungus flies (Diptera: Mycetophilidae) in beech woodland. *Ecol. Entomol.* **4**, 355–364.

Shorrocks, B., and Charlesworth, P. (1980). The distribution and abundance of the British fungal-breeding *Drosophila. Ecol. Entomol.* **5**, 61–78.

Shorrocks, B., and Charlesworth, P. (1982). A field study of the association between the stinkhorn *Phallus impudicus* Pers. and the British fungal-breeding *Drosophila. Biol. J. Linn. Soc.* **17**, 307–318.

Shorrocks, B., and Rosewell, J. (1986). Guild size in drosophilids: a simulation model. *J. Anim. Ecol.* **55**, 527–542.

Shorrocks, B., and Rosewell, J. (1988). Spatial patchiness and community structure: coexistence and guild size on ephemeral resources. *In* "Organization of Communities: Past and Present" (J. H. R. Gee and P. S. Giller, eds.), Blackwell, Oxford (in press).

Sivinski, J. (1981). Arthropods attracted to luminous fungi. *Psyche* **88**, 383–390.

Slatkin, M. (1974). Competition and regional coexistence. *Ecology* **55**, 128–134.

Snetsinger, R. (1972). Laboratory studies of mushroom infesting arthropods. *Mushroom Sci.* **8**, 199–208.

Southwood, T. R. E. (1977). Habitat, the templet for ecological strategies? *J. Anim. Ecol.* **46**, 337–365.

Spencer, K. A. (1972). "Handbooks for the Identification of British Insects. Diptera Agromyzidae". R. Ent. Soc. London.

Ståhls, G. (1987). Svamplevande Dipterer i Kilpisjärvitrakten. M.Sc. Thesis, Univ. Helskinki (unpublished).

Stephens, D. W., and Krebs, J. R. (1986). "Foraging Theory". Princeton Univ. Press, Princeton.

Strong, D. R., Lawton, J. H., and Southwood, T. R. E. (1984). "Insects on Plants". Blackwell, Oxford.

Terriere, L. C. (1984). Induction of detoxication enzymes in insects. *Annu. Rev. Entomol.* **29**, 71–88.

Tiensuu, L. (1935). Die bisher aus Finnland bekannten Musciden. *Acta Soc. Fauna Flora Fenn.* **58**, 1–56.

Ulvinen, T. (1976). "Suursieniopas". Suomen Sieniseura, Helsinki.

Väisänen, R. (1981). Is there more than 1 successional phase in the mycetophilid (Diptera) community feeding on a mushroom? *Ann. Zool. Fenn.* **18**, 199–202.

Van Valen, L. (1965). Morphological variation and the width of the ecological niche. *Am. Nat.* **100**, 377–389.

Vet, L. E. M. (1983). Host habitat location through olfactory cues by *Leptopilina clavipes* (Hymenoptera: Eucoilidae), a parasitoid of fungivorous *Drosophila*: the influence of conditioning. *Neth. J. Zool.* **33**, 225–248.

Vogt, K. A., and Edmonds, R. L. (1980). Patterns of nutrient concentration in basidiocarps in western Washington. *Can. J. Bot.* **58**, 694–698.

Welch, H. E. (1959). Taxonomy, life cycle, development, and habits of two new species of Allantonematidae (Nematoda) parasitic in drosophilid flies. *Parasitology* **49**, 83–103.

Wheeler, Q. D. (1984). Evolution of slime mold feeding in leiodid beetles. *In* "Fungus–Insect Relationships, Perspectives in Ecology and Evolution" (Q. D. Wheeler, and M. Blackwell, eds), pp. 446–478. Columbia Univ. Press, New York.

Wong, W. C., and Preece, T. F. (1980). *Pseudomonas tolaasi* in mushroom crops: a note on primary and secondary sources of the bacterium on a commercial farm in England. *J. Appl. Bacteriol.* **49**, 305–314.

Wyatt, I. J. (1959). Some aspects of cecid biology of importance in mushroom culture. *Mushroom Sci.* **4**, 271–279.

3

The Mutualistic Association between Macrotermitinae and *Termitomyces*

T. G. WOOD AND R. J. THOMAS

I. INTRODUCTION

Ever since König's (1779) observations on the fungi growing on comb-like structures in the nests of Macrotermitinae (H9), there have been many observations and much speculation on the relationship between the termites, the fungi observed on the combs and the occasional, often seasonal, appearance above ground of basidiocarps of the genus *Termitomyces* (F5.10) growing out of the combs. The combs are convoluted, greyish-brown and in the nest the most obvious manifestation of the fungus are the round, white bodies ("mycotêtes") on their surface. The relationship between the termites and *Termitomyces* has been confused by the common observation (e.g. Grassé, 1937) that when the combs are removed from the nest they are rapidly covered by a dense growth of hyphae of many fungi, dominantly *Xylaria* (F4.34). Bathellier (1927) confirmed Petch's (1906) suspicions that the mycotêtes on the comb were part of the life-cycle of the *Termitomyces* basidiocarp, but the relationships between the termites, *Termitomyces* and associated fungi on the comb are still only partially understood in spite of significant advances in the last twenty to thirty years.

Sands (1969) reviewed current knowledge of the relationship between *Termitomyces* and the termites and the function of the fungus combs, and in particular showed that the termites could not survive without the combs. Further observations and, more recently, specific investigations have increasingly indicated the strong mutual dependency of the termites and the fungus and the complexity of this relationship. In this paper we review this relationship in the light of recent research and indicate gaps in our knowledge and avenues for future research.

II. THE ORGANISMS—TERMITES AND FUNGUS

A. The Termites—Macrotermitinae

1. Distribution and Abundance

Macrotermitinae are distributed throughout tropical Africa and parts of Arabia and Indomalaya. Some species, particularly *Macrotermes* in Africa and *Macrotermes* and *Odontotermes* in Indomalaya, construct large mounds, up to 9 m high (Fig. 1A), which are very conspicuous and have been the subject of most investigations of the termite–fungus relationship. Most species, however, construct entirely subterranean nests. They are one of four subfamilies of the Termitidae, collectively known as "higher termites", which constitute approximately 75% of all termite species. All other termites belong to one or other of the six families collectively known as "lower termites". The latter are characterized by their symbiotic intestinal protozoa on which they depend for digestion of the complex polysaccharides (cellulose, lignin) in their food—wood, grass or various forms of plant debris. The Termitidae have no symbiotic protozoa; their digestive processes are less well understood, but depend partly on their own digestive enzymes, including cellulases, and symbiotic microorganisms in the gut (Breznak, 1983; Wood and Johnson, 1986).

Macrotermitinae often dominate the termite fauna in tropical savannas and forests (Abe and Matsumoto, 1979; Abe and Watanabe, 1983; Wood and Sands, 1978), reaching live weight biomasses of up to 10 $g\,m^{-2}$. They are general detritivores, feeding mainly on dead wood, dead grass, dung; some also feed on roots of dead and living plants. They often play a dominant role in litter removal (Abe, 1980; Buxton, 1981; Collins, 1981a; Josens, 1977; Lepage, 1981). For example in Southern Guinea savanna in West Africa the annual production of wood-, grass- and leaf-litter was estimated at 535 $g\,m^{-2}$; Macrotermitinae consumed approximately 26.5% and other termites 7.5% (Collins, 1981a; Wood and Sands, 1978). The factors contributing to these large consumption rates are their high numerical abundance, their high production : biomass ratios (2–4 times that of other termites, Wood, 1978; Collins, 1983) and the metabolic rate of the fungus combs which is up to 5.5 times that of the termite colony (Rohrmann, 1977; Wood and Sands, 1978; Collins, 1983).

The processing of these food resources is a vital function of the termite–fungus system (Section III, B). In addition to fungus combs some species, such as *Macrotermes bellicosus* (Smeathman), have food stores (Fig. 1B) which have the general appearance of moist sawdust and consist of comminuted food mixed with saliva. After approximately one week the food store is re-ingested and passed through the gut to be deposited as fresh faeces on the fungus comb (Collins, 1977; unpublished). It is not known whether all food is processed in this way or whether some food by-passes the food store.

2. *Fungus Combs*

The controversy as to whether fungus combs were constructed from food material or faeces was summarized by Sands (1969), who favoured a faecal origin based on his earlier observations (Sands, 1960) of *Ancistrotermes* behaviour. The staining techniques adopted by Josens (1971) comprehensively confirmed that combs are constructed from fresh faecal material. In some species they are located centrally in a hive or brood chamber, as in *Macrotermes* and some *Odontotermes*; in many subterranean species there is no central hive and the combs are located in small chambers distributed throughout the soil to depths of several metres. The combs are dynamic structures. As new faecal material is added, old parts of the comb are eaten by the termites resulting in a "turnover" time of five to eight weeks (Josens, 1971). Mycotêtes on the combs are also consumed by the termites. The net result is the almost complete decomposition of faecal material with no direct return to the ecosystem, except on death of the colony. Return of energy and nutrients to the ecosystem is largely via the diverse vertebrate and invertebrate predators that exploit alate swarms, foraging parties or colony brood centres (Wood and Sands, 1978).

The fungus combs vary in shape and size. Some, such as those of many

Fig. 1. *Macrotermes bellicosus.* (A) A large mound of a mature colony and (B) the central portion (hive) of a young colony, removed to show foodstore (fs) and fungus comb (fc). (Fig. 1B photograph by N. M. Collins.)

Microtermes, are small (2–4 cm diameter), subspherical and more or less solid. Many others have complex cellular and laminar patterns that obviously increase the surface area per unit weight of fungus comb. The adaptive advantages of comb architecture to the termites and fungus are unknown.

3. Colony Foundation

Colony foundation is an integral aspect of the termite-fungus relationship (Section III, A). Each year sexually mature, winged males and females fly from the nests in large swarms. They are usually produced at fairly specific seasons, often in response to rain. The flights for a given species are usually synchronous and can consist of 40% of the colony biomass (Wood and

Sands, 1978). Successfully paired alates have the potential for forming new colonies once they have burrowed into the soil. The first batch of eggs hatch into workers which forage for food and their faecal deposits are the beginning of the first fungus comb. Unless the comb is inoculated with *Termitomyces* the colony dies (Johnson, 1981).

B. The Fungus—Termitomyces

1. Termite–Fungus Associations

Termitomyces is a basidiomycete (family Amanitaceae (F5.10); Heim, 1942b) and occurs only in association with Macrotermitinae. It is less well known biologically and taxonomically than the termites. The mycotêtes on the comb have variously been called spherules (Batra and Batra, 1979), nodules and synnemata (Martin and Martin, 1978), conidial spheres (Coaton, 1961) and conidia (Abo-Khatwa, 1977). They consist of spherocysts bearing conidia of blastosporic ontogony and are the asexual stage of *Termitomyces*. The fungus is not found elsewhere in the nest structure (Thomas, 1987a, b). Outside the fungus combs it only occurs in the guts of workers (Thomas, 1987a, b) and in some species the guts of winged reproductives (Johnson *et al.*, 1981). The sexual stage of the fungus, the basidiocarp, arises as a pseudorhiza from a fungus comb and continues as a stipe, terminating in a cap with a prominent umbo or perforatorium. They usually grow directly from the comb and therefore have to penetrate the soil or the walls of mounds. Some species of *Odontotermes* bring fragments of comb onto the soil surface, giving rise to many small mushrooms of *Termitomyces microcarpus* (Berk. and Br.) Heim. The mushrooms usually appear at specific times in the rainy season, but as they last no more than a few days, are easily overlooked. Reviews by Sands (1969) and Thomas (1981, unpublished) indicated that there is some degree of specificity in the association between termite genera and species of fungus; apparently anomalous relationships are likely to be due to misidentification of either termites or fungus. The known termite–fungus associations are summarized in Table I. Basidiocarps are unknown in certain termite–*Termitomyces* associations (e.g. *Microtermes*) and therefore the identity of the fungus is not known.

2. *Other Fungi in the Nest*

A wide range of other fungi are associated with *Termitomyces* on the comb and are also found in other regions of the nest where *Termitomyces* does not occur (Das *et al.*, 1962; Batra and Batra, 1966; Mohindra and Mukerji, 1982; Ruyooka, 1979; Singh *et al.*, 1978; Thomas 1987a, b). These fungi consist of:

1. common soil fungi which are carried by termites on their bodies and collected food (e.g. *Aspergillus* (F6.2), *Penicillium* (F6.2)). In *Macrotermes bellicosus* comminuted food mixed with saliva is deposited in the food store, re-ingested after approximately one week, or passed directly through the gut to be deposited on the fungus comb (see Section II, A). The fate of these fungi is shown in Table II. The number of species declines from 21 in soil walls of the mound to 14 in the food store, 12 in fresh comb, 11 in middle comb and 8 in old comb. The large drop in the number of propagules from 230 in the food store to 9 in fresh fungus comb indicates that passage through the termite gut and/or conditions on the comb greatly reduced the viability and number of spores. *Termitomyces* does not occur in the

TABLE I. The association of termites with *Termitomyces* basidiocarps.

Termitomyces species	Termite species	Author	Comments
Termitomyces microcarpus (Berk. & Br.)	*Odontotermes badius* (Haviland)	Bottomley and Fuller (1921)	South Africa
		Coaton (1961)	South Africa
		Heim (1942a, 1958, 1977)	Africa
		Sands (1956)	Kenya
	O. kibarensis (Fuller)	Mukiibi (1973)	Uganda
		Piearce (pers. comm.)	Uganda
	O. latericius (Haviland)	Heim (1977)	
		Piearce (pers. comm.)	Zambia
	O. obesus (Rambur)	Bose (1923)	India
	O. transvaalensis (Sjöstedt)	Bottomley and Fuller (1921)	South Africa
		Coaton (1961)	South Africa
		Heim (1942a, 1958, 1977)	Africa
	O. vulgaris (Haviland)	Bottomley and Fuller (1921)	South Africa
		Coaton (1961)	South Africa
		Heim (1958, 1977)	Africa
	Odontotermes spp.	Heim (1951, 1952b, 1958)	Intertropical and Southern Africa, India, Sri Lanka, Indo-China
		Petch (1906, 1913)	Sri Lanka
	Macrotermes falciger (Gerstäcker)	Piearce (pers. comm.)	Zambia
	Macrotermes natalensis (Haviland)	Heim (1942a, 1977)	Not *natalensis* but *subhyalinus*; doubtful if association is with termite. Guinea Republic
	Microtermes obesi Holmgren	Batra and Batra (1979)	India. Circumstantial evidence
Termitomyces albuminosus (Berk.) Heim	*Odontotermes badius*	Coaton (1961)	Africa
	O. gurdaspurensis Holmgren K. & N.	Batra and Batra (1979)	India
	O. horni (Wasmann)	Bathellier (1927)	Vietnam
		Heim (1977)	Central Asia

(Continued)

TABLE I. Continued.

Termitomyces species	Termite species	Author	Comments
	O. obesus	Batra and Batra (1979)	India
		Bose (1923)	India
		Heim (1977)[a]	Central Asia
	O. obscuriceps (Wasmann)	Petch (1906)	Sri Lanka
	O. redemanni (Wasmann)	Petch (1906)	Sri Lanka
	O. sundaicus (Kemner)	Kemner (1934)	
	Microtermes insperatus (Kemner)	Kemner (1934)	Termite has morphological features not characteristic of *Microtermes*
T. cartilagineus (Berk) Heim	*O. horni*	Grassé and Heim (1950)	
		Heim (1942a)	Indo-China
T. clypeatus Heim	*O. lacustris* Harris	Piearce (pers. comm.)	Zambia
T. entolomoides Heim	*Protermes minutus* (Grassé)	Heim (1952b, 1977)	Zaire
T. eurhizus (Berk) Heim	*O. badius*	Pegler and Rayner (1969)	Kenya. Circumstantial evidence
	Pseudacanthotermes spiniger (Sjöstedt)	Piearce (pers. comm.)	Zambia
T. fuliginosus Heim	*Acanthotermes acanthothorax* (Sjöstedt)	Heim (1940, 1951, 1958, 1977)	Congo, Guinea Republic
		Van Ryn (1974)	French Guinea, Zaire
T. globulus Heim and Goossens	*O. badius*	Pegler and Rayner (1969)	Kenya. Circumstantial evidence only
T. letestui (Pat.) Heim	*Macrotermes natalensis*	Heim (1948)	Not *natalensis*, possibly *subhyalinus*. Cameroon. This is a very doubtful association as Heim only identified it as this species or *T. schimperi* from mycotêtes

T. mammiformis Heim	*Pseudacanthotermes* (probably *militaris* Sands 1969)	Heim (1940)	Guinea Republic
	Pseudacanthotermes and *Acanthotermes*	Heim (1977)	Congo, Guinea Republic
T. medius Heim & Grassé	*Ancistrotermes latinotus* (Holmgren)	Coaton (1961)	Zambia
		Grassé and Heim (1950)	Central African Republic
	Ancistrotermes sp.	Heim (1951)	Central African Republic
	Odontotermes patruus (Sjöstedt)	Piearce (pers. comm.)	Zambia
T. robustus (Beeli) Heim	*Acanthotermes acanthothorax*	Heim (1951, 1952b, 1958, 1977)	Congo and Zaire. (combs not this sp.; Sands, 1969)
	Macrotermes natalensis	Zoberi (1979)	Nigeria. Not clear from paper if this is the association
T. schimperi (Pat.) Heim	*Macrotermes natalensis*	Heim (1958)	Not *natalensis*, possibly *subhyalinus*. Zaire, Ivory Coast, Guinea Republic, Cameroon, Ethiopia
	M. natalensis	Heim (1951, 1952b)	Cameroon, Ethiopia
	M. natalensis	Heim (1977)	Cameroon, Ethiopia
	M. natalensis	Piearce (pers. comm.)	Zambia
	Odontotermes patruus perhaps others		
T. striatus (Beeli) Heim	*Pseudacanthotermes militaris* (Hagen)	Heim (1940, 1942a, 1948)	Guinea Republic, Cameroon, Liberia, Sierra Leone, Ivory Coast
	P. militaris and on nests of several other species	Heim (1952b, 1977)	As above and Congo
var. *aurantiacus* Heim	*P. militaris* and other species	Heim (1977)	Subtropical West Africa, Congo, equatorial Africa

[a] Natarajan (1979) considers the species to be *T. heimii* Natarajan.

TABLE II. Variety and abundance of fungi in nests of *Macrotermes bellicosus* (adapted from Thomas, 1981, unpublished, 1987c).

Location	Number of species (dp):[a] SEL + SF[b]	Number of propagules (dp) ($\times 10^3$) per g d.w. SF		Isolation index[c]			
				Termitomyces		Other fungi	
		Termitomyces	Other fungi	SEL	SF	SEL	SF
Soil walls	21	0	16				
Foodstore	14	0	230	0	0	30	121
Fresh comb	12	20	9	72	39	0	61
Middle comb	11	440	15 }				
			}	89	49	6	61
Old comb	8	30	8 }				

[a] dp, dilution plating.
[b] SF, soil fungus medium; SEL, selective medium for *Termitomyces* (Thomas, 1985).
[c] Isolation index, number of isolates × 100/number of isolations attempted.

food store, but totally dominates the fungal flora of the fungus comb. The evidence points to the fungus comb as being a very selective environment—favourable to *Termitomyces* but unfavourable to other fungi.

2. *Xylaria* (F4.34), which Grassé (1937) reported to exist as mycelium in the comb but which does not produce spores or stroma until the comb is removed from the nest or is abandoned by the termites. The commonest and most widely distributed is *X. nigripes* (Klotzsch) Cook which has only been recorded in association with nests of Macrotermitinae. The consensus of evidence (Sands, 1969; Batra and Batra, 1979; Thomas, 1987a, b, c) is that it is not found as active mycelium in the combs but is an obligate saprophyte which becomes active when combs die or are removed from the nest (see Section III, C).

III. THE TERMITE–FUNGUS SYSTEM

A. Reproductive Ecology

The question to be faced is: how do new termite colonies acquire an inoculum of *Termitomyces* and what are the roles of the asexual spores on the fungus comb and the basidiospores on the basidiocarp in inoculation? There appear to be at least two answers (Johnson *et al.*, 1981; Sieber, 1983). In all species of *Microtermes* that have been studied, the female alates ingest asexual spores on the combs prior to flying in swarms from the parent colony. The spores are carried as a bolus in the gut and pass through the alimentary canal onto the primordial fungus comb to inoculate the first faecal deposits of foraging workers. There is similar behaviour in *Macrotermes bellicosus*, except that it is the male that carries the bolus of spores. Basidiocarps are not known in association with these species, but have been seen in all other "non-carrying" species studied (except *Odontotermes smeathmani* (Fuller), Johnson *et al.*, 1981). In these species the appearance of basidiocarps coincides with the appearance of the first foraging workers from newly established colonies. Sieber (1983) showed that new colonies of *Macrotermes michaelseni* (Sjöstedt) and *Odontotermes montanus* Harris established fungus comb only after foraging workers had consumed basidiospores. It appears that in non-carrying species the asexual spores have no reproductive function. These relationships have only been studied in a handful of species in West and East Africa and the subject is ripe for further investigations.

These conclusions raise interesting questions with regard to the genotype of the fungus. Presumably in *Termitomyces* which do not have basidiocarps, within a given species of termite, the fungus is of uniform genotype. In species which produce basidiospores foraging termites from already existing colonies will transfer these to their own fungus combs where they will presumably germinate to monokaryotic hyphae. These hyphae will have the potential to fuse with existing monokaryotic hyphae to form dikaryotic hyphae and thus new basidiocarps. As far as we are aware this subject has not been investigated.

B. Utilization of Food

The digestive processes of termites are based largely on symbiotic relationships with various microorganisms (O'Brien and Slaytor, 1982; Breznak, 1983). Four groups can be recognized (Wood and Johnson, 1986): lower termites (all families except Termitidae), Macrotermitinae, soil-feeding termites, and other Termitidae. All groups produce various digestive enzymes. Macrotermitinae have been shown to produce in the gut, chitinase (Rohrmann and Rossman, 1980), cellobiase (Abo-Khatwa, 1978), amylase, invertase and β-galactosidase (Singh, 1976), and various cellulases (Abo-Khatwa, 1978; Martin and Martin, 1978; Mishra and Ranganathan, 1954). Some of these may be produced by gut bacteria (Singh, 1976).

Enzymes produced by *Termitomyces* include C_1 cellulase, C_x cellulase and β-glucosidase (all necessary for cellulose degradation, Martin and Martin, 1978) and lignases (Rohrmann and Rossman, 1980; Thomas, 1981, unpublished). The latter author showed that cultured mycelium produced few or no enzymes capable of degrading starch, chitin or pectins, although Gosh and Sengupta (1987) isolated a multisubstrate amylase from cultures and Mishra and Sen-Sarma (1985) detected sucrase, maltase, melibiase, cellobiase, lactase, amylase, xylanase, gallactanase, proteinase, polypeptidase and lipase in the mycotêtes.

Fresh faecal deposits on the fungus comb ("mylospheres", Grassé, 1978) consist of finely comminuted, largely structurally unaltered plant material (Rohrmann, 1978). It appears that the fungus breaks down lignin and cellulose and the breakdown products are utilised by the termites when they re-ingest "old" portions of the comb. The presence of cellulases in the termite gut partly results from the consumption of "mycotêtes" which contain C_1 cellulases (Martin and Martin, 1978, 1979). These authors showed that C_x cellulase and β-glucosidase in the gut were derived partly from ingested fungal material and partly from the midgut epithelium and salivary glands. The β-glucosidase produced by the fungus has different characteristics from that produced by the termites (Rouland *et al.*, 1986). Cellulolytic bacteria are present in the gut but their significance is unknown (Rohrmann and Rossman, 1980). Thus enzymes in the gut are acquired through ingestion of mycotêtes, while others are derived from gut bacteria or synthesized by the termites.

The key to the efficient digestion of plant material lies in the combined enzyme system of the termites and the *Termitomyces*, and the termites' behaviour of rapid recycling of the comb. The fungus is provided with a highly comminuted supply of food which it can rapidly invade and utilise and the termites are provided with partially "digested" food, relatively rich in nitrogen. Collins (1977, unpublished), Matsumoto (1976) and Thomas (1981, unpublished) showed that the net effect was to convert food of low N-content (0.3%) to fungus comb (0.8–2.0% N), mycotêtes (6–8% N) and termites (9–10% N). In quantitative terms it appears that 90–100% of the dietary nitrogen is used in the production of termite tissue (Collins, 1983).

This picture of food utilization is based on studies of some of the more advanced Macrotermitinae. A recent study of a more primitive species, *Macrotermes muelleri* (Sjöstedt) shows that although fungus combs are constructed from fresh faecal deposits ("mylospheres"), re-ingestion of comb does not result in near complete utilization of food, as small, but significant quantities of black faecal material ("final faeces") are found under the combs (Garnier-Sillam, 1987).

C. Competition Between *Termitomyces* and Other Fungi in the Fungus Combs

We have already noted the common observation that when fungus combs are removed from the nest they are rapidly covered by a dense growth of hyphae of fungi other than *Termitomyces*. In ordinary soil fungus medium *Termitomyces* grows very slowly and is rapidly swamped by other fungi. In order to culture it in the laboratory Thomas (1985) developed a selective medium that suppressed the growth of other microorganisms. These observations indicate that *Termitomyces* is a poor competitor. How, then, is it maintained on the combs where it has to compete for resources with other microorganisms?

Various mechanisms have been suggested for preventing the development of other fungi on combs in the nest. These are discussed below, and possibly a combination of factors are responsible, but whether or not this is the case, it now appears that inhibition by termite secretions is a major factor (Batra and Batra, 1966; Batra, 1971; Thomas, 1987c).

a. Mechanical activity of the termites. Within the nest combs are regularly re-ingested, comminuted and recycled but there is no evidence for physical "weeding" of fungi, although Batra and Batra (1979) observed workers removing mycelium from the comb and burying it in the soil.

b. Inhibition by termite secretions. Some Macrotermitinae have a "foodstore" which has the general appearance of moist sawdust and consists of comminuted food mixed with saliva. In *Macrotermes bellicosus* it contains spores of a wide variety of fungi, but very few germinate when foodstore is incubated, although many germinate when spores are washed out of the foodstore (Thomas, 1987c). This suggests that some factor in the foodstore (e.g. saliva) inhibits spore germination and growth. Batra and Batra (1966) showed that soil that has recently been manipulated by worker termites, and therefore was mixed with saliva, was fungistatic. Thomas (1987c) showed that benzene and methanol extracts of whole *Macrotermes bellicosus* workers and worker guts and methanol extracts of foodstore significantly reduced spore germination. Batra and Batra (1966) indicated that the defensive oral secretion of *Odontotermes gurdaspurensis* Holmgren K. and N. soldiers was fungistatic. They also showed that saliva of workers promoted the growth of mycotêtes. Exudates of queens have also been reported as inhibiting fungal germination (Sannasi and Sundara Rajulu, 1967).

c. Nest microclimate. The temperature around fungus combs in *Macrotermes* mounds (see Section III, D below) is fairly constant at around

30°C, which is close to the optimum for growth of *Termitomyces* (Thomas, 1987c). High CO_2 concentrations (1.2–5.2%) have been found in *Macrotermes* mounds (Matsumoto, 1977). These levels may inhibit growth and germination of some fungi, such as *Penicillium* (F6.2), whereas many others would be unaffected (Burges and Fenton, 1953). There can be little or no control of microclimate around the fungus combs of *Microtermes* (see Section III, D below).

d. Antibiotic production by Termitomyces. Heim (1952a) noted antagonism between *Termitomyces* (F5.10) and *Xylaria* (F4.34) mycelium and Thomas (1987c) noted that growth of *Xylaria* was slightly retarded in the presence of *Termitomyces*. Otherwise there is no evidence for antibiotic production by *Termitomyces*, and as has been noted earlier, it is rapidly overgrown by other fungi in culture and when combs are removed from the nest.

e. Chemical composition of the comb. The low pH (4.1–4.6) of the fungus comb would tend to prevent the development of bacteria; otherwise the comb appears to be a favourable substrate for microbial activity (Thomas, 1987c).

D. Environmental Regulation Within the Nest

There is a vast literature on the size, shape and architecture of termite nests (particularly mounds) in relation to environmental conditions inside the nest (Lüscher, 1961; Noirot, 1970; Lee and Wood, 1971). Conditions within mounds of Macrotermitinae are remarkably constant: in *Odontotermes obesus* (Rambur) there are no diurnal fluctuations, and seasonal variations are within 4°C and 4% relative humidity (Agarwal, 1980). In mounds of *M. bellicosus* the temperature of the habitacle is maintained at around 30°C, which Thomas (1981, unpublished) showed to be the optimum for growth of *Termitomyces*. We have already referred to the fact that most of the collected food is metabolized by the fungus combs. The complex internal architecture of the *Macrotermes* nests described by Lüscher (1961), Ruelle (1964) and Collins (1979) is designed to dissipate excess heat. In regions with more variable climates the nests are designed to dissipate heat in the hot season and to retain it in the cool season (Rajagopal, 1982). Without the large production of heat from centrally located fungus combs, this regulation would not be possible.

A stable microclimate within the nest throughout the year has obvious advantages for growth and development of the colony. In contrast, the termite–fungus relationship can have very little impact on environmental regulation in *Microtermes*, where the fungus combs occur singly in small chambers 2–4 cm in diameter, distributed throughout the soil.

E. Evolution of the Termite–Fungus Relationship

Heim's (1952a) view that *Termitomyces* was a commensal tolerated by the termites led him to propose (Heim, 1977) that the ancestral *Eutermitomyces* grew through the soil from underground combs, and that the next step was the *Praetermitomyces* which grew on combs which were expelled from the nest by the termites in much the same way as some present day *Odontotermes* (see Section II, B). This was followed by the non-basidiocarp forming *Termitomyces* of *Microtermes* and finally with the appearance of *Termitomyces*—free combs of *Sphaerotermes sphaerothorax* (Sjöstedt) which rid themselves of the mycotêtes and mycelium as well.

There are several objections to this theory. Firstly *Sphaerotermes*, which builds comb-like structures without *Termitomyces*, is a primitive member of the subfamily as opposed to being at the peak of its evolution. Secondly, the evidence presented in this paper indicates a very close, interdependent relationship between termites and the fungus. The fungus has no active existence outside the fungus combs, the termite and the fungus have complementary, interactive enzyme systems which produce a relatively nitrogen-rich diet for the termites, and termite secretions are a major factor in suppressing other fungi on the fungus combs that could compete with *Termitomyces*. It is likely that the relationship has evolved gradually with mutual adaptations causing the relationship to become even closer, rather than by a process of gradually eliminating the fungus. Many species of termites construct carton-like structures in their nest from faeces, often in concentrated form (Lee and Wood, 1971). These structures are re-ingested as nests are enlarged and, in a macrotermitine ancestor with an inefficient digestive system, they would present an unexploited niche which could be invaded by a saprophytic fungus tolerant of the environment in the nest. The co-evolution of complementary enzyme systems should, theoretically, produce species–species associations of differing efficiencies. This is a virtually untouched subject but there is some evidence to show that its study could be rewarding. Thomas (1981, unpublished) showed that polyphenol oxidase production (which is correlated with lignin degradation) by *Termitomyces* associated with *Macrotermes bellicosus* was lower than in the *Termitomyces* associated with the more advanced *Microtermes*. The net effect of differing efficiencies in enzyme systems may explain the fact that abandoned, old combs can be found in *Macrotermes* nests (Collins, 1977, unpublished), possibly due to accumulation of undigestable components during the recycling of comb material, whereas in *Microtermes* the combs are almost totally metabolized (Wood and Johnson, 1978). The loss of the perfect stage of the fungus would appear to have occurred at least twice during the evolution of the relationship—once leading to the *Macrotermes bellicosus* association and again leading to the *Microtermes* association. Further studies of the efficiency of species–species enzyme systems and reproductive ecology should help to elucidate the evolution of the multi-faceted termite–fungus relationships which have led to the great variety of nest and comb architectures.

IV. UTILIZATION OF THE TERMITE–FUNGUS ASSOCIATION

Possible outlets for utilizing our knowledge of the mechanisms involved in sustaining these highly effective mutualistic associations include:

1. Control of termites using fungicides to depress fungal activity.
2. Enzymatic hydrolysis of lignocellulose to yield fermentable sugars for conversion to liquid fuels (Osore, 1985).
3. Development of fungal growth regulators by analysis of the fungistatic properties of termite secretions.

Research on (1) is being carried out at the UK's Overseas Development Natural Resources Institute (ODNRI) (ex-Tropical Development and Research Institute) in collaboration with the University of Khartoum, Sudan. Research on (2) is being carried out at the International Centre for Insect Physiology and Ecology in Nairobi, Kenya. As far as we are aware, with the exception of the works quoted in this paper, there has been no research on (3) and none is being pursued.

To complete this paper we will outline the ODNRI's research on developing new, selective methods of termite control using fungicides.

Macrotermitinae cause widespread and severe losses to field crops, tree crops, forestry trees, domestic dwellings and wooden structures in Africa and Indomalaya. They are effectively controlled using cyclodiene insecticides which give the necessary toxicity and persistence. Worldwide there are ever increasing restrictions placed on the use of these compounds, due to greater public and governmental awareness of their health and environmental hazards. There are also economic reasons in developing countries, as exportable food and food products (e.g. groundnut oil) contaminated with detectable residues of these compounds are refused by many importing countries. Currently, there are no commercially available alternatives. Basic research on the termite–fungus association (Thomas, 1981, unpublished; Thomas, 1985, 1987a, b, c) opened up the possibility of developing a new control strategy that would be specific to the Macrotermitinae, whereby foraging termites could be presented with fungicide-impregnated food, resulting in the deposition of fungicidal faeces on the combs. The research has now progressed through the following stages:

1. Bioassay of fungicides against *Termitomyces* in culture.
2. Bioassay of compounds toxic to *Termitomyces* against Macrotermitinae for repellency, toxicity and palatability.
3. Bioassay of faecal residues for fungicidal activity.
4. Bioassay of compounds, after passing (1), (2) and (3), against young laboratory-reared termite colonies.
5. Durability of fungicide-impregnated baits in a tropical environment.
6. Pilot field trials.

The research is currently at (6), where there has been little success. Trials with ^{14}C-labelled fungicide showed that subterranean combs of *Microtermes* contained fungicide which is undoubtedly being diluted by faecal deposits from natural sources of food.

V. DISCUSSION

Insects having mutualistic relationships with fungi include certain midges (Diptera, Itonididae (H28.9)), scale insects (certain Diaspididae (H20)), wood wasps (Siricidae (H33.1)) of the genera *Sirex, Tremex* and *Urocercus*, ship timber worms (Coleoptera, Lymexylidae (H25.24)), bark and ambrosia beetles (H25.63) (Beaver, this volume) and two groups of social insects, Macrotermitinae (H9) and the leaf-cutting ants (H33.9) (Cherrett *et al.*, this volume).

If "success" of a group of animals can be evaluated in terms of their ability to exploit abundant natural resources then ants and termites are undoubtedly successful (Brian, 1978). In discussing reasons for the existence of the ant–fungus mutualism, Cherrett (1980) posed the fundamental question, "what are the selective advantages of a fungus-cultivating habit which have caused it to evolve in competition with the normal habit of direct consumption of the leaf by the insect?". A similar question could be asked of the Macrotermitinae, as other groups of termites form one of the dominant groups of decomposers in tropical regions, such as Australasia and South America, where Macrotermitinae are absent. In both ants and termites the fungus has the ability to degrade cellulose, and one could reason that this enables both groups to tap a virtually inexhaustible supply of food. This was in fact suggested by Hodgson (1955) as the reason for the success of the leaf-cutting ants. However, this was discounted by Cherrett (1980) as only 5% of the energy requirements of the colony come from the fungus, the rest being supplied by plant sap. The main reason for the ant–fungus mutualism appears to be the increased polyphagy made possible for both fungus and ant through their alliance to break through the plants' chemical defence mechanisms. On the other hand termites evolved from roach-like ancestors that fed on decaying plant material; throughout their evolution fungi have been involved in their diet, and fungus-rotted wood is a preferred food source for many termites (Sands, 1969).

Thus polyphagous detritivores, like the Macrotermitinae, have never been involved in the insect–plant "arms race" of plant chemical defences versus insect detoxification, digestion or behaviour, although some dead plants retain compounds that make them distasteful or repellent (Wood, 1978). Dead plant material has high C : N and lignin : cellulose ratios and is difficult to digest. Termites accomplish this digestion through the combined enzyme system of their gut and their gut symbionts. Their ability to exploit dead plant material is partly limited by the number of symbionts that can be accommodated in the gut. The Macrotermitinae have by-passed this limitation by cultivating *Termitomyces* outside the gut. Rohrmann and Rossman (1980) estimated that 27–33% of the comb consisted of *Termitomyces*. If we take a figure of 30% and apply it to the biomass of *Macrotermes bellicosus* of 0.5 $g\,m^{-2}$ (Collins, 1981b) and a mass of fungus combs of 2.6 $g\,m^{-2}$ (Collins, 1977, unpublished) we find that the biomass of *Termitomyces*, 0.78 $g\,m^{-2}$, is of a similar biomass to that of the termites. The net result is efficient conversion of energy and nitrogen in dead plant material into termite tissue at a much greater rate and efficiency than is achieved by other termites.

Similarly, the fungus is presented with large quantities (up to 25% of available dead plant material) of comminuted food through which hyphae can rapidly spread, in a favourable microclimate in which competitors are suppressed. In contrast "free-living", wood-rotting basidiomycetes have to colonize their resources of dead wood, their hyphae have to penetrate solid material, and they live in a variable climate and in competition with other fungi.

One can speculate on the selective advantages of these processes. In termites most of the colony biomass and production is invested in the neuter castes (Wood and Sands, 1978; Collins, 1983). This production compensates for the heavy predation, largely by ants, on foraging parties (Longhurst *et al.*, 1978, 1979). Investment in neuter castes has no selective advantage unless it results in a higher chance of successfully founding new colonies. One expression of this is the production of alate reproductives, and the little evidence available (Table III) suggests that the alate production : colony biomass ratio may be greater in Macrotermitinae than other termites.

TABLE III. Production (P): biomass (B) ratios in Macrotermitinae and other termites (adapted from Collins, 1983).

	Total P : total B	Alate reproductive P : total B
Macrotermitinae	5.38–13.23	0.76–2.02
Other Termites	1.47–3.97	0.04–2.33

VI. SUMMARY

The mutualistic relationship between Macrotermitinae (H9) and *Termitomyces* (F5.10) is multifaceted, but is largely based on their complementary enzyme and adaptive reproductive systems. The termites forage on dead plant material, which is passed rapidly and relatively unchanged through the gut and their faecal deposits used to construct fungus combs. *Termitomyces* mycelium grows on the comb and degrades lignin and cellulose. After 5–8 weeks "old" comb is re-ingested by the termites which also consume nitrogen-rich groups of asexual spores (mycotêtes) on the comb surface. The food is further degraded in the termite's gut by enzymes which digest cellulose, starch, soluble carbohydrates, pectins and chitin. Some enzymes are produced by the termites, while others are acquired by ingestion of mycotêtes or are produced by gut bacteria. In the few species studied, approximately 80% of the food is metabolized by the fungus, thereby concentrating nitrogen. The net effect of the association enables the termites to commandeer a major resource—dead vegetation—and convert it into termite tissue to a degree not accomplished by other termites. In turn the fungus grows in a favourable, competitor-free environment. Some termites construct nests with a complex internal architecture, thereby utilizing heat, mostly resulting from fungal metabolism, to regulate the nest temperature at a favourable level throughout the year. In turn the environment on the comb is made favourable for the *Termitomyces* and unfavourable for other fungal propagules on the comb which are brought in on the termites' bodies and their food. This is largely achieved by termite secretions, principally saliva, which suppress the germination and growth of other fungi. Passage through the termite gut also lowers the number and variety of microorganisms associated with their food. The termites and *Termitomyces* have mutually adapted life-cycles to ensure that new termite colonies receive an inoculum of the fungus and that the fungus is implanted into a suitable substrate. In the majority of associations the fungus produces basidiocarps which arise from the fungus combs and appear above ground, briefly, at specific times of the year. These times coincide with the first foraging forays by workers from new colonies. The workers ingest basidiospores with their food and the spores inoculate the faecal deposits which are the beginnings of the first fungus comb. In a few associations the fungus has no sexual stage. Transmission is achieved by winged reproductive termites which ingest asexual spores from the fungus combs before they fly from the parent colonies. The spores are retained in the gut and deposited on the first faecal deposits of the young foraging workers.

REFERENCES

Abe, T. (1980). Studies on the distribution and ecological role of termites in a lowland rain forest of W. Malaysia: (4). The role of termites in the process of wood decomposition in Pasoh. *Rev. Ecol. Biol. Sol.* **17**, 23–24.

Abe, T., and Matsumoto, T. (1979). Studies on the distribution and ecological role of termites in a lowland rain forest of west Malaysia: (3) Distribution and abundance of termites in Pasoh Forest Reserve. *Jpn. J. Ecol.* **29**, 337–351.

Abe, T., and Watanabe, H. (1983). Soil macrofauna in a subtropical rainforest and its adjacent cassava plantation in Okinawa—with special reference to the activity of termites. *Physiol. Ecol. Japan* **20**, 101–114.

Abo-Khatwa, N. (1977). Natural products from the tropical termite *Macrotermes subhyalinus*: chemical composition and function of "fungus-gardens". *Scr. Varia Pontif. Acad. Sci.* **41**, 447–467.

Abo-Khatwa, N. (1978). Cellulase of fungus-growing termites: A new hypothesis on its origins. *Experientia* **34**, 559–560.

Agarwal, V. B. (1980). Temperature and relative humidity inside the mound of *Odontotermes obesus* (Rambur) (Isoptera, Termitidae). *Proc. Indian Acad. Sci. Sect.B* **89**, 91–99.

Bathellier, J. (1927). Contribution a l'étude systematique et biologique des termites de l'Indochine. *Faune Colon. Fr.* **1**, 125–365.

Batra, L. R., and Batra, S. W. T. (1966). Fungus growing termites of tropical India and associated fungi. *J. Kans. Entomol. Soc.* **39**, 725–738.

Batra, L. R., and Batra, S. W. T. (1979). Termite-fungus mutualism. *In* "Insect Fungus Symbiosis" (L. R. Batra, ed.), pp. 117–163. Allanheld, Osmun, Montclair, N.J.

Batra, S. W. T. (1971). The behaviour and ecology of the fungus growing termites (Termitidae, Macrotermitinae). (Abstract) *Am. Zool.* **11**, 642.

Bose, S. R. (1923). The fungi cultivated by the termites of Barkuda. *Rec. Ind. Mus.* **25**, 253–258.

Bottomley, A. M., and Fuller, C. (1921). The fungus food of certain termites. *S. Afr. J. Nat. Hist.* **3**, 139–144, addendum 223.

Breznak, J. A. (1983). Biochemical aspects of symbiosis between termites and their intestinal microbiota. *In* "Invertebrate–microbial Interactions" (J. M. Anderson, A. D. Rayner and D. H. Walton, eds), pp. 173–203. Cambridge Univ. Press, Cambridge.

Brian, M. V., ed. (1978). "Production Ecology of Ants and Termites". Cambridge Univ. Press, Cambridge.

Burges, A., and Fenton, E. (1953). The effect of carbon dioxide on the growth of certain soil fungi. *Trans. Br. Mycol. Soc.* **36**, 104–108.

Buxton, R. D. (1981). Changes in the composition and activities of termite communities in relation to changing rainfall. *Oecologia* **51**, 371–378.

Cherrett, J. M. (1980). Possible reasons for the mutualism between leafcutting ants (Hymenoptera: Formicidae) and their fungus. *Biol. & Ecol. Méditerr.* **7**, 113–122.

Coaton, W. G. H. (1961). Association of termites and fungi. *Afr. Wild Life* **15**, 39–54.

Collins, N. M. (1977). The population ecology and energetics of *Macrotermes bellicosus* (Smeathman), Isoptera. Ph.D. Thesis, Univ. of London.

Collins, N. M. (1979). The nests of *Macrotermes bellicosus* (Smeathman) from Mokwa, Nigeria. *Insectes Soc.* **26**, 240–246.

Collins, N. M. (1981a). Consumption of wood by artificially isolated colonies of the fungus-growing termite *Macrotermes bellicosus. Entomol. Exp. Appl.* **29**, 313–320.

Collins, N. M. (1981b). Populations, age structure and survivorship of colonies of *Macrotermes bellicosus* (Smeathman) (Isoptera: Macrotermitinae). *J. Anim. Ecol.* **50**, 293–311.

Collins, N. M. (1983). The utilisation of nitrogen resources by termites (Isoptera). *In* "Nitrogen as an Ecological Factor" (A. Lee, S. McNeill and I. H. Rorison, eds), pp. 381–412. Blackwell Scientific, Oxford.

Das, S. R., Maheshwari, K. L., Nigam, S. S., Shukla, R. K., and Tandon, R. N. (1962). Micro-organisms from the fungus garden of the termite *Odontotermes obesus* (Rambur). *In* "Termites in the Humid Tropics". Proc. Symp. New Delhi, 1960, pp. 163–166. UNESCO, Paris.

Garnier-Sillam, E. (1987). Biologie et role des Termites dans les processes d'humification des sol forestiers tropicaux du Congo. Thèse de Docteur ès Sciences, Université Paris Val de Marne.

Gosh, A. K., and Sengupta, S. (1987). Multisubstrate specific amylase from mushroom *Termitomyces clypeatus. J. Biosci.* **11**, 275–285.

Grassé, P. P. (1937). Recherches sur la systematique et la biologie des termites de l'Afrique occidentale Française. *Ann. Soc. Entomol. Fr.* **106**, 1–100.

Grassé, P. P. (1978). Sur la véritable nature et le rôle des meules à champignons construites par les Termites Macrotermitinae (Isoptera, Termitidae). *C. R., Hebd. Seances Acad. Sci., Sér.D* **287**, 1223–1226.

Grassé, P. P., and Heim, R. (1950). Un *Termitomyces* sur meules d'un *Ancistrotermes* Africain. *Rev. Sci.* **88**, 3–13.

Heim, R. (1940). Études descriptives et expérimentales sur les agarics termitophiles d'afrique tropicale. *Mém. Acad. Sci. Inst. Fr.* **64**, 1–74.

Heim, R. (1942a). Les champignons des termitieres. Nouveaux aspects d'un probleme de biologie et de systematique generales. *Rev. Sci.* **80**, 69–86.

Heim, R. (1942b). Nouvelles études descriptives sur les agarics termitophiles d'Afrique tropicale. *Archs. Mus. Natn. Hist. Nat. (Paris)* **18**, 107–166.

Heim, R. (1948). Nouvelles réussites culturales sur les *Termitomyces. C. R. Hebd. Seances Acad. Sci., Ser.D* **226**, 1488–1491.

Heim, R. (1951). Les *Termitomyces* due Congo Belge recueillis par Madame M. Goossens-Fontana. *Bull. Jard. Bot. État Brux.* **21**, 205–222.

Heim, R. (1952a). Classement raisonné des parasites, symbiotes, commensaux et saprophytes d'origine fongique associés aux termites. *6th Congr. Int. Patol. Comp., Madrid*, pp. 15–21.

Heim, R. (1952b). Les *Termitomyces* du Cameroun et du Congo Français. *Denkschr. Schweiz. Naturforsch. Ges.* **80**, 1–29.

Heim, R. (1958). Fasc. 7: *Termitomyces. In* "Flore Iconographique des Champignons du Congo", pp. 139–151. Jardin Bot. État, Bruxelles.

Heim, R. (1977). "Termites et Champignons". Société nouvelle des éditions Boubée, Paris.

Hodgson, E. S. (1955). An ecological study of the behaviour of the leaf-cutting ant *Atta cephalotes. Ecology* **36**, 293–304.

Johnson, R. A. (1981). Colony development and establishment of the fungus comb in *Microtermes* sp. nr. *usambaricus* (Sjöst) (Isoptera, Macrotermitinae) from Nigeria. *Insectes Soc.* **28**, 3–12.

Johnson, R. A., Thomas, R. J., Wood, T. G., and Swift, M. J. (1981). The inoculation of the fungus comb in newly founded colonies of some species of the Macrotermitinae (Isoptera) from Nigeria. *J. Nat. Hist. (London)* **15**, 751–756.

Josens, G. (1971). Le renouvellement des meules à champignons construites par quatre Macrotermitinae, (Isoptères) des savanes de Lamto-Pacobo (Côte-d'Ivoire). *C. R. Hebd. Seances Acad. Sci., Ser.D* **272**, 3329–3332.

Josens, G. (1977). Recherchez sur la structure of fonctionnement des nids hypoges de quatre especes de Macrotermitinae (Termitidae) communes dans les savanes de Lamto (Côte d'Ivoire). *Mém. Acad. R. Belg. Cl. Sci. 2nd Serie.* **42**, 1–123.

Kemner, N. A. (1934). Systematische und biologische Stüdien über die Termiten Javas und Celebes. *K. Sven. Vetenskapsakad. Handl.* **13**, 4–241.

König, J. G. (1779). Naturgeschichte der sogenannten weissen Ameisen. *Beschäft. Berlin. Ges. Naturf. Fr.* **14**, 1–28.

Lee, K. E., and Wood, T. G. (1971). "Termites and Soils". Academic Press, New York and London.

Lepage, M. G. (1981). The impact of foraging populations of *Macrotermes michaelseni* on a semi-arid ecosystem. II. Food offtake, comparison with large herbivores. *Insectes Soc.* **28**, 309–319.

Longhurst, C., Johnson, R. A., and Wood, T. G. (1978). Predation by *Megaponera foetens* (Fabr.) (Hymenoptera: Formicidae) on termites in the Nigerian southern Guinea savanna. *Oecologia* **32**, 101–107.

Longhurst, C., Johnson, R. A., and Wood, T. G. (1979). Foraging, recruitment and predation by *Decamorium uelense* (Santschi) (Formicidae: Myrmicinae) on termites in southern Guinea savanna, Nigeria. *Oecologia* **38**, 83–91.

Lüscher, M. (1961). Air conditioned termite nests. *Sci. Am.* **205**, 138–145.

Martin, M. M., and Martin, J. S. (1978). Cellulose digestion in the midgut of the fungus-growing termite *Macrotermes natalensis*: the role of acquired digestive enzymes. *Science* **199**, 1453–1455.

Martin, M. M., and Martin, J. S. (1979). The distribution of origins of the cellulolytic enzymes of the higher termite, *Macrotermes natalensis. Physiol. Zool.* **52**, 11–21.

Matsumoto, T. (1976). The role of termites in an equatorial rain forest eco-system of West Malaysia. I. Population density, biomass, carbon, nitrogen and calorific content and respiration rate. *Oecologia* **22**, 153–178.

Matsumoto, T. (1977). Respiration of fungus combs and carbon dioxide concentration in the centre of mounds of some termites. *Proc., Int. Congr. Int. Union Study Soc. Insects, 8th, 1977*, 104–106.

Mishra, J. N., and Ranganathan, V. (1954). Digestion of cellulose by the mound building termite, *Termes (Cyclotermes) obesus* (Rambur). *Proc. Indian Acad. Sci.* **39**, 100–113.

Mishra, S. C., and Sen-Sarma, P. K. (1985). Nutritional significance of fungus comb and *Termitomyces albuminosus* in *Odontotermes obesus* (Isoptera, Termitidae). *Mater. Org.* **20**, 205–214.

Mohindra, P., and Mukerji, K. G. (1982). Fungal ecology of termite mounds. *Rev. Ecol. Biol. Sol.* **19**, 351–361.

Mukiibi, J. (1973). The nutritional value of some Uganda mushrooms. *Acta Hortic.* **33**, 171–175.

Noirot, C. (1970). The nests of termites. *In* "Biology of Termites" (K. Krishna and F. M. Weesner, eds), Vol. II, pp. 73–125. Academic Press, New York and London.

O'Brien, R. W., and Slaytor, M. (1982). Role of microorganisms in the metabolism of termites. *Aust. J. Biol. Sci.* **35**, 239–262.

Osore, H. (1985). The role of micro-organisms isolated from fungus-comb-constructing African termites in the degradation of lignocellulose. *In* "Energy from Biomass" (W. Palz, J. Coombs, and D. O. Hall, eds) *3rd E.C. Conf., Venice, 25–29 March 1985*, pp. 999–1000. Elsevier Applied Science Publishers, London.

Pegler, D. N., and Rayner, R. W. (1969). A contribution to the agaric flora of Kenya. *Kew Bull.* **23**, 347–412.

Petch, T. (1906). The fungi of certain termite nests. *Ann. R. Bot. Gdns Peradeniya* **3**, 185–270.

Petch, T. (1913). White ants and fungi. *Ann. R. Bot. Gdns Peradeniya* **5**, 389–393.

Rajagopal, D. (1982). Mound building behaviour of *Odontotermes wallonensis* (Isoptera, Termitidae). *Sociobiol.* **7**, 289–304.

Rohrmann, G. F. (1977). Biomass, distribution and respiration of colony components of *Macrotermes ukuzii* Fuller (Isoptera: Termitidae: Macrotermitinae). *Sociobiol.* **2**, 283–295.

Rohrmann, G. F. (1978). The origin, structure and nutritional importance of the comb in two species of Macrotermitinae (Insecta, Isoptera). *Pedobiologia* **18**, 89–98.

Rohrmann, G. F., and Rossman, A. Y. (1980). Nutrient strategies of *Macrotermes ukuzii* (Isoptera: Termitidae). *Pedobiologia* **20**, 61–73.

Rouland, C., Mora, P., Matoub, M., Renoux, J., and Petek, F. (1986). Comparative study of two β-glucosidases from *Macrotermes mülleri* and its symbiotic fungus *Termitomyces* sp. *Actes des Colloques Insectes Sociaux* **3**, 109–118.

Ruelle, J. E. (1964). L'architecture du nid de *Macrotermes natalensis* et son sens fonctionnel. *In* "Etudes sur les Termites Africains" (A. Bouillon, ed.), pp. 327–362. UNESCO, Paris.

Ruyooka, D. B. A. (1979). Associations of *Nasutitermes exitiosus* (Hill) (Termitidae) and woodrotting fungi in *Eucalyptus regnans* F. Muell. and *Eucalyptus grandia* W. Hill ex Maiden: Choice-feeding, laboratory study. *Z. Angew, Entomol.* **87**, 377–388.

Sands, W. A. (1956). Some factors affecting the survival of *Odontotermes badius. Insectes Soc.* **3**, 531–536.

Sands, W. A. (1960). Initiation of fungus comb construction in laboratory colonies of *Ancistrotermes guineensis* (Silvestri). *Insectes Soc.* **7**, 251–259.

Sands, W. A. (1969). The association of termites and fungi. *In* "Biology of Termites" (K. Krishna and F. M. Weesner, eds), Vol. I, pp. 495–524. Academic Press, New York and London.

Sannasi, A., and Sundara Rajulu, G. (1967). Occurrence of antimicrobial substance in the exudate of physogastric queen termite, *Termes redemanni* Wasmann. *Curr. Sci.* **16**, 436–437.

Sieber, R. (1983). Establishment of fungus comb in laboratory colonies of *Macrotermes michaelseni* and *Odontotermes montanus* (Isoptera, Macrotermitinae). *Insectes Soc.* **30**, 204–209.

Singh, N. B. (1976). Studies on certain digestive enzymes in the alimentary canal of *Odontotermes obesus* (Isoptera: Termitidae). *Entomol. Exp. Appl.* **20**, 113–122.

Singh, U. R., Singh, J., and Singh I. D. (1978). Microbial association with the termites in a tropical deciduous forest at Varansi. *Trop. Ecol.* **19**, 163–173.

Thomas, R. J. (1981). Ecological studies on the symbiosis of *Termitomyces* Heim with Nigerian Macrotermitinae. Ph.D. Thesis, Univ. of London.

Thomas, R. J. (1985). Selective medium for isolation of *Termitomyces* from termite nests. *Trans. Br. Mycol. Soc.* **84**, 519–526.

Thomas, R. J. (1987a). Distribution of *Termitomyces* Heim and other fungi in the nests and major workers of *Macrotermes bellicosus* (Smeathman) in Nigeria. *Soil Biol. Biochem.* **19**, 329–333.

Thomas, R. J. (1987b). Distribution of *Termitomyces* and other fungi in the nests and major workers of several Nigerian Macrotermitinae. *Soil Biol. Biochem.* **19**, 335–341.

Thomas, R. J. (1987c). Factors affecting the distribution and activity of fungi in the nests of *Macrotermitinae* (Isoptera). *Soil Biol. Biochem.* **19**, 343–349.

Van Ryn, R. (1974). Corrélation entre la poussée des carpophores de *Termitomyces* (Heim) et l'essaimage de Macrotermitinae (Isoptera) dans la région de Kinshasa. *Revue Zool. Afr.* **88**, 703–705.

Wood, T. G. (1978). Food and feeding habits of termites. *In* "Production Ecology of Ants and Termites" (M. V. Brian, ed.), pp. 55–80. Cambridge Univ. Press, Cambridge.

Wood, T. G., and Johnson, R. A. (1978). Abundance and vertical distribution in soil of *Microtermes* (Isoptera, Termitidae) in savanna woodland and agricultural ecosystems at Mokwa, Nigeria. *Memorab. Zool.* **29**, 203–213.

Wood, T. G., and Johnson, R. A. (1986). The biology, physiology and ecology of termites. *In* "Economic Impact and Control of Social Insects" (S. B. Vinson, ed.), pp. 1–68. Praeger, New York.

Wood, T. G., and Sands, W. A. (1978). The role of termites in ecosystems. *In* "Production Ecology of Ants and Termites" (M. V. Brian, ed.), pp. 245–292. Cambridge Univ. Press, Cambridge.

Zoberi, M. H. (1979). The ecology of some fungi in a termite hill. *Mycologia* **71**, 537–545.

4

The Mutualism between Leaf-Cutting Ants and their Fungus

J. M. CHERRETT, R. J. POWELL AND D. J. STRADLING

I. BIOLOGY OF THE ANTS

A. Taxonomic Position

The subfamily Myrmicinae within the family Formicidae (the ants (H33.9)) includes a taxonomically compact group of 12 genera which constitute the tribe Attini and contains approximately 190 species (Weber, 1972). These ants, whose distribution is confined to the nearctic and neotropical biogeographic regions, share a unique feature among the Formicidae in their obligate dependence on symbiotic fungi as a source of larval food. Their closest ecological equivalent appears to be the fungus-growing Macrotermitinae (H9) of the old world (see Wood and Thomas, this volume).

Wilson (1971) has suggested dividing the 12 attine genera into primitive, transitional and advanced groups on the basis of colony size, worker polymorphism and the types of substrates utilized in the cultivation of the fungi (Table I). The six genera which constitute the primitive group are characterized by small- to medium-sized colonies of no more than a few hundred monomorphic workers. Their fungi are cultivated on substrates of insect faeces (especially caterpillar frass) and dead vegetable matter.

The three transitional genera also develop small- to medium-sized colonies and, with the exception of a slight tendency to worker polymorphism in some species of *Trachymyrmex*, are generally characterized by a monomorphic worker caste. The fungal substrates, however, include a significant proportion of fallen flowers and fruit among other dead vegetable matter.

Acromyrmex and *Atta*, two of the three genera in the advanced group, differ from the other two groups in developing large colonies of strongly and functionally polymorphic workers, in addition to distinctive soldiers in *Atta*. Furthermore they use fresh leaves, stems, fruit and flowers cut from living plants as fungal substrates, and are consequently known as leaf-cutting ants.

There are 24 species of *Acromyrmex* and 15 of *Atta*. Species of the former genus are characterized by colonies occupying few nest chambers and attaining worker populations of between 10 and 20×10^3. *Atta* colonies are generally much larger, sometimes occupying more than 1000 nest chambers (Jonkman, 1977). Worker populations are concomitantly large,

TABLE I. Genera of the tribe Attini, the fungus-growing ants (number of species). (After Wilson, 1971 and Weber, 1972.)

Primitive genera
Cyphomyrmex (30), *Mycetosoritis* (5), *Mycetophylax* (7), *Mycocepurus* (4) *Mycetarotes* (2), *Myrmicocrypta* (20)
Transitional genera
Apterostigma (27), *Sericomyrmex* (19), *Trachymyrmex* (37)
Advanced genera
Acromyrmex (24), *Pseudoatta* (1) (a worker-less social parasite), *Atta* (15)

and estimates run to 2.2×10^6 for *A. sexdens sexdens* (Weber, 1966) and even to 7.0×10^6 for *A. vollenweideri* (Jonkman, 1978).

B. Ecological Dominance

Leaf-cutting ants, as competitors with man for plant material, are important pests (Cramer, 1967)—some evidence for this is the large number of local names given to them (60 for *Acromyrmex* spp. and 71 for *Atta* spp.: Cherrett, 1986a). Losses are difficult to compile as many crops suffer, but Cherrett (1986b) has suggested that potential losses (expected losses if no control measures are adopted) could well exceed the figure of US $1000 million first suggested in 1923 by Townsend.

According to a postal survey of Latin American countries (Cherrett and Peregrine, 1976), 47 agricultural and horticultural crops and 13 species of pasture plants are attacked, so we can readily appreciate why Belt (1874) described them as "... one of the greatest scourges of tropical America ...".

Their impact on the natural vegetation of tropical America, with which they co-evolved, is less clear, and has been the subject of fewer studies. Cherrett (in press), using mean figures from several studies, has calculated that in tropical rain forest leaf-cutting ants may be harvesting 0.8% of the gross plant productivity, or, more significantly, 17% of total leaf production. This supports the observations of Wint (1983) in Panamanian rain forest who found that as much as 80% of the leaf damage observed was caused by *Atta* spp. For grasslands, Fowler *et al.* (1986) have collated the results of studies on nest densities and rates of defoliation and conclude that a colony of the grass-cutting ant *Atta capiguara* takes 30–150 kg dry matter $ha^{-1}\ yr^{-1}$, *A. vollenweideri* 90–250 $kg\ ha^{-1}\ yr^{-1}$, and a colony of *Acromyrmex landolti* 0.4–2.2 $kg\ ha^{-1}\ yr^{-1}$. At the densities observed in South America, the dry weight of grass harvested ranged between 84 and 8775 $kg\ ha^{-1}\ yr^{-1}$. The significance of this can be seen when we compare these rates of consumption with those of domestic cattle. Cherrett *et al.* (1974) suggested a consumption rate of 5400 $kg\ ha^{-1}\ yr^{-1}$ for the cattle in an improved pasture at Ebini, Guyana, whilst Fowler *et al.* (1986) considered that the number of head of cattle which a pasture can carry is reduced by 10–30% by *Atta capiguara*. In natural, unimproved grasslands, leaf-cutting ant populations may be somewhat lower.

These studies suggest that in both man-made and natural vegetation, leaf-cutting ants can become dominant exploiters of living vegetation, so justifying their title of "dominant invertebrates" (Wheeler, 1907) and "prevalent herbivores" (Wilson, 1982). How have a few species of insects achieved such ecological dominance?

C. Polyphagy

In general, the total number of species of trees per hectare for a given type of community increases from the poles to the equator. As an example, coniferous forest in Northern Canada will have 1–5 spp. ha^{-1}, deciduous forest in North America 10–30 and tropical rain forest in South America 40–100 spp. ha^{-1}. How this situation has arisen is controversial (Begon *et al.*, 1986), but to achieve ecological dominance in tropical rain forest it would be necessary for a herbivore to have a broad diet breadth (be polyphagous). However, as Strong *et al.* (1984) point out "one consequence of the tremendous chemical and physical diversity among plants is the narrow diet tolerances of phytophagous insects". This has been illustrated by Lawton and McNeill (1979), who showed that the vast majority of British Agromyzids (Diptera) were monophagous, very few being polyphagous. Figure 1A illustrates the typical right-skewed frequency distribution of diet breadth which results.

The leaf-cutting ant species found in tropical rain forest have, by contrast, a very wide diet breadth. Cherrett (1968) recorded a nest of *Atta cephalotes* cutting 50% of the plant species growing in the nest area during 8 weeks observation, whilst in Costa Rica, Rockwood (1976) found one particularly active colony which cut 77%, and a colony of *A. colombica* which cut 67%. This is a most impressive degree of polyphagy, unusual for an insect, and comparable with that of howler monkeys which Rockwood and Glander (1979) observed utilizing 64% of the tree species. The diet breadth of grass-cutting ants in natural grasslands has been much less studied.

Within the range of plants cut, however, leaf-cutting ants exhibit preferences, and although in laboratory trials *A. cephalotes* picked up some leaf discs from 92% (34 from 37) of a random sample of plant species from Costa Rica, Hubbell and Wiemer (1983) demonstrated that some were greatly preferred, the form of the diet breadth frequency distribution for the single ant species again being highly skewed to the right (Fig. 1B). A series of studies has shown this to be due to deterrents which inhibit cutting, pick-up or feeding, and include toughness (Cherrett, 1972; Waller, 1982a), the presence of sticky latex (Stradling, 1978), and a wide range of secondary defensive chemicals (Littledyke and Cherrett, 1978; Waller, 1982b; Hubbell *et al.*, 1983; Febvay *et al.*, 1985). The latter have recently been subject to intense study by Wiemer and his associates (see for example Hubert *et al.*, 1987). However, as it is generally believed that narrow diet breadth is imposed on most insects by the toxic, repellent or physical unsuitability of the vegetation they do not eat, selectivity on the part of leaf-cutting ants, and the role of secondary plant chemicals in it comes as no surprise. The

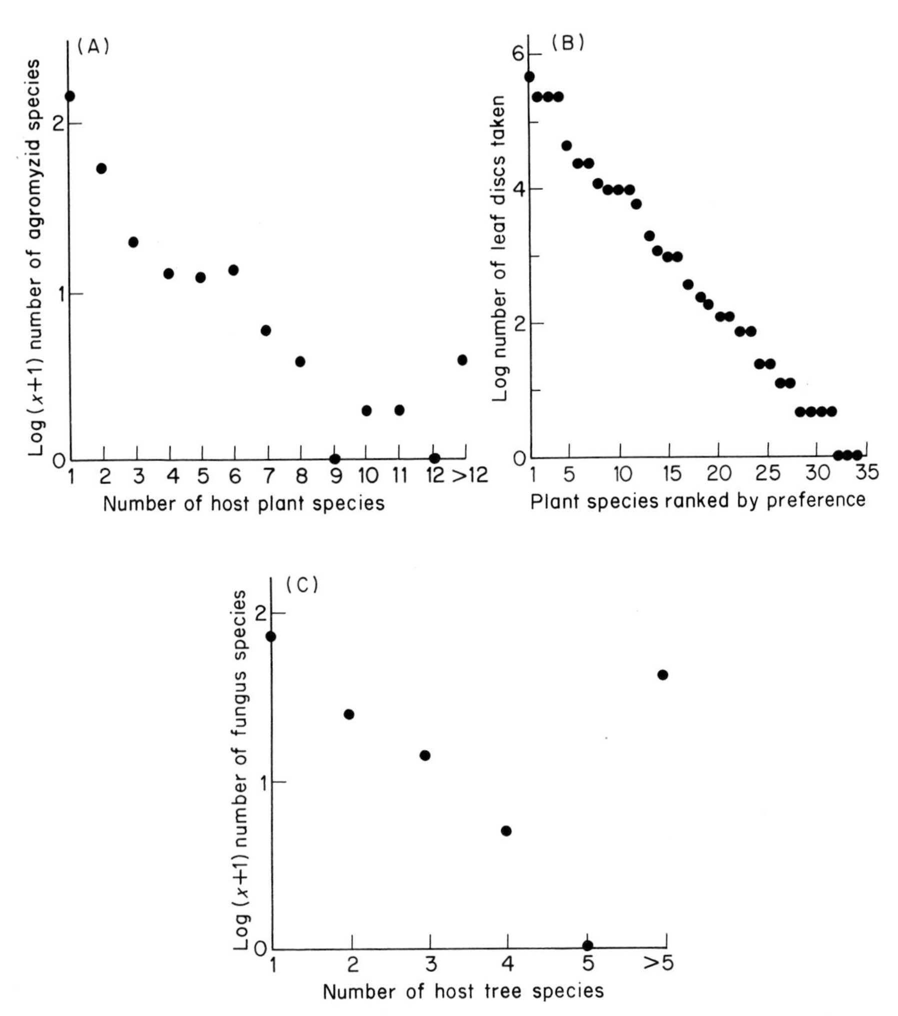

Fig. 1. A. Diet breadth of British Agromyzidae (after Lawton and McNeill, 1979). B. Dietary composition of a laboratory culture of *Atta cephalotes* (after Hubbel and Wiemer, 1983). C. Diet breadth of foliar fungi of Western trees (after Funk, 1985).

interesting question is how they maintain the high degree of polyphagy upon which their ecological dominance depends. At this point we should remember that the all-important growth stage of leaf-cutting ants, the larvae, are probably monophagous, being restricted to the nutritive bodies (gongylidia) of their mutualistic fungus *Attamyces bromatificus* (F6.2) (Quinlan and Cherrett, 1979). It is only the adults which obtain the majority of their energy requirements directly from plant sap, and it is likely that they drink plant sap from a much narrower range of plant species than the ones they cut (Littledyke and Cherrett, 1976). Ecologically, leaf-cutting ants have a broader diet than they have physiologically.

D. Specialized Physiology and Behaviour for Fungus Culture

Among the few physiological adaptations to fungiculture that attines possess is the ability to concentrate a cocktail of enzymes in their hindgut which appear in their faeces and are used to initiate the digestion of the freshly collected plant fragments. An adaptation to mycophagy is the production of certain enzymes, such as chitinase by the labial glands and lipase, maltases, trehalases and proteases by the gut, to digest the gongylidia (Febvay and Kermarrec, 1981a).

The ants' fungus gardens are continuously subjected to contamination by spores of a wide range of potential competitors brought in on the substrate materials collected by foragers. If the ants are removed, fungus gardens are soon overwhelmed by the germination of these contaminant spores. How the ants maintain their monoculture free from competing fungi is not yet fully understood. By physically removing contaminants and by planting large inocula of the symbiotic fungus on new substrate, so that it colonizes it quickly, the worker ants must help its competitive ability. It seems likely that the fungus itself produces some antibiotics (Hervey and Nair, 1979; Angeli-Papa, 1984).

There has been speculation about a fungicultural role for secretions of the metathoracic glands of attines. The major components are β-hydroxy-decanoic acid (myrmicacin) and phenylacetic acid, both of which are known to show antibiotic activity (Schildknecht and Koob, 1971). However, acidic secretions are not unusual among myrmicine ants. Myrmicacin has received most attention, as bees also secrete a homologue with similar properties. Iwanami (1978) found that myrmicacin was the most potent member of a homologous series of "myrmic acids" which reversibly inhibited mitosis in pollen grains. Its potency was governed by acidity and it was effective in very low concentrations (Iwanami and Iwadare, 1979). Powell (1984) discovered that it had the same effect upon both the germination and production of spores in fungi such as *Aspergillus* (F6.2) and *Penicillium* (F6.2) species, both common contaminants of attine fungus-gardens, but had little effect upon yeasts. This probably explains the observations of Schildknecht *et al.* (1973) that it suppressed the growth of moulds in fruit and that the fungus gardens of attines were quickly overrun by contaminant fungi soon after the ants were removed. As Powell (1984) found that the concentration of myrmicacin that inhibited fungal spore germination was much higher than that recorded in fungus gardens, he suggested that other attine secretions may enhance its properties. He found that indolylacetic acid also secreted by the metathoracic glands, has this effect but apparently only through increasing the acidity of the mixture. In commercial preparations, the antifungal properties of myrmicacin are enhanced by the addition of ethanol at pH values lower than 5 (Osberghaus *et al.*, 1974). Also, Koppensteiner and Bansemir (1975) have demonstrated that it possesses antibacterial properties if mixed with methanol under acidic conditions. It is therefore possible that chemical synergists of myrmicacin are to be found in the fungus-gardens. Beattie (1986) has speculated that myrmicacin evolved as a protection from soil pathogens for naked ant larvae and it may explain why ants make poor pollinators. Macrotermitinae also secrete a selective fungistat to enhance the competitiveness of their fungal symbiont (Thomas, 1981).

It has been suggested that the phenylacetic acid may inhibit bacterial multiplication in the fungus gardens, but Papa and Papa (1982a) found large populations of bacteria in *Acromyrmex* fungus gardens without any evidence of a major effect.

The acidity of the fungus garden is limiting to bacterial growth (Papa and Papa, 1982b), and the ants' acidic secretions contribute to this and modify the pH of its constituent plant tissues. It falls consistently from 4.8 in the younger zones to about 4.5 in the older parts irrespective of the plant tissues used, the pH of which may range from 5 to 8 (Powell, 1984). This suggests that the ants can regulate the acidity to levels normally outside those tolerated by the majority of bacteria. In fungus combs of Macrotermitinae the pH drops from 4.6 to 4.1 in older zones, and the optimum in laboratory culture is 5.2 (Thomas, 1981).

Fungus-garden pH is not only thought to be crucial in the control of bacterial contamination, but also in optimizing the growth of the attine symbionts. Cultures of attine symbionts on buffered potato dextrose agar showed maximum yields of gongylidia biomass between pH 4.5 and 5.0. (Powell and Stradling, 1986). For the ants the gongylidia constitute the most important component of fungal productivity, and represent the yield directly utilized by the larvae.

Behavioural adaptations of attines to fungiculture are more numerous as the ants must provide suitable conditions for their fungal symbionts to grow. Incubation temperature depends upon the depth at which the nest chambers are excavated in the soil, and Powell and Stradling (1986) demonstrated a temperature optimum close to 25°C, the fungus being killed by temperatures higher than 30°C at which workers evacuate the fungus-garden. At lower temperatures the size and number of gongylidia in fungal isolates from *Atta cephalotes* were reduced. There is evidence that, in the field, *A. sexdens* keeps its fungus-gardens at depths which vary according to season.

The ants control nest humidity by adjusting the air flow. This is accomplished by opening and closing entrance holes, and by regurgitating water onto the gardens. The fungus is also kept in darkness, and Powell and Stradling (1986) showed that light slowed down its growth rate, whilst ultraviolet light was lethal.

Other behavioural adaptations to fungiculture must include the selection of suitable substrate, although the extent to which the workers can adjust their diet to the needs of their fungus is a controversial issue. It is conceivable that the ants could monitor the performance of the fungus on fragments of vegetation, and then subsequently cease cutting those plant species on which performance was poor. Alternatively, preferences for suitable plant material could reflect past selection pressures when colonies taking in material poisonous to the fungus compromised their ability to produce offspring. However, the ants show marked preferences between crops recently introduced to South America, and it is an open question if there has been enough time, or enough contact with the ants, for the observed preferences to have evolved in this way. "Nutritional wisdom" is still a controversial topic in the feeding behaviour of large animals (Hafez, 1969) and is unresolved for leaf-cutting ants. The large number of plant species cut, the wide dispersion of the fragments over many fungus-gardens, and the time factor which results in parts of the garden of different ages containing different materials, probably provides a very wide safety margin against poisoning of the entire fungus culture of a colony. Perhaps the ants simply reject leaves which are distasteful to them as a result of the physiological responses which the defensive secondary compounds induce. If the ants' biochemistry is not affected, perhaps the more robust biochemistry of the fungus will not be fatally affected either.

There is evidence that the acidity of the plant tissues is important and that ants tend to avoid material more acidic than the fungus-garden. In doing so they probably obtain plant tissues which are low in tannins and consequently possess a high "available protein" content (Powell, 1984). Neutral or slightly alkaline tissues can always be acidified to the fungal optimum by means of acidic secretions, whilst the reverse adjustment cannot apparently be achieved.

The treatment and preparation of plant fragments prior to incorporation into the fungus-garden has been well documented by Quinlan and Cherrett (1977). This involves assiduous licking which removes the waxy cuticle that acts as a barrier to fungus penetration, and may help to decontaminate the leaf surface. The leaf fragments are crimped between the mandibles to express juices and also aid hyphal penetration. The ants then defaecate on the fragments, faecal enzymes digesting cell walls and possibly destroying phytoalexins. Finally, the ants pick hyphal fragments from more mature parts of the garden, and plant them at high density amongst the newly prepared substrate, so providing a large initial inoculum.

E. Fungus Transmission

As the mutualistic fungus *Attamyces bromatificus* (F6.2) has never been found outside leaf-cutting ant nests and is unlikely to produce sporophores, and as the ants are completely dependent upon it, they have evolved specialist behaviour to transmit the fungus to new colonies. These are normally founded by an individual queen which unaided digs a chamber in the ground after its mating flight. Soon the queen is found to be tending a minute fungus garden which she frequently fertilizes with her own liquid faeces whilst raising the first workers. When these emerge they begin to cut and prepare leaves as substrate. Von Ihering (1898) was the first to discover how the fungus was transmitted, and his observations were repeated and amplified by Huber (1905). Gynes accumulate fragments of fungal mycelia and other amorphous material which they compress into a pellet about 0.6 mm in diameter and lodge in the infrabuccal pocket, a small cavity beneath the opening of the oesophagus. On completing the excavation of her first small underground chamber, each foundress queen expels the pellet and tends it assiduously. Actively growing hyphae soon develop from it. If at this stage the queen loses her inoculum, as perhaps 25% of them do (Quinlan and Cherrett, 1978a), then she dies. This is a remarkable piece of behaviour and several writers have implied that gynes deliberately pick fungal hyphae from their parent garden and pack them into their infrabuccal pockets before issuing from the nest on their mating flight (Mariconi, 1970; Wilson, 1971). There seems to be no direct evidence for this and, in colonies large enough to produce sexuals, it would be very difficult to observe the behaviour of gynes immediately prior to flight. Wheeler (1910) is probably more nearly correct in suggesting that "This pellet is the unexpelled refuse of her last meal".

Other Hymenoptera (H33) as well as the ants possess an infrabuccal pocket (Janet, 1905), and it serves as a receptacle for fine particles licked up during grooming and taken in during the consumption of food. Ants only take liquids into their gut, and the leaf-cutting ant *Acromyrmex octospinosus* is able to filter particles as small as 10 μm from the liquid it drinks. These particles collect in the infrabuccal pocket, and workers expel them at regular intervals (Quinlan and Cherrett, 1978a; Febvay and Kermarrec, 1981b). Although cultures of the mutualistic fungus have been obtained from viable hyphae in worker pellets, in nature they are never allowed to remain on the fungus-garden, being immediately removed to middens. This may be because they also contain contaminants due to the workers' habit of cleaning themselves, the queen, the leaf fragments brought in and the nest chamber itself by assiduous licking. Febvay and Kermarrec (1986) have shown that the labial glands of *A. octospinosus* secrete a chitinase which erodes the cell walls of fungal material collected in the infrabuccal pocket, and this may reduce the viability of contaminant spores and hyphae. Quinlan and Cherrett (1978a) have suggested that the use of the infrabuccal pocket by leaf-cutting ant queens to transmit the fungus to newly founded colonies results from a small modification to the general behaviour pattern found in the workers and probably in most ants. This behaviour contrasts markedly with the transmission process in the Macrotermitinae, in which spores of *Termitomyces* have been reported in the guts of *Macrotermes* alates, and workers also collect spores when foraging (Johnson *et al.*, 1981).

II. BIOLOGY OF THE FUNGUS

A. Taxonomic Position and Identity

Unlike the clearly identifiable termite symbiont *Termitomyces*, the identity of the attine symbiotic fungi has long been the subject of controversy due to doubts over sporophores. After Belt (1874) realized the true significance of fungi to attines, it was Möller (1893) who first provided detailed descriptions of fungi isolated from the fungus-gardens of various attine species, including members of the genera *Acromyrmex, Apterostigma* and *Cyphomyrmex*. Besides isolating the symbiotic fungi from fragments of fungus-garden he also repeatedly found and described fruiting bodies occurring on the surface of leaf-cutting ant nests. He named these *Rozites gongylophora* (F5.10), a Basidiomycete (Cortinariaceae). To confirm that this was the symbiotic fungus, basidiospores from the fruit bodies were propagated and gave rise to a mycelium bearing what Möller described as "kohlrabi bodies", now known as gongylidia (Wheeler, 1907). These swollen hyphal tips were taken by the ants to feed to their larvae as though they had been cultivated in their own fungus-gardens. The name *Rozites* is still used by many modern authors to describe the food fungi of both *Atta* and *Acromyrmex* species.

These reports were rejected as unproven by Wheeler (1910, 1913), who claimed that the fungus was an Ascomycete (F4). The acceptability of the fungus to the ants as evidence that it was their own fungus was also questioned by Weber (1938), because he had found the ants to accept other fungi including *Agaricus* (F5.10) and *Phallus* (F5.18) species. This prompted further searches for the elusive fruit bodies of the ants' symbiotic fungus. Many of the subsequent reports must be treated with caution since there is a high probability of encountering the sporophores of opportunist fungi in and around attine nests. For example, many of the fruit bodies described came from nest middens (Spegazinni, 1899, 1921; Bruch, 1921), abandoned fungus-gardens (Weber, 1938, 1966; Heim, 1957) or nests which had been disturbed, for example by digging (Stahel and Geijkes, 1941).

Contemporary with attempts to find sporophores, other German mycologists (Goetsch, 1939; Goetsch and Stoppel, 1940; Stoppel, 1940) had attempted to isolate the vegetative mycelium from Brazilian *Atta cephalotes* nests. They described various Ascomycete "competing contaminants" but were unable to find any Basidiomycetes at all. On this basis, they criticized Möller's work and regarded *Rozites* as erroneous. Unable to find any gongylidia in their isolates, they considered that these structures were only found in the presence of the ants and resulted from their chewing the hyphae.

Many attempts have also been made to propagate fruit bodies from cultivated samples of the symbiotic mycelium. Apparent success was achieved by Weber (1957, 1966) using the fungi isolated from nests of the primitive attines *Cyphomyrmex costatus* and *Myrmicocrypta buenzlii*. Hervey *et al.* (1977) also succeeded in obtaining sporophores from the symbiont of *M. auriculatum*. The sporophores were identified as a species of *Lepiota* (F5.10) hitherto undescribed (Smith and Locquin in Weber, 1957). However, in no case did they report the recovery of gongylidia-bearing mycelium from spores obtained from these sporophores.

Heim (1957) compared Weber's *Lepiota* sporophores with Möller's *R. gongylophora* and considered them to be identical, calling them *Leucocoprinus gonglyophorus, Leucoagaricus gongylophora* or *Agaricus gongylophora* (Heim, 1958). Another widely used name, *Attamyces bromatificus*, was designated by Kreisel (1972, 1975) for the symbiont of *Atta insularis* found in Cuba. He recognized Möller's original classification of the fungus as a Basidiomycete (F5.10) but put the vegetative mycelium in the Mycelia Sterilia (F6.2) because of the absence of sporophores.

The validity of each one of these names is questionable, not only because confirmatory spore germination tests are lacking in the majority of cases, but also because no type material exists. We propose to follow Kreisel's "cautious opinion" (Kermarrec *et al.*, 1986) which does not depend on sporophores of dubious origin and because the name he uses is comparable with the ecologically equivalent *Termitomyces*. We feel that the name *Attamyces bromatificus* avoids difficulties.

The symbiotic fungi of *Atta* and *Acromyrmex* species are now known to be Basidiomycetes due to the presence of dolipore septation in the mycelium. This was described by Powell (1984) using electron microscopy and also a modified fixative for light microscopy. Angeli-Papa and Eyme (1979) showed electron microscope pictures of the *Acromyrmex* symbiont but failed to realize the significance of the dolipore shown in one of their sections.

In the absence of sporophores, morphological comparisons of the fungal symbionts of different species of higher attines are limited. On the basis of hyphal structure, isolates from different ant species are indistinguishable although Powell (1984) found evidence of minor differences in the sizes of gongylidia. The compatibility of different isolates determined by paired-culture tests was reported by Stradling and Powell (1986). In a total of fifteen tests involving six different isolates no interactions were observed. This implies that there are no major differences and that some species of *Atta, Acromyrmex* and *Trachymyrmex* cultivate the same species of fungus.

B. Ecological Dominance

Fungi which exploit the bodies of higher plants (as distinct from their liquid exudates) can adopt one of two broadly different life styles. They can become saprophytes, breaking down the plant tissue once it has died. Fungi dominate this process with some 6000 m of hyphae per gramme of soil being recorded in tropical forests (Swift *et al.*, 1979). However, as Quinlan and Cherrett (1977) have pointed out, dead plant material which falls as litter to the forest floor is nutritionally inferior to living tissue. Fungi attempting to exploit the latter parasitically have however to cope with a battery of plant defence mechanisms. By "domesticating" an ant to pick and comminute selected fresh leaf material for it, *Attamyces bromatificus* can function as a saprophyte in a competitor-free environment, decomposing dying plant matter which has the nutritional content of living material.

It is difficult to obtain figures for the amount of leaf tissue normally consumed in tropical forests and grasslands by parasitic leaf fungi, or the number of fungal species involved. Burdon and Chilvers (1974) studied fungal parasites in mixed stands of *Eucalyptus* saplings in Australia, and on the basis of 10 tree species and site combinations, they estimated that the percentage effective leaf area lost through death and disease, the greater part of which was due to parasitic fungi, was 11.6% ± 3.1 standard errors of the mean. By contrast, the figure of 17% of total leaf production in tropical rain forests harvested by leaf-cutting ants for their single mutualistic fungus is high, suggesting that in such situations, *A. bromatificus* might well be the dominant species of fungus exploiting fresh leaf material. As with the ants themselves, this provokes the question how has this one species of fungus achieved such an ecologically dominant position in its group?

C. Polyphagy

Using a similar argument to that used for the ants, the ecological dominance exhibited by the fungus can only be achieved by polyphagy. As with phytophagous insects, plant defence mechanisms have tended to impose a narrow diet breadth on parasitic fungi. Thus, Funk (1985) in his catalogue of the foliar fungi of western trees which covers needle casts, needle blights, leaf spots, tar spots, sooty moulds and mildews, but excludes the epiphytic fungi, records nearly 150 species of which 70 (47%) were monophagous, whilst 71% exploited three or fewer host tree species (Fig. 1c). In this context, the ability of *Attamyces bromatificus* to utilize the 50–77% of tropical rain forest species cut for it by the ants appears remarkably polyphagous.

D. Specialized Anatomy and Physiology

The fungus produces swollen hyphae (gongylidia) which are clustered into groups like bunches of grapes and which the ants harvest, eat themselves and feed to their larvae. These structures seem to have no functional significance for *Attamyces* except as a food source for the mutualistic ant. They represent a considerable investment in metabolic products not used directly by the fungus for its own growth. It is not clear what is the evolutionary origin of these structures, nor how the individual gongylidia have become aggregated into "bite sized" staphylae which the ants can crop. The whole topic of the number and size of gongylidia in a staphyla, how these change with the age of the garden and the species of ant, and the preferences of different sized workers for different staphylae merits further study.

Much light was shed on the biochemical relationships between attine ants and their symbiotic fungi by Martin and co-workers from 1969 onwards. Significant enzyme activity was discovered in the faecal droplets which attine workers apply to the fungus garden substrate during its preparation. These enzymes, which included proteases (Martin and Martin, 1970), α-amylase and chitinase (Martin *et al.*, 1973), appear to accelerate the digestion of the fresh substrate and thus the rate of growth of the fungus on it. The chitinase was thought to improve the competitiveness of *Attamyces* by lysing some of the fungal contaminants. Later (Martin, 1974; Martin and Boyd, 1974), it became clear that these enzymes together with those which digest pectin, sodium pectate, xylan and carboxymethylcellulose (Martin *et al.*, 1973, 1975) originated from the fungal symbiont and were not secreted in the ant guts. Thus, Boyd and Martin, (1975b) suggested that the fungus was capable of synthesizing the enzymes but not of secreting them. The ants were thought to have taken over this role, removing fungal enzymes from vigorous, well-established mycelium to newly inoculated parts of the fungus garden, concentrating the enzymes in their guts in the interim.

Powell (1984) has shown by using defined media and by estimating the productivity of the fungi in terms of the gongylidia biomass, that the fungal symbionts of *Atta cephalotes*, *Acromyrmex octospinosus* and *Trachymyrmex urichi* are capable of utilizing glucose, fructose, galactose, cellobiose, maltose, sucrose, starch, cellulose and pectin. This means that the fungi at least possess the following enzymes: maltase, sucrase, β-D-galactosidase, α-amylase, endopolygalacturonase and cellulase. The ability to clear starch and pectin media strongly implies that the fungi can secrete the necessary digestive enzymes. Even though the ants defaecate enzymes onto plant fragments, the fungal symbiont seems quite capable of digesting the structural components of plant tissues unaided. This was seen as a zone of clearing around the fungus in solid media containing macerated plant tissues.

Boyd and Martin (1975a, b) had assumed the fungal symbiont to be the major dietary item of attines. Their work was done, however, before the importance of plant sap in the diet of leaf-cutting ant workers was demonstrated by Littledyke and Cherrett (1976) and Stradling (1978). Drinking plant sap requires various lipases, carbohydrases, and proteases in the midgut for digestion, and Febvay and Kermarrec (1981a, 1986) have shown that the midguts of adult workers do secrete some enzymes including a lipase, maltases, trehalase and proteases (exopeptidases but not endopeptidases). In addition, the labial glands secrete chitinolytic enzymes which they considered to be an adaptation to mycophagy.

Since plant sap not only contains nutrients but also some toxins, the ants' guts should also contain detoxifying enzymes.

The most ubiquitous of all plant defences, especially among highly apparent species such as rain forest trees, are the phenolics, the majority of which occur as soluble, inactive glycosides. Active aglycones, however, are released upon rupture of plant cells by herbivores or pathogens. Oxidation of aglycones by the plant then produces quinones which are more reactive and bind strongly with proteins and amino acids, interfering with enzyme systems. Quinone activity can only be overcome by further oxidation and, to this end, some phytophagous insects and pathogenic fungi have evolved polyphenol oxidase as a detoxifying agent (Miles, 1968).

Phenolic compounds have long been considered a biochemical hurdle for the ant–fungus symbiosis. Powell (1984), however, has shown that the fungal symbionts of at least three attine genera (*Atta, Acromyrmex* and *Trachymyrmex*) secrete enzymes capable of breaking down a range of phenolic compounds. One example is polyphenol oxidase, which oxidizes various phenolic substances possessing *ortho*-hydroxyl groups, and is typical of many white-rot basidiomycete fungi (F5). These attack wood by digesting phenolic substances including lignin, before they digest the white cellulose. That the attine fungal symbionts are white-rotters was confirmed by Powell (1984) who demonstrated their dephenolization of lignin. Evidence that the degradation of phenolics occurs in the fungus-gardens was obtained using two-dimensional chromatograms of privet leaves made before and after passage through an *Atta* nest. These showed a marked decrease in the quantities of many of these substances and the disappearance of others.

Tannins are phenolic compounds which represent an important barrier to herbivory (Harborne, 1977) and also provide protection against fungal attack (Levin, 1976). They exhibit wide structural divergence but fall into two main groups: the hydrolysable tannins and proanthocyanidin (condensed) tannins which are more resistant to hydrolysis and tend to be found in mature and more persistent plant tissues. Tannins afford protection to the plant by their toxicity and, under certain conditions, by reducing the nutritional availability of its proteins, polysaccharides and nucleic acids.

Powell (1984) showed that all attine fungi tested were capable of growing normally on media containing up to 0.1% tannic acid (a hydrolysable tannin). A ring of coloured products (the Bavendamm reaction), also observed by Bottrell (1980), indicated that tannase (an esterase) and polyphenol oxidase had been secreted. Tannase breaks the ester bonds between the carbohydrate core of tannic acid and its attached phenolic acids. Cochrane (1958) suggested that the final breakdown products may be assimilated by the fungus. Thus attine symbionts may actually derive nutrients from the breakdown of tannins and other phenolic substances. A further demonstration of tannase secretion was the clearing of protein flocculated by tannic acid in agar plates. The characteristic Bavendamm reaction was also seen in plates containing macerated plant tissues, indicating that naturally occurring phenolics were degraded in the same way. Seaman (1984) criticized the earlier studies on the effects of tannins on the growth of the ant symbiont (Quinlan and Cherrett, 1987b) on the valid grounds that the effects of the tannins are reduced, as they interact with the agar medium itself. He used a liquid medium, and obtained a much greater inhibitory action of the tannins on fungus growth. However, a liquid medium is open to the opposite criticism, namely that detoxification enzymes secreted by the fungus will now be swept away by the mass flow of the liquid and will no longer exert a local effect immediately adjacent to the hyphae.

Although the fungi could tolerate up to 0.1% tannic acid, they were inhibited by at least one-tenth as much condensed tannin (Powell, 1984). This was not surprising since only a few microorganisms, such as *Penicillium adametzi* (F6.2) (Grant, 1976), are known to be capable of degrading it.

Having demonstrated that detoxifying enzymes are secreted by attine fungal symbionts, Powell (1984) went on to investigate whether, like the fungal proteases, these enzymes survive passage through worker guts. Tannase and polyphenol oxidase, having the same substrate specificity, optimum temperature and pH as those secreted by the fungi, were discovered in the faeces of *Atta, Acromyrmex* and *Trachymyrmex*. The temperature and pH optima of these enzymes were found to coincide with the conditions maintained in the fungus-gardens of the ants.

Plants possess various biochemical defences other than phenolics, including alkaloids, terpenoids, mustard oils and saponins. The ants cut and drink the sap of some leaves known to contain toxins and are apparently unaffected by them. Perhaps the workers acquire other detoxifying enzymes from their fungal symbionts in addition to those involved in the breakdown of phenolic substances.

Now that it appears that the fungus does secrete enzymes directly onto its substrate, and that other enzymes are added to the substrate via the ants' faeces, we can expect a good deal of extracellular digestion of plant material to occur in the fungus-garden. Quinlan and Cherrett (1978a) estimated that at any one time, 30% of the worker ants found on the surface of their fungus-garden were licking it, and there is no entirely satisfactory explanation of what they are doing. Littledyke and Cherrett (1976) induced colonies of both *Acromyrmex octospinosus* and *Atta cephalotes* to produce a radioactive fungus-garden, by giving them leaves internally labelled with ^{32}P to use as substrate. The ants engaged in preparing the garden were then removed, and clean ants were introduced. These latter progressively picked up radioactive label during the next 4 days, presumably before appreciable numbers of radioactive gongylidia had been produced. No explanation was offered for how this occurred and it is interesting to speculate that by repeatedly licking the substrate and fungus, the workers may be directly ingesting the breakdown products of extracellular digestion before they can all be absorbed by the fungus. This possibility should be investigated.

E. Life-cycle

The fungus gardens of advanced attines are monocultures of an obligate symbiotic basidiomycete (F5) which has lost the complex of sporulating sexual and asexual reproductive stages that characterize the life-cycles of *Termitomyces* (F5.10) and of free-living basidiomycete fungi.

The selective advantage of a monoculture is a matter for speculation, but the efficiency of conversion of substrate to a nutritious food supply at a rate commensurate with the requirements of the developing ant larvae provides a plausible explanation.

The establishment of new fungus cultures by incipient attine colonies is a continuous vegetative process equivalent to cloning which lacks both sexual and asexual reproductive stages, the mycelium being periodically subcultured from the isolates carried by gynes. This mechanism implies complete isolation of clones, always inherited from the parental colony, but we do not know how true this is. In *Atta texana* several newly fertilized queens will cooperate in colony foundation (primary pleometrosis) (Mintzer and Vinson, 1985) and, if these come from different parental colonies, the fungus-garden which results could contain elements from several infrabuccal pellets, with the possibility of subsequent genetic exchange by heterokariosis. However, such behaviour would only lead to heterokaryosis between clones within an ant species. Primary pleometrosis has not been reported in other attine species. Also, if young queens settle within the territory of an existing colony, the young colonies they establish are eventually discovered and destroyed (Jutsum *et al.*, 1979). During this process it is possible that individual workers will take back portions of the destroyed fungus-gardens to their own colony. Aggression between leaf-cutting ant species has also been recorded in the field (Fowler, 1977) but whether or not this leads to fungus interchange is again unknown.

The cloning of *Attamyces* by attines in the presence of the spore germination inhibitor myrmicacin over evolutionary time would militate against, and is probably responsible for the loss of any sporulating reproductive and dispersal stages in the life-cycle.

F. Productivity

Chemical analysis of fungal material from attine colonies (Martin *et al.*, 1969) has shown it to be a rich source of amino acids and carbohydrates but lacking in lipids. The rate of conversion of substrate nutrients to fungal biomass therefore constitutes one determinant of the larval food supply and may limit colony success.

In order to compare the possible influences of plant secondary compounds on the growth of *Attamyces*, Powell and Stradling (1986) performed tests to determine the optimum culture conditions and most suitable growth indices. These were carried out using isolates from six different species and subspecies of attines. All grew best at 25°C on PDA plates with a substrate close to pH 5.0. Growth performance was measured most satisfactorily in terms of total or gongylidial biomass. Under these conditions increases in fungal fresh weight shown by cultures of the six isolates (Fig. 2) indicated significant differences in productivity after six weeks of incubation (Stradling and Powell, 1986). Total fungal production was

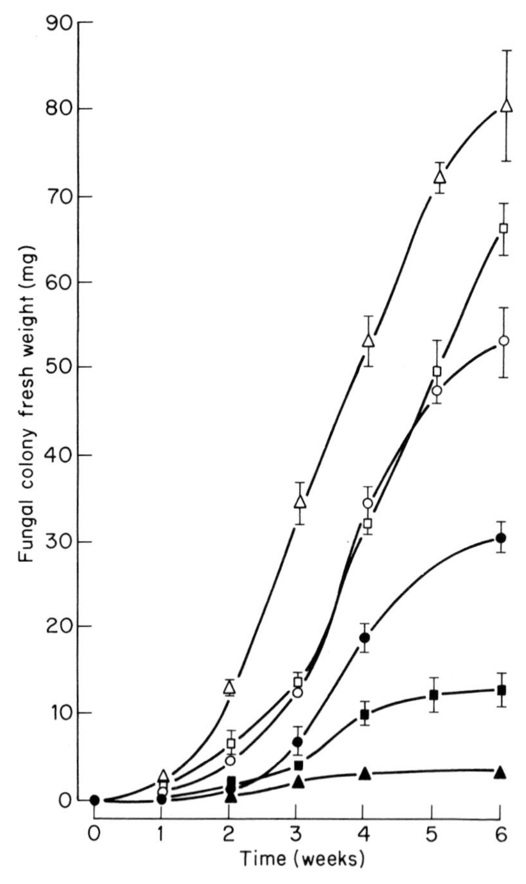

Fig. 2. Total production (± SE) by PDA cultures of fungal isolates from different attines during a period of six weeks. (▲) *Trachymyrmex urichi*, (■) *Acromyrmex octospinosus*, (•) *Atta cephalotes*, (○) *Atta sexdens sexdens*, (□) Brazilian *Atta sexdens rubropilosa*, (△) Paraguayan *Atta sexdens rubropilosa*. (After Stradling and Powell, 1986.)

reflected in the yield of gongylidia biomass (Table I). It is therefore evident that despite the lack of morphological differences and interactions between different fungal isolates, each attine species cultivates a strain of fungus characterized by a different productivity.

Of particular interest is the strong correlation ($r = 0.99$, $p < 0.001$) between fungal productivity and estimates of mature colony worker populations for the respective ant species (Table II). This relationship is described by the linear regression shown in Fig. 3. Thus Brazilian *Atta sexdens rubropilosa*, whose colonies reach worker populations of 3×10^6, cultivates a strain of fungus whose productivity is 3.17 times that of *A. cephalotes*, a species with colonies of up to about 0.7×10^6 workers, and 6.39 times that of *Acromyrmex octospinosus* whose colonies are of the order of 14×10^3.

The suppression of spore germination by ant secretions and the absence of sporulation in the life-cycle of *Attamyces* leaves somatic mutation as the only source of differences in fungal productivity. Stradling and Powell (1986) suggest that the selection of more highly productive mutant strains has been an important factor in the speciation of the higher attines. Since the gongylidia constitute the larval diet, their rate of production might be expected to influence colony expansion, sexual production and the success of foundress queens. Larger colony worker populations would lead to an extended foraging range, thus facilitating an expansion of the trophophoric field and niche separation. Such an improved food supply would in turn confer a selective advantage on increased queen fecundity and colony population (Stradling and Powell, 1986). Some support for this hypothesis is gained from the observation that a weak laboratory culture of *Trachymyrmex urichi*, when provided with the more highly productive fungal strain of *Atta cephalotes*, expanded rapidly and produced sexuals.

By analogy with the high yielding cereal clones used in developed and mechanized systems of agriculture however, it is to be expected that there would be penalties to be paid for high productivity. A rich substrate lacking fungal toxins may be required, growing conditions may be stringent, competitive ability may be poor, and susceptibility to pathogens high. The green revolution in cereal production can only be adopted by those farmers willing and able to provide the enhanced agricultural inputs required. It is an open question whether the low productivity of some attines results from the historic lack of high yielding fungus clones, or from the ecological conditions of the niche they occupy which dictates a lower yielding, less demanding, more hardy symbiont.

If indeed the selection and cloning of more highly productive fungal mutants underlies the speciation of the attines, the relatedness of fungal strains from different ant species might be expected to throw light on their

TABLE II. Mean estimated gongylidia production (mg) after six weeks of incubation by PDA cultures of fungal isolates from attine ant colonies[a] and estimated worker populations of mature colonies.

Ant species	Mean gongylidia biomass ± SE	Source	Maximum worker population	Source
Trachymyrmex urichi	1.72 ± 0.16	Trinidad	763 ± 76	Weber (1972)
Acromyrmex octospinosus	6.49 ± 0.13	Trinidad	14,278	Lewis (1975)
Atta cephalotes	13.10 ± 0.24	Trinidad	0.7×10^6	Based on Weber (1972)[b]
Atta sexdens sexdens	33.49 ± 0.14	Guyana	2.2×10^6	Weber (1966)
A. sexdens rubropilosa	41.47 ± 0.51	Brazil	3.0×10^6	Dias[c]
A. sexdens rubropilosa	50.12 ± 0.40	Paraguay		

[a] After Stradling and Powell (1986).
[b] Derived by applying Weber's direct count of 8762 workers in a representative fungus-garden of this species to our own observation of a mature field colony containing 80 full-sized fungus-gardens.
[c] Dias (pers. comm.).

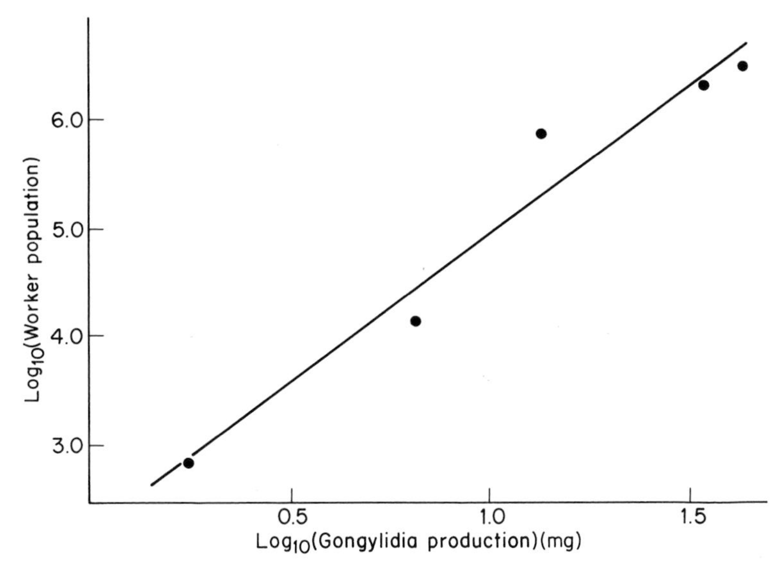

Fig. 3. Relationship between estimated maximum worker populations of attine colonies and gongylidia production by PDA cultures of their symbiotic fungi over a period of six weeks.

phylogeny. Preliminary restriction mapping studies of fungal DNA by Stradling (unpublished) have recently revealed detectable differences between fungal isolates from *Atta sexdens rubropilosa* and *Acromyrmex octospinosus*.

III. SELECTIVE ADVANTAGE OF THE MUTUALISM

It is evident that the mutualism has involved evolutionary adaptations to the life-cycle, physiology, behaviour and ecology of both the ant and the fungus. The intriguing question is what are the selective advantages of the mutualism which have driven this co-evolutionary process? Phytophagous insects make up approximately 25% of all living species found on earth (Strong *et al.*, 1984), but the overwhelming majority of these eat green plant material directly without having to culture a fungus to aid in its breakdown. The use of microorganisms by insects which feed on the less nutritious dead plant material is of course much more common. Similarly, other fungi which feed on living leaves have done so without domesticating an ant to aid them in the process.

As we have seen, for both the ant and the fungus, the outcome of their cooperation is a most unusual degree of ecological dominance in diverse tropical vegetation, brought about by unprecedented polyphagy. Any theory about the selective advantages of the mutualism should take these two facts into account. Cherrett (1980) has reviewed the principal theories, which may be summarized as follows.

A. The Provision of Energy from the Breakdown of Cellulose

Martin and Weber (1969), after demonstrating that a minimum of 45% of the cellulose originally present in the fresh leaves cut and placed on the fungus-garden was consumed before the spent substrate of the garden was cast out as refuse, concluded. "The contribution of the fungus to the ants is clearly the metabolic capability of converting cellulose into carbohydrates which can be metabolized by the ants." However, Quinlan and Cherrett (1979) subsequently calculated, on the basis of the rate of gongylidia harvested, that only about 4% of the respiratory requirements of the colony was coming from direct fungus consumption in this way, the bulk being obtained by adult workers directly drinking plant sap. This theory does not provide a sufficient explanation.

B. The Provision and Augmentation of Essential Nutrients

Quinlan and Cherrett (1979) demonstrated that the fungus is much more important in the diet of the growing larvae than it is in the diet of the adult workers, which mainly require a carbohydrate energy source. This implies a particular significance for the fungus in the provision of proteins and essential nutrients during growth. Martin *et al.* (1969) listed 20 amino acids found in the fungus together with ergosterol, and estimated that it contained 17% protein (dry weight) compared with a 13% average for trees and shrubs. As a consequence, they described the fungus as an excellent diet "... high in protein which is rich in essential amino acids." Whilst the argument goes some way to explaining the success of the ant through acquiring a microorganism to enrich its food supply, it does not explain the success of the fungus, nor how both partners achieved polyphagy.

C. The Creation of an Unholy Alliance to Break Down Plant Defences to Herbivore Exploitation

Cherrett (1980) pointed out that living plants have to defend themselves against a very wide range of potential herbivores, from bacteria and fungi to insect larvae and vertebrate grazers. This involves an equally wide range of defences, from simple physical barriers to penetration by micro-organisms to sophisticated hormone mimics against insects and vertebrates. This permits two organisms with very different physiologies and morphologies to cooperate to break through plant defences. Thus, the heavily chitinized mandibles of an insect would quickly break down a plant's physical barriers to hyphal penetration, whilst the enzymic virtuosity of a fungus could well denature an insect juvenile hormone analogue synthesized by a plant. This unholy alliance enables the insect to breach the plant's defences against fungi, and the fungus to breach the defences against insects, so that both can utilize a far wider range of host plants than either could alone. It is this which bestows the selective advantage of polyphagy on both ant and fungus. In the production of tannase and polyphenol oxidase, in the provision of high protein gongylidia from low protein substrate already discussed, and in the ability to denature the insect juvenile hormone analogue reported by Little *et al.* (1977) we see some evidence of the biochemical role of the fungus. In the selection of young susceptible leaves by the ants, in their cutting and comminuting them to an easily invaded pulp, in the licking off of inhibitory leaf waxes, and in the provision of liquid faeces containing active enzymes which begin to break down cell walls, we have evidence of the largely physical contribution of the ants to the destruction of plant defences. In passing, it is interesting to note that the soil fauna is often credited with the physical comminution of dead plant material, so facilitating its subsequent biochemical degradation by microorganisms (Swift *et al.*, 1979). Certainly as Janzen (1975) has pointed out, comparable polyphagy is usually only achieved by vertebrates such as ruminants, which support a complex microbial fauna capable of detoxifying plant secondary compounds.

Two weaknesses remain in this theory of the selective advantages of the ant–fungus symbiosis:

1. Comparable studies to those described for leaf-cutting ants have not yet been performed on the grass-cutting species, so we do not know how polyphagous they are, nor how an unholy alliance might operate with the very different secondary compounds of grasses.
2. It is not clear how the leaf-cutting ant mutualism evolved from the fungus-growing habits of the lower attines. Why do ants which collect caterpillar droppings (an odd diet by any standards) have to culture a fungus? Investigations in this fascinating field are still at a very early stage.

REFERENCES

Angeli-Papa, J. (1984). La culture d'un champignon par les fourmis attines; mise en evidence de pheromones d'antibioses dans le nid. *Cryptogam. Mycol.* **5**, 147–154.

Angeli-Papa, J., and Eyme, J. (1979). Le champignon cultive par la "fourmi-manioc" *Acromyrmex octospinosus* Reich en Guadeloupe: resultats preliminaires sur le mycelium en culture pure et sur l'infrastructure des hyphes. *C. R. Hebd. Seances Acad. Sci., Ser.D* **289**, 21–24.

Beattie, A. J. (1986). "The Evolutionary Ecology of Ant–Plant Mutualisms." Cambridge Univ. Press, Cambridge.

Begon, M., Harper, J. L., and Townsend, C. R. (1986). "Ecology. Individuals, Populations and Communities." Blackwell Scientific, Oxford.

Belt, T. (1874). "The Naturalist in Nicaragua." John Murray, London.

Bottrell, C. M. (1980). Studies on plant defences against attack by leaf-cutting ants of the genera *Atta* and *Acromyrmex*. M.Sc. Thesis, University College of North Wales, UK. (unpublished).

Boyd, N. D., and Martin, M. M. (1975a). Faecal proteinases of the fungus-growing ant, *Atta texana* (Hym, Formicidae): properties, significance and possible origin. *Insect Biochem.* **5**, 619–635.

Boyd, N. D., and Martin, M. M. (1975b). Faecal proteinases of the fungus-growing ant, *Atta texana* (Hym, Formicidae): their fungal origin and ecological significance. *J. Insect Physiol.* **21**, 1815–1820.

Bruch, C. (1921). Estudios Mirmecologicos. *Rev. Mus. La Plata* **26**, 175–211.

Burdon, J. J., and Chilvers, G. A. (1974). Fungal and insect parasites contributing to niche differentiation in mixed species stands of *Eucalyptus* saplings. *Aust. J. Bot.* **22**, 103–114.

Cherrett, J. M. (1968). The foraging behaviour of *Atta cephalotes* L. (Hymenoptera, Formicidae). 1. Foraging pattern and plant species attacked in tropical rain forest. *J. Anim. Ecol.* **37**, 387–403.

Cherrett, J. M. (1972). Some factors involved in the selection of vegetable substrate by *Atta cephalotes* (L.) (Hymenoptera: Formicidae) in tropical rain forest. *J. Anim. Ecol.* **41**, 647–660.

Cherrett, J. M. (1980). Possible reasons for the mutualism between leaf-cutting ants (Hymenoptera, Formicidae) and their fungus. *Biol.-Ecol. Mediterr.* **7**, 113–122.

Cherrett, J. M. (1986a). The economic importance and control of leaf-cutting ants. *In* "Economic Impact and Control of Social Insects" (S. B. Vinson, ed.), pp. 165–192. Praeger, New York.

Cherrett, J. M. (1986b). The biology, pest status and control of leaf-cutting ants. *Agric. Zool. Rev.* **1**, 1–37.

Cherrett, J. M. (in press). Leaf-cutting ants: their ecological role, diversity and zoogeography. *In* "Tropical Rainforest Ecosystems: Ecosystems of the World" (H. Lieth and M. J. A. Werger, eds). Elsevier, Amsterdam.

Cherrett, J. M., and Peregrine, D. J. (1976). A review of the status of leaf-cutting ants and their control. *Ann. Appl. Biol.* **84**, 124–128.

Cherrett, J. M., Pollard, G. V., and Turner, J. A. (1974). Preliminary observations on *Acromyrmex landolti* (For.) and *Atta laevigata* (Fr. Smith) as pasture pests in Guyana. *Trop. Agric. (Trinidad)* **51**, 69–74.

Cochrane, V. W. (1958). "Physiology of Fungi." Wiley Chapman and Hall, New York and London.

Cramer, H. H. (1967). "Plant Protection and World Crop Production." 'Bayer' Pflanzenschutz, Leverkusen.

Febvay, G., and Kermarrec, A. (1981a). Activités enzymatiques des glandes salivaires et de l'intestin moyen d'une fourmi attine (adultes et larves): *Acromyrmex octospinosus* (Reich) (Formicidae, Attini). *Arch. Biol.* **92**, 299–316.

Febvay, G., and Kermarrec, A. (1981b). Morphologie et fonctionnement du filtre intrabuccal chez une attine *Acromyrmex octospinosus* (Reich) (Hymenoptera, Formicidae). Role de la poche infrabuccale. *Int. J. Insect Morphol. Embryol.* **10**, 441–449.

Febvay, G., and Kermarrec, A. (1986). Digestive physiology of leaf-cutting ants. *In* "Fire Ants and Leaf-Cutting Ants. Biology and Management" (C. S. Lofgren and R. K. Vandermeer, eds), pp. 274–288. Westview Press, Boulder, Colorado and London.

Febvay, G., Bourgeois, P., and Kermarrec, A. (1985). Antiappetants pour la fourmi attinae, *Acromyrmex octospinosus* (Reich) (Hymenoptera, Formicidae), chez certaines espèces d'igname (Dioscoreaceae) cultive aux Antilles. *Agron. (Paris)* **5** (5), 439–444.

Fowler, H. G. (1977). Field response of *Acromyrmex crassispinosus* (Forel) to aggression by *Atta sexdens* (Linn.) and predation by *Labidus praedator* (Fr. Smith) (Hymenoptera: Formicidae). *Aggressive Behav.* **3**, 385–391.

Fowler, H. G., Forti, L. C., Pereira-da-Silva, V., and Saes, N. B. (1986). Economics of grass-cutting ants. *In* "Fire Ants and Leaf-Cutting Ants. Biology and Management" (C. S. Lofgren and R. K. Vandermeer, eds), pp. 18–35. Westview Press, Boulder, Colorado and London.

Funk, A. (1985). "Foliar Fungi of Western Trees." Canadian Forest Service, Pacific Forest Research Centre.

Goetsch, W. (1939). Pilzzuchtende ameisen. *Umschau* **43**, 157–159.

Goetsch, W., and Stoppel, R. (1940). Die pilze der blattschneider-ameisen. *Biol. Zentralbl.* **60**, 393–398.

Grant, W. D. (1976). Microbial degradation of condensed tannins. *Science* **193**, 1137–1139.

Hafez, E. S. E., ed. (1969). "The Behaviour of Cattle." Baillière, Tindall and Cassell, London. 2nd Edition.

Harborne, J. B. (1977). "Introduction to Ecological Biochemistry." Academic Press, London.

Heim, R. (1957). A propos du *Rozites gongylophora. Rev. Mycol.* **22**, 293–299.

Heim, R. (1958). *Leucoagaricus gongylophorus. In* "Index of Fungi," Vol. 2, Part 17. C.A.B. C.M.I., Kew, Surrey.

Hervey, A., and Nair, M. S. R. (1979). Antibiotic metabolite of a fungus cultivated by gardening ants. *Mycologia* **71**, 1064–1066.

Hervey, A., Rogerson, C. T., and Leong, I. (1977). Studies in fungi cultivated by ants. *Brittonia* **29**, 226–236.

Hubbell, S. P., and Wiemer, D. F. (1983). Host plant selection by an Attine ant. *In* "Social Insects in the Tropics" (P. Jaisson, ed.), Vol. 2, pp. 133–154. Proceedings of the First International Symposium, organised by the International Union for the Study of Social Insects and the Sociedad Mexicana Entomologia, Université Paris-Nord.

Hubbell, S. P., Wiemer, D. F., and Adeboye, A. (1983). An antifungal terpenoid defends a neotropical tree (*Hymenaea*) against attack by fungus-growing ants (*Atta*). *Oecologia* **60**, 321–327.

Huber, J. (1905). Ueber di Koloniengrundung bei *Atta sexdens* L. *Biol. Zentralbl.* **25**, 606–619, 625–635.

Hubert, T. D., Okunade, A. L., and Wiemer, D. F. (1987). Quadrangolide, a heliangolide from *Eupatorium quadrangulate. Phytochemistry* **26**, 1751–1753.

Ihering, H. von (1898). Die Anlage neuer Colonien und Pilzgarten bei *Atta sexdens. Zool. Anz.* **21**, 238–245.

Iwanami, Y. (1978). Myrmicacin, a new inhibitor for mitotic progression after metaphase. *Protoplasma* **95**, 267–271.

Iwanami, Y., and Iwadare, T. (1979). Myrmic acids: a group of new inhibitors analagous to myrmicacin (β-hydroxydecanoic acid). *Bot. Gaz. (Chicago)* **140**, 1–4.

Janet, C. (1905). "Anatomie de la tête du *Lasius niger*." Ducourtieux and Gout, Limoges.

Janzen, D. H. (1975). "Ecology of Plants in the Tropics." Edward Arnold, London.

Johnson, R. A., Thomas, R. J., Wood, T. G., and Swift, M. J. (1981). The inoculation of the fungus comb in newly founded colonies of some species of Macrotermitinae (Isoptera) from Nigeria. *J. Nat. Hist.* **15**, 751–756.

Jonkman, J. C. M. (1977). Biology and ecology of *Atta vollenweideri* Forel 1893 and its impact on Paraguayan pastures. Thesis, Universiteitsbibliothick, Leiden.

Jonkman, J. C. M. (1978). Nests of the leaf-cutting ant, *Atta vollenweideri*, as accelerators of succession in pastures. *Z. Angew. Entomol.* **86**, 25–34.

Jutsum, A. R., Saunders, T. S., and Cherrett, J. M. (1979). Intraspecific aggression in the leaf-cutting ant *Acromyrmex octospinosus*. *Anim. Behav.* **27**, 839–844.

Kermarrec, A., Febvay, G., and Decharme, M. (1986). Leaf-cutting ant symbiotic fungi: a synthesis of recent research. *In* "Fire Ants and Leaf-Cutting Ants. Biology and Management" (C. S. Lofgren and R. K. Vandermeer, eds), pp. 339–356. Westview Press, Boulder, Colorado and London.

Koppensteiner, G., and Bansemir, K. (1975). *Use of B-hydroxycarboxylic acids in combination with formaldehyde as antimicrobial composition*. Patent: Ger Offen 2349058.

Kreisel, H. (1972). Pilze aus Pilzgarten von *Atta insularis* in Kuba. *Z. Allg. Mikrobiol.* **12**, 643–654.

Kreisel, H. (1975). *Attamyces*. *In* "Index of Fungi," Vol. 4, Part 10. C.A.B. C.M.I., Kew, Surrey.

Lawton, J. H., and McNeill, S. (1979). Between the devil and the deep blue sea: on the problems of being a herbivore. *In* "Population Dynamics" (R. M. Anderson, B. D. Turner and L. R. Taylor, eds), pp. 223–244. Blackwell Scientific, Oxford.

Levin, D. A. (1976). The chemical defenses of plants to pathogens and herbivores. *Annu. Rev. Ecol. Syst.* **7**, 121–159.

Lewis, T. (1975). Colony size, density and distribution of the leaf-cutting ant, *Acromyrmex octospinosus* (Reich) in cultivated fields. *Trans. R. Entomol. Soc. London* **127**, 51–64.

Little, C. H., Jutsum, A. R., and Cherrett, J. M. (1977). Leaf-cutting ant control. The possible use of growth-regulating chemicals. Proceedings of the VIII Congress of the International Union for the Study of Social Insects, Wageningen, pp. 89–90.

Littledyke, M., and Cherrett, J. M. (1976). Direct ingestion of plant sap from cut leaves by the leaf-cutting ants *Atta cephalotes* (L.) and *Acromyrmex octospinosus* (Reich) (Formicidae, Attini). *Bull. Entomol. Res.* **66**, 205–217.

Littledyke, M., and Cherrett, J. M. (1978). Defence mechanisms in young and old leaves against cutting by the leaf-cutting ants *Atta cephalotes* and *Acromyrmex octospinosus* (Reich) (Hymenoptera: Formicidae). *Bull. Entomol. Res.* **68**, 263–271.

Mariconi, F. A. M. (1970). "As Sauvas." Editora Agronomica Ceres, São Paulo.

Martin, J. S., and Martin, M. M. (1970). The presence of protease activity in the rectal fluid of attine ants. *J. Insect Physiol.* **16**, 227–232.

Martin, M. M. (1974). Biochemical ecology of attine ants. *Acc. Chem. Res.* **7**, 1–5.

Martin, M. M., and Boyd, N. D. (1974). Properties, origin and significance of the fecal proteases of the fungus-growing ants. *Am. Zool.* **14**, 1291.

Martin, M. M., and Weber, N. A. (1969). The cellulose utilising capacity of the fungus cultured by the attine ant *Atta colombica tonsipes*. *Ann. Entomol. Soc. Am.* **62**, 1386–1387.

Martin, M. M., Carman R. M., and MacConnell J. G. (1969). Nutrients derived from the fungus cultured by the fungus-growing ant *Atta colombica tonsipes*. *Ann. Entomol Soc. Am.* **62**, 11–13.

Martin, M. M., Gieselmann, M. J., and Martin, J. S. (1973). Rectal enzymes of attine ants: alpha-amylase and chitinase. *J. Insect Physiol.* **19**, 1409–1416.

Martin, M. M., Boyd, N. D., Gieselmann, M. J., and Silver, R. G. (1975). Activity of faecal fluid of a leaf-cutting ant toward plant cell wall polysaccharides. *J. Insect Physiol.* **21**, 1887–1892.

Miles, P. W. (1968). Insect secretions in plants. *Ann. R. Phyto.* **6**, 137–164.

Mintzer, A., and Vinson, S. B. (1985). Cooperative colony foundation by females of the leaf-cutting ant *Atta texana* in the laboratory. *J. N.Y. Entomol. Soc.* **93**, 1047–1051.

Möller, A. (1893). Die Pilzgarten einiger sudamerikanischer Ameisen. *Bot. Mitt. Trop.* **6**, 1–27.

Osberghaus, R., Krauch, C. H., Kolaczinski, G., and Koppensteiner, G. (1974). *Antimicrobial β-hydroxyalkanoic acids.* Patent: Ger Offen 2312280.

Papa, F., and Papa, J. (1982a). Etude de l'activité microbiologique dans les nids d'*Acromyrmex octospinosus* Reich en Guadeloupe. *Bull. Soc. Pathol. Exot.* **75**, 404–414.

Papa, J., and Papa, F. (1982b). Inhibition des bacteries dans les nids d'*Acromyrmex octospinosus* Reich. *Bull. Soc. Pathol. Exot.* **75**, 415–425.

Powell, R. J. (1984). The influence of substrate quality on fungus cultivation by some attine ants. Ph.D. Thesis, University of Exeter, UK (unpublished).

Powell, R. J., and Stradling. D. J. (1986). Factors influencing the growth of *Attamyces bromatificus*, a symbiont of attine ants. *Trans. Br. Mycol. Soc.* **87**, 205–213.

Quinlan, R. J., and Cherrett, J. M. (1977). The role of substrate preparation in the symbiosis between the leaf-cutting ant *Acromyrmex octospinosus* (Reich) and its food fungus. *Ecol. Entomol.* **2**, 161–170.

Quinlan, R. J., and Cherrett, J. M. (1978a). Studies on the role of the infrabuccal pocket of the leaf-cutting ant *Acromyrmex octospinosus* (Reich) (Hym., Formicidae). *Insectes Soc.* **25**, 237–245.

Quinlan, R. J., and Cherrett, J. M. (1978b). Aspects of the symbiosis of the leaf-cutting ant *Acromyrmex octospinosus* (Reich) and its food fungus. *Ecol. Entomol.* **3**, 221–230.

Quinlan, R. J., and Cherrett, J. M. (1979). The role of fungus in the diet of the leaf-cutting ant *Atta cephalotes* (L.). *Ecol. Entomol.* **4**, 151–160.

Rockwood, L. L. (1976). Plant selection and foraging patterns in two species of leaf-cutting ants (*Atta*). *Ecology* **57**, 48–61.

Rockwood, L. L., and Glander, K. E. (1979). Howling monkeys and leaf-cutting ants: comparative foraging in a tropical deciduous forest. *Biotropica* **11**, 1–10.

Schildknecht, H., and Koob, K. (1971). Myrmicacin, the first insect herbicide. *Angew. Chem. Int. Ed. Engl.* **9**, 173.

Schildknecht, H., Reed, P. B., Reed, F. D., and Koob, K. (1973). Auxin activity in the symbiosis of leaf cutting ants and their fungus. *Insect Biochem.* **3**, 439–442.

Seaman, F. C. (1984) The effects of tannic acid and other phenolics on the growth of the fungus cultivated by the leaf-cutting ant, *Myrmicocrypta buenzlii*. *Biochem. Syst. Ecol.* **12**, 155–158.

Spegazinni, C. (1899). Fungi Argentini novi v critici. *Ann. Mus. Nat. Hist., N.Y.* **3**, 81–365.

Spegazinni, C. (1921). Description de hongos mirmecofilos. *Rev. Mus. La Plata* **26**, 166–174.

Stahel, G. and Geijkes, D. C. (1941). Weitere Untersuchungen uber Nestbau und Gartenpilz von *Atta cephalotes* L. und *Atta sexdens* L. (Hymenoptera: Formicidae). *Rev. Entomol., Rio de J.* **12**, 243–268.

Stoppel, R. (1940). Pilze der pilzgarten von *Atta sexdens*. *Planta* **31**, 406–413.

Stradling, D. J. (1978). The influence of size on foraging in the ant *Atta cephalotes* and the effect of some plant defense mechanisms. *J. Anim. Ecol.* **47**, 173–188.

Stradling, D. J., and Powell, R. J. (1986). The cloning of more highly productive fungal strains: a factor in the speciation of fungus growing ants. *Experientia* **42**, 962–964.

Strong, D. R. Jr., Lawton, J. H., and Southwood, T. R. E. (1984). "Insects on Plants: Community Patterns and Mechanisms." Blackwell Scientific, Oxford.

Swift, M. J., Heal, O. W., and Anderson, J. M. (1979). "Decomposition in Terrestrial Ecosystems." Blackwell Scientific, Oxford.

Thomas, R. J. (1981). Ecological studies on the symbiosis of *Termitomyces* with Nigerian Macrotermitinae. Ph.D. Thesis, University of London, UK (unpublished).

Townsend, C. H. T. (1923). Um inseto de um bilhao de dollares e sua eliminacao. A formiga sauva. *Almanak Agric. Brasil.* **12**, 253–254.

Waller, D. A. (1982a). Leaf-cutting ants and live oak: the role of leaf toughness in seasonal and intraspecific host choice. *Entomol. Exp. Appl.* **32**, 146–150.

Waller, D. A. (1982b). Leaf-cutting ants and avoided plants: Defences against *Atta texana* attack. *Oecologia* **52**, 400–403.

Weber, N. A. (1938). The biology of the fungus-growing ants (3). The sporophore of the fungus grown by *Atta cephalotes* and a review of other reported sporophores. *Rev. Entomol., Rio de J.* **8**, 265–272.

Weber, N. A. (1957). Weeding as a factor in fungus culture by ants. *Anat. Rec.* **28**, 638.

Weber, N. A. (1966). Fungus-growing ants. *Science* **153**, 587–604.

Weber, N. A. (1972). 'Gardening Ants: the Attines.' *Mem. Am. Philos. Soc.* **92**, 1–146.

Wheeler, W. M. (1907). The fungus-growing ants of North America. *Bull. Am. Mus. Nat. Hist.* **23**, 669–807.

Wheeler, W. M. (1910). "Ants, their Structure, Development and Behaviour." Columbia Univ. Press, New York and London.

Wheeler, W. M. (1913). "Ants." Columbia Univ. Press, New York and London.

Wilson, E. O. (1971). "The Insect Societies." Harvard Univ. Press (Belknap), Cambridge, Massachusetts.

Wilson, E. O. (1982). Of insects and man. *In* "The Biology of Social Insects: Proceedings of the Ninth Congress of the International Union for the Study of Social Insects" (M. D. Breed, C. D. Michener and H. E. Evans, eds), pp. 1–3. Westview Press, Boulder, Colorado.

Wint, G. R. W. (1983). Leaf damage in tropical rain forest canopies. *In* "Tropical Rain Forest: Ecology and Management" (S. L. Sutton, T. C. Whitmore and A. C. Chadwick, eds), pp. 229–239. Special Publication No. 2, British Ecological Society, Blackwell Scientific, Oxford.

5

Insect–Fungus Relationships in the Bark and Ambrosia Beetles

R.A. BEAVER

I. INTRODUCTION

Bark and ambrosia beetles (H25.63) are well known as forest insects which often cause economic damage to trees and timber. They are commonly, perhaps generally, associated with fungi. The fungi frequently play an important, sometimes an essential, role in the lives of the beetles. Various aspects of the beetle–fungus interaction have been reviewed by Baker (1963), Francke-Grosmann (1966, 1967), Graham (1967), Norris (1979) and Whitney (1982). An annotated bibliography for the period 1965–1974 is given by Barras and Perry (1975), and most of the earlier literature is covered by the reviews cited. This paper looks at the fungus–beetle association, primarily from the point of view of the beetles. It is particularly concerned with the mutual relationships that have evolved several times within the bark and ambrosia beetles, the benefits to the partners, and the effects that the mutualism has had on their morphology, biology, ecology and geographical distribution.

II. THE ORGANISMS INVOLVED

A. The Beetles

The bark and ambrosia beetles have been considered either as two families (Scolytidae and Platypodidae) within the weevils (Curculionoidea) (e.g. Wood, 1973, 1982), or as subfamilies of Curculionidae (H25.63) (e.g. Crowson, 1967, 1981; and in other papers in this volume). Most workers have taken the first option, and this tradition is followed here. Typically the beetles breed in woody plants, making characteristic gallery systems either in the phloem (bark beetles) or in the wood (ambrosia beetles). Other habitats used include large seeds and fruits, the pith of twigs and the petioles of large fallen leaves. The vast majority of species breed only in dead or dying plants or parts of plants such as shaded-out branches. A very small proportion are able to attack and kill healthy trees. Of the latter group, species of the bark beetle genera *Dendroctonus* and *Ips* have been extensively studied because of their economic importance, but the arboricidal habits are not typical.

Usually only a single generation of beetles can breed in a particular habitat unit, e.g. a branch or a seed, before the unit is either exhausted as a food source, or too altered to be of further use for breeding. Each generation of adult beetles has to disperse to find new units. This has obvious implications for the development of any fixed association between particular beetles and particular fungi.

Feeding habits may be classified into six categories. Phloeophagy involves feeding on the phloem or inner bark, the nutritive value of which is high, particularly when fresh. This habit is widespread in the Scolytidae. Less common feeding habits include herbiphagy (feeding on the soft tissues of herbaceous plants), myelophagy (feeding on the pith of twigs), xylophagy (feeding on the xylem of woody stems) and spermatophagy (feeding on fruits and seeds). The other major feeding habit is that of the ambrosia beetles—xylomycetophagy—feeding on wood-inhabiting ectosymbiotic fungi introduced into the gallery system by the beetle. The term xylomycetophagy is used for those species for which the ectosymbiotic "ambrosia" fungi form the major part of the food of both larvae and adults, and are essential for the completion of the beetles' life-cycle. The

TABLE I. Occurrence of xylomycetophagy in Scolytidae and Platypodidae.

Scolytidae	
Hylesinini	*Dactylipalpus* Chapuis[a], *Phloeoborus* Erichson[a]
Hyorrhynchini	Whole tribe[c]
Bothrosternini	*Bothrosternus* Eichhoff
Phloeosinini	*Hyleops* Schedl
Scolytini	*Camptocerus* Latreille
Scolytoplatypodini	*Scolytoplatypus* Blandford[b]
Xyloterini	Whole tribe[c]
Xyleborini	Whole tribe[c]
Cryphalini	*Hypothenemus curtipennis* Schedl
Corthylini	All Corthylina[c], *Pityoborus* Blackman, probably *Dacnophthorus* Wood
Platypodidae	All except *Mecopelmus* Blackman, *Schedlarius* Wood, *Protoplatypus* Wood[d]

[a] Inferred from presence of mycangia.
[b] Only genus in tribe.
[c] For included genera, see Wood (1986).
[d] Taxonomic position of genus controversial.

great majority of ambrosia beetles breed in the trunks and branches of trees. A few breed in twigs or in the pith of small stems. The major part of the wood excavated by the beetles is not ingested but pushed out of the gallery, and there appears to be little digestion of the fragments that are ingested. The majority of tropical Scolytidae and almost all species of Platypodidae are xylomycetophagous. Ambrosia beetles occur in at least 10 tribes of Scolytidae (Table I), indicating the parallel evolution of the habit a number of times (see Section IV), often in a single genus within a tribe.

The classification of the Scolytidae and Platypodidae according to their feeding habits is useful, but it must be realized that the beetles "do not cooperate very readily in tidy classifications" (Browne, 1961a). Thus the larvae of some bark beetles are primarily phloeophagous, but feed partly on fungi growing in the larval galleries, e.g. some species of *Dendroctonus, Ips* and *Tomicus*. There is no absolute distinction between phloeophagous or myelophagous, and xylomycetophagous species. All are normally associated to some extent with fungi.

B. The Fungi

Fungi associated with ambrosia beetles have been listed by Baker (1963), Batra (1967), Francke-Grosmann (1967), Kok (1979) and Norris (1979). Those associated with bark beetles are listed by Francke-Grosmann (1967) and Whitney (1982). Most of the fungi involved in mutualistic associations are only known in the imperfect state (Deuteromycotina). Where the perfect state is known, the majority belong to the Ascomycotina. Blue-stain fungi associated mutualistically with bark beetles are mostly members of the plectomycete genus *Ceratocystis* (F4.24), although several other genera may be involved (Francke-Grosmann, 1967; Whitney, 1982). Some species of *Ceratocystis* are also believed to occur as ambrosia fungi (Francke-Grosmann 1967). Several genera of yeasts (e.g. *Pichia* (F4.11), *Hansenula* (F4.11)) show mutualistic associations with bark and ambrosia beetles (e.g. Baker, 1963; Bridges *et al.*, 1984; Callaham and Shifrine, 1960; Francke-Grosmann, 1967; Gusteleva, 1982; Whitney, 1982). Other important genera of ambrosia fungi include the hyphomycetes (F6.2) *Fusarium, Cephalosporium* and *Ambrosiella*, and filamentous hemiascomycetes such as *Endomycopsis* (F4.11). Basidiomycotinous ambrosia fungi (F5) are known only from *Dendroctonus* spp. (Happ *et al.*, 1976a), *Xyleborus dispar* (Happ *et al.*, 1976b), and *Pityoborus* spp. (Furniss *et al.*, 1987).

An important characteristic of the ambrosia fungi is their pleomorphism. Most can grow either as hyphae forming an extended mycelium, or in "ambrosia" form (see below), or in yeast-like form. The mycelium ramifies in the xylem and phloem, but usually sporulates only in the beetle galleries. There is usually one species of fungus, the primary ambrosia fungus, dominant in the galleries, but other secondary or auxiliary ambrosia fungi occur, together with bacteria. Together they form a symbiotic microbial complex (Haanstad and Norris, 1985). The relative dominance of the species may differ in different parts of the gallery system, and at different times during its occupation by the beetles and their offspring.

Many other fungi and bacteria have been isolated from the beetles and their galleries (Baker, 1963; Whitney, 1982). Many of them, e.g. *Penicillium* spp. (F6.2) and *Trichoderma* spp. (F6.2), seem to be "weed" fungi with no more than a commensal relationship with the beetles. They may sometimes be used as a supplementary food source, but tend to occur in old or abandoned galleries. The pathogenic fungi, such as *Beauveria* (F6.2) and *Metarhizium* (F6.2), while occasionally important in the life and population dynamics of the beetles, fall outside the scope of this paper, as do the endosymbiotic bacteria found in the gut, Malpighian tubules and elsewhere (Büchner, 1965), and transmitted via the eggs (Peleg and Norris, 1972).

III. BENEFITS OF MUTUALISM

By definition, mutualism involves a benefit to both partners in the association. The primary benefit to the fungi is probably the reliable dispersal (relative to wind dispersal) of the spores, and their inoculation into new habitats suitable for their growth. The beetles' tunnelling helps rapid penetration of the mycelium. Once established, the fungus also benefits from the activities of the beetles. These help, in ways which are poorly understood, to maintain the dominance of the ambrosia fungi in the galleries, and to hinder or prevent the growth of other fungi. The direct benefits to the beetles are more varied. The fungi form the only real source of food for ambrosia beetle adults and larvae. They may provide a supplementary source of food for some bark beetle larvae and teneral adults. Fungi may be involved in the conversion of host tree chemicals into beetle pheromones. The range of host trees and host tree conditions in which successful breeding is possible may be widened. The inoculation of plant-pathogenic fungi into a living host tree by the beetle may provide both direct and indirect benefits. Thus certain *Dendroctonus* and *Scolytus* species which attack living conifers may gain direct benefit, because the fungus helps to reduce host defence mechanisms in the vicinity of the beetle attacks. The chances of successful beetle colonization are improved (Berryman, this volume). For beetles which breed in dead trees, the benefits may be indirect. Thus maturation feeding by the reproductively immature adults of various species of *Scolytus* on the twigs of living elm trees can result in the introduction of the pathogen, *Ceratocystis ulmi* (F4.24). The fungus spreads through the tree, gradually killing it. The beetles can then breed in branches which have been killed by the fungus.

These examples indicate that the beetle–fungus association should not be considered apart from the host tree in which both partners live (Crowson, 1981). For the great majority of bark and ambrosia beetles, the host tree is an essentially passive partner, for it is already dead when they attack; but if alive, it is an active defendant against beetle and fungus attacks. The relationship between tree and fungus, or tree and beetle, is then antagonistic rather than mutualistic (Webber and Gibbs, this volume).

The direct mutual benefits of the beetle–fungus association will now be examined in more detail, particularly in relation to ambrosia beetles.

A. Fungal Transmission and Beetle Mycangia

Neger (1911) considered that transmission of ambrosia fungi from the old to the new host tree occurred in the gut of the ambrosia beetle. Later, Doane and Gilliland (1929) suggested transmission by the adhesion of fungal spores to the exoskeleton of the beetle. The spores of blue-stain fungi (*Ceratocystis* (F4.24)) are adapted to epizoic transport by arthropods (Dowding, 1969, 1979). They produce a sticky mucilage which helps to hold together spore masses and aids adhesion to the arthropod exoskeleton (in this case that of the ambrosia beetle) or to the bark. An outer shell of dried

TABLE II. Types of Mycangia found in Scolytidae and Platypodidae.

Type	Species	Reference
Head		
Paired mandibular pouches	♀ *Xyleborus* (many spp.)	Schedl (1962), Francke-Grosmann (1967), Takagi (1967), Baker and Norris (1968), Nakashima (1982)
	♀ *Euwallacea fornicatus*	Fernando (1960)
	♀ *Ips acuminatus*	Francke-Grosmann (1963)
	♂ ♀ *Dryocoetes confusus*	Farris (1969)
	♂ ♀ *Xyloterinus politus*	Abrahamson and Norris (1966)
Maxillary pouches	♂ ♀ *Dendroctonus ponderosae*	Whitney and Farris (1970)
Single pouch at rear of preoral cavity	♀ *Crossotarsus niponicus*	Nakashima (1971, 1975, 1982)
Paired pharyngeal pouches	♀ *Premnobius cavipennis*	Schedl (1962)
Mentum-pregular pouch	♀ *Monarthrum bicallosum*	Schedl (1962)
Pits on vertex, cheeks	♂ ♀ *Scolytus ventralis*	Livingston and Berryman (1972)
Pronotum		
Below anterior margin	♀ *Dendroctonus brevicomis*	Francke-Grosmann (1966, 1967)
	♀ *D. adjunctus*	Francke-Grosmann (1966), Barras and Perry (1971)
	♀ *D. frontalis*	Francke-Grosmann (1965), Barras and Perry (1972)
	♀ *D. approximatus* (probable in *D. vitei, D. mexicanus*)	Whitney (1982)
Median pronotal pit	♀ *Scolytoplatypus* spp.[a]	Schedl (1962)
2–many dorsal pits (often fewer or 0 in ♂)	♀ *Platypus* (many spp.), ♀ *Carchesiopygus* spp., ♂ ♀ *Genyocerus* spp.[a], ♀ *Neotrachyostus* spp., ♀ *Periommatus* spp.	Farris and Funk (1965), Francke-Grosmann (1967), Schedl (1972), Nakashima (1975, pers. obs.)

Anterior transverse crevice	♀ *Dactylipalpus* spp.[a]	Nunberg (1951, pers. obs.)
Posterior transverse crevice (sometimes in 2 arcuate halves or with associated pores)	♂ ♀ *Diapus* spp.[a]	Nakashima (1975, pers. obs.)
Shallow lateral depressions	♀ *Hypothenemus curtipennis*	Beaver (1986)
	♀ *Pityoborus* spp.[a]	Furniss *et al.* (1987)
Paired tubes opening into precoxal cavity	♂ *Corthylus columbianus*	Giese (1967)
	♂ *C. punctatissimus*	Finnegan (1963)
	♂ *C. schaufussi*	Schedl (1962)
	♂ *Microcorthylus castaneus*	Schedl (1962)
Paired tubes opening on proepimeron	♀ *Trypodendron* spp.[a]	Francke-Grosmann (1956), Abrahamson *et al.* (1967), French and Roeper (1972)
	♀ *Xyloterinus politus*	Abrahamson and Norris (1966)
Paired tubes opening on proepisternum	♀ *Phloeoborus* spp.[a]	Wood (1982)
	♀ *Bothrosternus* spp.[a]	Wood (1982)
Pro/Mesonotal		
Pouch extended below pronotum	♀ *Xylosandrus* spp.[a]	Francke-Grosmann (1956), Takagi and Kaneko (1965)
Pouch extended below mesonotum	♀ *Xyleborus dispar*	Francke-Grosmann (1956), Happ *et al.* (1976b)
Mesonotum involuted in spiral	♀ *Eccoptopterus* spp.[a]	Francke-Grosmann 1958
Elytra		
Pouches in elytral bases	♀ *Cryptoxyleborus* spp.	Pers. obs.
	♀ *Xyleborinus* spp.[a]	Francke-Grosmann (1956), Schedl (1962)
Enlarged Coxal Cavities		
Procoxae	♂ *Gnathotrichus* spp.[a]	Farris (1963), Schneider and Rudinsky (1969)
	♀ *Monarthrum* spp.	Schedl (1962), Francke-Grosmann (1963), Farris (1965), Lowe *et al.* (1967), Roeper and French (1978)
Pro- and meso-coxae	♂ *Platypus* spp.	Nakashima (1972, 1975, 1982)

[a] Probably occurring in all species in genus.

spores and mucilage protects the inner spore mass against desiccation and irradiation (Dowding, 1969). The mucilage also protects the spores from digestion in the gut of the beetle (Francke-Grosmann, 1963), so that both epizoic and endozoic dispersal may be possible. However, the conidia of some ambrosia fungi are unable to survive even short periods of slight desiccation (Zimmermann and Butin, 1973), and would be unable to survive dispersal on the outside of the beetle. Ambrosia beetles have evolved special structures for fungal transmission which protect such spores against desiccation.

In 1951, Nunberg suggested that the tubelike glands found in the pronotum of females of *Trypodendron* spp. could play a role in fungal transmission, and that glandular pits and cavities found in the pronotum of the scolytid genera *Dactylipalpus, Phloeoborus* and *Scolytoplatypus*, and the platypodid genera *Platypus* and *Diapus* might also have a similar function. Nunberg's suggestion was confirmed by Francke-Grosmann (1956) for *Trypodendron* and various other Scolytidae and Platypodidae, and the term "mycangia" for such organs was suggested (Batra, 1963). Since then, a great variety of mycangia have been described in various parts of scolytid and platypodid beetles (Table II). They occur not only in ambrosia beetles, but also in some bark beetles (e.g. some species of *Dendroctonus*).

The mycangia have a similar basic structure and consist of tubes, pouches, cavities or pits of various sizes, associated with glandular cells. (If these cells are absent, as in male *Dendroctonus adjunctus*, the structure is termed a pseudomycangium (Barras and Perry, 1971), and is probably not regularly involved in fungal transport.) Flattened setae or bunches of hairs often help to retain the contents of the mycangium.

Mycangia may have evolved both as pocket-like internal expansions of normal parts of the body, and as newly developed hollow glands (Francke-Grosmann, 1966). Whitney (1982) suggests that the mycangia of certain *Dendroctonus* spp. evolved as defensive glands which were later colonized by microorganisms able to tolerate the defensive secretions. Another suggestion (Francke-Grosmann, 1966) is that the glands originally secreted an oily substance used as a water-repellent against excessive sap flow, and as a lubricant while boring. Fungi which were able to utilize the secretions and grow there in yeast-like form were dispersed more regularly, and a specialized symbiotic relationship gradually evolved. Whatever the truth of these speculations, it is clear that mycangia have evolved on numerous occasions within the Scolytidae and Platypodidae in conjunction with the evolution of mutualistic relationships with fungi.

The type of mycangium is usually specific to the genus, but there can be variation between species within a genus, e.g. *Monarthrum, Platypus* and *Xyleborus*. Mycangia may apparently not be present in all members of a genus, e.g. *Dendroctonus*. The mycangia may occur in females only, males only or in both sexes, depending on the species. If occurring in both sexes, the mycangia are often reduced, and sometimes non-functional in the males, e.g. *Dendroctonus, Xyleborus*. If occurring in males only (*Corthylus, Microcorthylus, Gnathotrichus*, some *Platypus*), it is the male which initiates the gallery system. However, the converse is not true. There are some Platypodidae, for example, with mycangia present in the female only, but the male still starts the gallery. Sometimes, two types of mycangia can be present in a single species, e.g. males of *Platypus severini* Blandford and several other *Platypus* spp. have small numbers of pronotal pits, and enlarged pro- and meso-coxal cavities (Nakashima, 1975). Both sexes of *Xyloterinus politus* have paired mandibular mycangia: the females also have pronotal mycangia (Abrahamson and Norris, 1966).

Mycangia are not always essential for the successful transmission of ambrosia fungi, and recent work on them has tended to overshadow the occurrence of other less specialized methods of fungal transport. The old and rather discredited suggestion that ambrosia fungi are transmitted through the gut has recently been shown to be correct in *Xyleborinus saxeseni* (Francke-Grosmann, 1975). She found that the principal ambrosia fungus, *Ambrosiella sulphurea* (F6.2), is transmitted in the form of micromycelium in the hind gut. The mycangia in the elytra only occasionally contained *A. sulphurea*, but regularly contained auxiliary ambrosia fungi (including yeasts (F4.11, etc.), *Ceratocystis* (F4.24) and *Penicillium* (F6.2)). Baker (1963) also found that yeasts (*Candida* sp. (F6.2)) and chlamydospores of the chief ambrosia fungus (*Raffaelea ambrosiae*—F6.2 (= *Sporothrix* sp.)) were carried in the gut of *Platypus cylindrus*.

No mycangia have been found in a number of Scolytidae and Platypodidae which are closely and regularly associated with particular fungi, or are known to be ambrosia beetles. *Dendroctonus pseudotsugae, D. rufipennis, Ips avulsus* and *Tomicus minor*, for example, are all bark beetles known to have consistent associations with particular fungi (Francke-Grosmann, 1952; Whitney, 1982), but none have mycangia. In *D. pseudotsugae*, fungal transmission probably occurs by the adhesion of mycelial fragments and spores to the exoskeleton of the beetle (Castello *et al.*, 1976). The same may be true of *D. rufipennis* and *I. avulsus*, or there may be transmission via the gut (Gouger, 1972). In *T. minor*, fungal spores are carried in the sutural groove of the left elytron and on the metanotum (Francke-Grosmann, 1956). Similar methods of transmission of fungal spores probably occur in those Platypodidae (all ambrosia beetles) in which no mycangia have been found, e.g. *Doliopygus* spp. (Lhoste and Roche, 1961; Francke-Grosmann, 1956). In some of the platypodid genera not known to have mycangia, the female frons may be impressed (sometimes deeply excavated), and often carries brushes of long hairs (e.g. *Chaetastus, Doliopygus, Mitosoma*). These structures may carry spores of the ambrosia fungi, although this seems unlikely to be their main function (Browne, 1961a, b).

Entry of the spores into the mycangia seems to be passive, either from the walls of the pupal cell, or from the walls of the gallery as the teneral adult moves around the gallery system, although Batra and Batra (1967) suggest that rocking movements of the teneral adults in the parental gallery system force spores into the mycangia. The acquisition of the fungi is non-specific, but only certain species will grow within the mycangia (e.g. Barras and Perry, 1971; Paine and Birch, 1983). The mycangial glands may secrete chemicals that inhibit the growth of "contaminating" fungi and/or favour the growth of the symbiotic fungi (Happ *et al.*, 1971; Schneider, 1976). Antibiotics produced by the fungi themselves could also be responsible (Nakashima *et al.*, 1982; Whitney, 1982).

The growth form of the fungi in the mycangia usually differs from that in the galleries. The fungi appear to be yeast-like, and grow by budding and fission, producing distinct cells, although sometimes yeast-mycelial "transition" cells and mycelioid forms are found (Happ *et al.*, 1975; Paine and Birch, 1983). Growth is slow. Although the gland cells of the mycangia produce an oily or waxy secretion, the lipids are not major nutrients for fungal growth (Kok, 1979); they probably help to protect the spores from desiccation. A major determinant of the yeast-like growth form appears to be free amino acids, which are present in large amounts in the haemolymph and body secretions (Abrahamson and Norris, 1970; Norris, 1979), and are utilized by the mycangial fungi (French and Roeper, 1973). The gland cells are active during the flight period of the beetles, and it seems to be during this period that fungal growth and the selective elimination of certain species occurs (Barras and Perry, 1971; Paine and Birch, 1983; Schneider and Rudinsky, 1969). The net result is that the number of ambrosial fungus cells available for inoculation into the new habitat unit is increased.

After beetle dispersal, the method of release of the fungi from the mycangia into the gallery system probably depends in part on the type and position of the mycangia. In most species, release is probably passive. The contents of mycangia associated with intersegmental membranes or coxal joints may be released by body movements (Schedl, 1962). Whitney (1982) suggests that spores may be squeezed out of the pronotal mycangia of *Dendroctonus* spp. by pressure on the thin mycangial wall during tunnelling.

B. Fungi as Food for the Beetles

Wood is an unpromising food source compared to most other organic substrates. It consists mainly of lignin and cellulose, both of which are resistant to degradation by enzymes found in most beetles (Haack and Slansky, 1987). It is deficient in essential B-vitamins and sterols which the beetles cannot synthesize themselves (Baker, 1963) and often rich in secondary compounds (Haack and Slansky, 1987). Wood that has been partly decayed by fungi provides a higher quality food resource, and the fungi themselves an even better one. Fungi break down the cellulose and lignin, producing organic molecules which can be assimilated by the beetles, and also synthesize chemicals essential for beetle development. In the case of the ambrosia beetles, this relatively nitrogen-rich food can be obtained with a minimum of effort, because the nutrients are translocated through the mycelium to the fungal layer lining the galleries, and little active tunnelling is required. Xylophagous species can acquire the necessary nutrients directly from the wood given time, but xylomycetophagous species obtain them faster and with far less effort. Their life-cycles tend to be shorter than those of xylophagous species (Haack and Slansky, 1987).

In the wood and in the beetle galleries, the fungal growth is mycelial, not yeast-like. The hyphae penetrate the phloem and wood, and can spread widely through the host plant's tissues. Within the galleries of ambrosia beetles, the fungal hyphae form an "ambrosial" layer covering the walls. The ambrosia consists of a layer of fungal hyphae, the ends of which may form a regular, erect palisade bearing conidia and/or chlamydospores, or may form tangled chains of moniliform cells (Baker, 1963; Batra, 1967). This layer is browsed by the beetles and their larvae, and this browsing may promote further sporulation (Batra, 1966).

The development of the ambrosia form of the fungus seems to depend in part on physical contact between beetle and fungus (Whitney, 1971; French and Roeper, 1972). It may also involve secretions of the beetles and their larvae (Francke-Grosmann, 1956, 1967; French and Roeper, 1972). However, Whitney (1982) points out that fungi can show pleomorphism in the absence of beetles, and that there is no need to involve mycangial or other secretions. Further investigations are needed to resolve the matter.

Contrary to early ideas, the ambrosia fungi are not species-specific to the beetles. The natural food of the beetles and their larvae is probably the whole mutualistic microbial complex (Haanstad and Norris, 1985), although the proportions of the various fungi and bacteria in the diet may vary with the stage of development of the beetles and the age of gallery system (Batra, 1966).

Ambrosia beetles normally disperse with empty guts (or with small amounts of material in the hind gut only), and with immature gonads (e.g. Baker, 1963; Brader, 1964; Browne, 1961b; Francke-Grosmann, 1967; French and Roeper, 1975; Roberts, 1961). They do not feed during initial gallery construction. Most of the wood that is excavated is pushed directly out of the gallery entrance. Feeding only begins after the ambrosia fungi have become established. This feeding is essential for gonadal maturation. There is a nutritional threshold for oviposition, and for the associated behavioural changes (Kingsolver and Norris, 1977). If the fungus does not become established, no offspring are produced, and the parent beetles eventually die.

Using *Xyleborus ferrugineus*, Norris (1979) found that the number of offspring produced depended on the species of fungus used as food. Brood production was maximal on the primary ambrosia fungus, *Fusarium solani* (F6.2), but reduced by 50% when the female fed on the secondary ambrosia fungus, *Cephalosporium* sp. (F6.2), and by 70% when fed on *Graphium* sp. (F6.2). A similar dependence of fecundity on the associated species of ambrosial fungus is indicated by Bridges (1983) for *Dendroctonus frontalis*. Bridges and Norris (1977) suggest that the initiation of reproduction depends on a particularly rich meal of certain essential amino acids provided by the symbiotes. Thus ovarian maturation has become symbiote-dependent rather than sperm-dependent. (In many of the scolytid ambrosia beetles, sib-mating occurs before the adults emerge from the gallery system in which they developed.)

In certain Platypodidae (Baker, 1963; Browne, 1961b; Cachan, 1957; Roberts, 1961), and possibly also in some Scolytidae, ingested thick-walled spores may pass through the gut unchanged, and germinate on the walls of the tunnels, but the hyphae are digested by the beetles and their larvae. The rate of growth of the fungi in the gallery system determines the rate of growth of the brood in ambrosia beetles (Brader, 1964; Hosking, 1973; Kaneko, 1965), and fungal growth rates depend primarily on the condition, moisture content and temperature of the wood. The presence of an active parent female is essential for the healthy development of the brood (Beaver, 1972; Norris, 1979). If the female dies, the brood dies too, apparently unable to control the fungal growth within the gallery. As noted by Norris (1979), we still do not understand how the female exerts her control over the ambrosia fungus.

There has been some progress in understanding the chemical basis of the dependence of the ambrosia beetles on their fungal symbionts for food. Fungal hyphae provide a richer source of protein than dead wood, and fungi are able to concentrate nitrogen from substrates in which it occurs in very low concentration (Martin, 1979; Swift and Boddy, 1984). There is probably some concentration of nitrogen in the ambrosial layer (P. J.

Whitney, pers. comm.). We may also note that nitrogen-fixing bacteria have been found associated with certain bark beetles (*Dendroctonus, Ips*), and it is suggested that they could be nutritionally important in certain conditions (Bridges, 1981).

In addition to the nitrogen compounds, the fungi also produce steroids (in particular Δ^7 steroids such as ergosterol), which have been shown by Norris and his co-workers, summarized by Norris (1976), to be essential for egg-hatching, complete larval development, pupation and oocyte maturation in several species of *Xyleborus*. A peculiarity of these species, relative to almost all other insects, is that cholesterol alone will not produce pupation (Kok *et al.*, 1970; Norris, 1966). Ergosterol, produced by the symbiotic fungi is essential, and is a critical nutrient which Kok (1979) suggests may form the chemical basis of the mutualistic relationship. Fungally-produced vitamins may also be involved. Gusteleva (1982) found that certain B-group vitamins synthesized by the symbiotic yeast, *Pichia* sp., were apparently essential for the development of *Ips cembrae* (= *I. subelongatus*).

Fungal growth in the gallery systems of bark beetles may provide a supplementary source of food for larvae and teneral adults. In *Dendroctonus frontalis*, it has been shown that the absence of associated symbiotic fungi results in decreased brood survival and a longer development time (Barras, 1973; Bridges, 1983). However, there is also evidence that the associated blue-stain fungus, *Ceratocystis minor* (F4.24), can be detrimental to the development of *D. frontalis* (Barras, 1970; Franklin, 1970). It appears that ascigerous *C. minor* can make the phloem unsuitable for *D. frontalis*, but that in the larval galleries, the formation of asci is prevented or delayed by the presence of the beetle larvae. The fungus develops an ambrosial character, with prolific formation of conidia, which are fed on by the larvae (Barras and Taylor, 1973).

C. Fungi and Beetle Pheromone Production

Brand *et al.* (1976) found that the mycangial fungi of *Dendroctonus frontalis* could convert *trans*-verbenol to verbenone. The latter compound is a multifunctional pheromone important in the aggregation of the adult beetles on host trees (Borden, 1985). They also found that the fungi could convert 3-methyl-2-cyclohexen-1-ol (seudenol) to the corresponding ketone, which is a multifunctional pheromone in *Dendroctonus pseudotsugae* (Borden, 1985). Brand *et al.* (1976) suggested that the development of the fungus in the plant host could play a role in influencing the behaviour of *D. frontalis* towards the host tree. Leufvén *et al.* (1984) have since shown that yeasts in the gut of *Ips typographus* can also interconvert verbenols and verbenone. Bacteria have also been shown to synthesize bark beetle pheromone components and attractive chemicals (Brand *et al.*, 1975; Chararas *et al.*, 1980). However, all this work has been done in the laboratory, and the importance of bacteria and fungi in the production of pheromone components in the field is unknown. It is perhaps significant that Conn *et al.* (1984) have shown that *Ips paraconfusus* and *Dendroctonus ponderosae* can produce their aggregative pheromones in the complete absence of normal symbiotic microorganisms.

Bacteria and fungi could also be involved in the primary (i.e. non-pheromone-induced) attraction of beetles to their host trees. Bark and ambrosia beetles are often attracted to stressed or dying trees, many of which will have fungal or bacterial infections. It is known that ethyl alcohol is attractive to many of these beetles (Moeck, 1981; Elliott *et al.*, 1983, pers. obs.), and ethyl alcohol is a well-known by-product of the metabolism of various yeasts and other fungi.

D. Fungi and Host Specificity of the Beetles

Most ambrosia beetles are polyphagous, and breed in a wide range of host trees; most bark beetles are much more restricted in their host range (Atkinson, 1988; Beaver, 1977, 1979a). One important reason for this is that the fungi on which the ambrosia beetles feed are polyphagous. So far as is known, the beetle feeds on the same fungi whatever host it is attacking. Thus the symbiotic association allows lower host specificity. However, it does not necessarily cause it, because it is the beetle that selects the host (Beaver, 1979a). Another reason for polyphagy may be the floristic diversity and high decomposition rates of the humid tropics, where ambrosia beetles make up the majority of the fauna. This makes it more difficult for host-specific beetles to find suitable host material (Beaver, 1979b). This situation would tend to lead to selection for greater polyphagy by all species whether xylomycetophagous or not. The published figures give some indication of this (Atkinson, 1988; Beaver, 1979a). However, Atkinson (1988) suggests that most phloeophagous species do not have the flexibility to widen their host ranges significantly.

Not all ambrosia beetles are polyphagous. There are various reasons for this. In Mexico, Atkinson (1988) found that the majority of host-specific ambrosia beetles were found in environments with few tree species, and attacked the most abundant trees. In Malaysia, most host-specific ambrosia beetles are restricted to the Dipterocarpaceae, the most abundant tree family in the forests, rather than to other plant families (Beaver and Browne, 1979). Thus abundance of a particular host group may permit host specificity (cf. Stevens, 1986).

The few ambrosia beetles that habitually attack living trees, and breed successfully without killing their hosts, also tend to show high host specificity. Thus *Dendroplatypus impar* is restricted to a group of *Shorea* species in Malaysia (Browne, 1961a), *Trachyostus ghanaensis* to *Triplochiton* in West Africa (Roberts, 1968), *Austroplatypus incompertus* to *Eucalyptus* in Australia (Harris *et al.*, 1976). In North America, *Corthylus columbianus* attacks a greater range of deciduous trees in several families (Kabir and Giese, 1966a). High host specificity in these cases is presumably related to the necessity for the beetle–fungus combination to overcome active host defences, and the ability of the particular fungus associates to grow in living trees. Unfortunately, with the exception of *C. columbianus*, the symbiotic fungi do not seem to have been investigated. In *C. columbianus*, the chief ambrosia fungi were reported to be *Ambrosiella xylebori* (F6.2) and *Pichia* sp. (F4.11), together with associated *Fusarium* spp. (F6.2) (Kabir and Giese, 1966b; Nord, 1972). *A. xylebori* is possibly a pleomorphic form of *Fusarium solani*, which is well known as a plant pathogen (Trujillo in Hara and Beardsley, 1979). It is noteworthy that *F. solani* is the primary ambrosia fungus of three other ambrosia beetles (*Xylosandrus compactus, X. germanus* and *Xyleborus ferrugineus*), which sometimes attack living trees (Weber and McPherson, 1984).

From an evolutionary point of view, it seems likely that the majority of ambrosia beetles are primitively polyphagous, and that monophagy has developed only secondarily. However, there may be exceptions, e.g. *Camptocerus, Pityoborus*, which are primitively host-specific (Atkinson, 1988).

IV. EVOLUTION OF THE MUTUALISM AND ITS RELATION TO CLIMATE

Batra (1966) has suggested that the association between the beetles and fungi was initially fortuitous. Fungi growing below the bark ramified into the frass-filled larval galleries of bark beetles, and were occasionally eaten along with the phloem. Phloeophagy is certainly the primitive feeding habit in Scolytidae and Platypodidae (H25.63) (Kirdendall, 1983; Wood, 1982). Once the beetles began to depend on the fungi for food, they were able to colonize the less nutritious, but less well defended xylem, which the fungi had already penetrated and weakened (Atkinson, 1988). Eventually the advantages of eating fungus alone (see Section III, B), rather than wood and fungus, led to xylomycetophagy and the development of a mutualistic relationship (Batra, 1966). Within the wood, often living in narrow tunnels, the beetles and their offspring are also better protected against predators and parasites. Mortality of ambrosia beetle broods seems to be generally lower than that of phloeophagous species (Beaver, 1977). The exploitation of a new ecological niche led in turn to extensive species radiations in some groups, particularly the Xyleborini, Corthylini and Platypodidae (Atkinson, 1988).

It seems probable that this evolutionary process occurred in tropical rain forests, in which the high temperatures and humidity are particularly favourable for fungal growth, and where associations between beetles and fungi would frequently have occurred. Atkinson and Equihua-Martinez (1986) suggest that the utilization of fungi as a primary source of food may have arisen from competition between beetles and fungi for the same substrate. Such competition might frequently have occurred in different beetle–fungus associations, leading to the multiple evolution of the ambrosia habit.

If the ambrosia-feeding habit is basically a response to high humidities and temperatures (Atkinson and Equihua-Martinez, 1986), it might be expected that it would occur more frequently in geographical regions with these climatic conditions. A general correlation with latitude was shown by Beaver (1979a). Tropical areas have a higher percentage of xylomyceto-phagous species than temperate areas. Recent studies by Atkinson (1988) of eight areas within Mexico have shown a more detailed correlation with climate. The percentage of xylomycetophagous species in the local fauna varies from zero in xeric scrub, and 13–16% in the relatively cool and dry, higher altitude tropical deciduous forest, to 60% in the hot and humid, lowland rain forest. Floristic composition and diversity have a modifying effect, but the importance of climate is clear.

V. COSTS OF MUTUALISM

Discussions of mutualism usually concentrate on the benefits that each of the partners gains from the association. However, there are also costs following from the loss of independence of each partner. The spores of the ambrosia fungi are not exposed in fruiting bodies on the surface of the bark. For dispersal and inoculation into new sites, they are dependent on the survival of the new generation of ambrosia beetles, and it is likely that mortality of the beetles during dispersal is high. The fungi are no longer able to determine fully their vegetative growth form, and they normally forego sexual reproduction. They lose a great deal of vegetative and reproductive production as a result of the feeding of the beetles and their larvae.

The production of brood by the ambrosia beetles, and the growth of that brood, is dependent on the growth of ambrosia fungi in the galleries, and not on food obtained directly from the wood, or from the phloem in which the beetles are tunnelling. Successful development of the progeny is dependent on the survival and brood-tending behaviour of the parent female. The offspring cannot be left to their own devices as in bark beetles. Pupation, in at least some of the ambrosia beetles, is dependent on fungus-produced steroids. There is an energy cost to the adult in the production of the specialized mycangia. Thus the evolution of an obligatory mutualism involves both costs and benefits. Even a partial failure by one partner to carry out its "obligations" can lead to the death of both.

Mutualistic relationships are also potentially open to "parasitism", either by one of the partners, or by organisms outside the association (Janzen, 1985). The fungus *Fusarium solani* (F6.2) is the primary ambrosia fungus of several ambrosia beetles, but it is also known to be weakly pathogenic to bark beetle larvae (Barson, 1976; Moore, 1971). It is possible that in certain conditions it may be similarly pathogenic to ambrosia beetle larvae.

Ambrosia beetles which are the first colonists of a dead tree must carry their ambrosia fungi with them to survive. Species which arrive later could make use of these introduced fungi if their own galleries are close to those of the first colonists. The late arrivals would not necessarily have to bear the cost of mycangia production, and they would have to wait a much shorter time for the fungus to grow in their own galleries. Such a relationship has been suggested by Kalshoven (1960) for certain *Xyleborus* species, but there is as yet no clear-cut evidence for it. The fungus-feeding mites and nematodes that are phoretic on the adult beetles (as hypopi or dauer-larvae respectively), may also be parasites of the beetle–fungus mutualism, but are not allowed to thrive in gallery systems containing active beetles. The ambrosia beetles and their symbiotic fungi form a small part of much larger food webs, the complexities of which we have barely started to understand.

REFERENCES

Abrahamson, L. P., and Norris, D. M. (1966). Symbiontic relationships between microbes and ambrosia beetles. I. The organs of microbial transport and perpetuation of *Xyloterinus politus*. *Ann. Entomol. Soc. Am.* **59**, 887–890.

Abrahamson, L. P., and Norris, D. M. (1970). Symbiontic relationships between microbes and ambrosia beetles (Coleoptera: Scolytidae). V. Amino acids as a source of nitrogen to the fungi in the beetle. *Ann. Entomol. Soc. Am.* **63**, 177–180.

Abrahamson, L. P., Chu, H.-M., and Norris, D. M. (1967). Symbiontic relationships between microbes and ambrosia beetles. II. The organs of microbial transport and perpetuation in

Trypodendron betulae and *T. retusum* (Coleoptera: Scolytidae). *Ann. Entomol. Soc. Am.* **60**, 1107–1110.

Atkinson, T. H. (1988). Effects of climate and vegetation on patterns of host use by bark and ambrosia beetles (Coleoptera: Scolytidae and Platypodidae). *Environ. Entomol.* (in press).

Atkinson, T. H., and Equihua-Martinez, A. (1986). Biology of bark and ambrosia beetles (Coleoptera: Scolytidae and Platypodidae) of a tropical rainforest in Southeastern Mexico with an annotated checklist of species. *Ann. Entomol. Soc. Am.* **79**, 414–423.

Baker, J. M. (1963). Ambrosia beetles and their fungi, with particular reference to *Platypus cylindrus* Fab. *Symp. Soc. Gen. Microbiol.* **13**, 232–265.

Baker, J. M., and Norris, D. M. (1968). A complex of fungi mutualistically involved in the nutrition of the ambrosia beetle *Xyleborus ferrugineus*. *J. Invertebr. Pathol.* **11**, 246–250.

Barras, S. J. (1970). Antagonism between *Dendroctonus frontalis* and the fungus *Ceratocystis minor*. *Ann. Entomol. Soc. Am.* **63**, 1187–1190.

Barras, S. J. (1973). Reduction of progeny and development in the Southern pine beetle following removal of symbiotic fungi. *Can. Entomol.* **105**, 1295–1299.

Barras, S. J., and Perry, T. (1971). Gland cells and fungi associated with prothoracic mycangium of *Dendroctonus adjunctus* (Coleoptera: Scolytidae). *Ann. Entomol. Soc. Am.* **64**, 123–126.

Barras, S. J., and Perry, T. (1972). Fungal symbionts in the prothoracic mycangium of *Dendroctonus frontalis* (Coleopt.: Scolytidae). *Z. Angew. Entomol.* **71**, 95–104.

Barras, S. J., and Perry, T. (1975). Interrelationships among microorganisms, bark or ambrosia beetles, and woody host tissue: an annotated bibliography, 1965–1974. *For. Serv., U. S. Dep. Agric., Gen. Tech. Rep.* SO-10, 34 pp.

Barras, S. J., and Taylor, J. J. (1973). Varietal *Ceratocystis minor* identified from mycangium of *Dendroctonus frontalis*. *Mycopathol. Mycol. Appl.* **50**, 293–305.

Barson, G. (1976). *Fusarium solani*, a weak pathogen of the larval stages of the large elm bark beetle, *Scolytus scolytus* (Coleoptera: Scolytidae). *J. Invertebr. Pathol.* **27**, 307–309.

Batra, L. (1963). Ecology of ambrosia fungi and their dissemination by beetles. *Trans. Kans. Acad. Sci.* **66**, 213–236.

Batra, L. R. (1966). Ambrosia fungi: extent of specificity to ambrosia beetles. *Science* **153**, 193–195.

Batra, L. R. (1967). Ambrosia fungi: a taxonomic revision and nutritional studies of some species. *Mycologia* **59**, 976–1017.

Batra, S. W. T., and Batra, L. R. (1967). The fungus gardens of insects. *Sci. Am.* **217**(5), 112–120.

Beaver, R. A. (1972). Biological studies of Brazilian Scolytidae and Platypodidae (Col.). I. *Camptocerus* Dejean. *Bull. Entomol. Res.* **62**, 247–256.

Beaver, R. A. (1977). Bark and ambrosia beetles in tropical forests. *Proc. Symp. Forest Pests and Diseases S. E. Asia, Bogor, 1976. BIOTROP* Spec. Publ. No. 2, 133–147.

Beaver, R. A. (1979a). Host specificity of temperate and tropical animals. *Nature (London)* **281**, 139–141.

Beaver, R. A. (1979b). Non-equilibrium "island" communities: a guild of tropical bark beetles. *J. Anim. Ecol.* **48**, 987–1002.

Beaver, R. A. (1986). The taxonomy, mycangia and biology of the first known cryphaline ambrosia beetle, *Hypothenemus curtipennis* (Schedl) (Coleoptera: Scolytidae). *Entomol. Scand.* **17**, 131–135.

Beaver, R. A., and Browne, F. G. (1979). The Scolytidae and Platypodidae (Coleoptera) of Penang, Malaysia. *Orient. Insects* **12**, 575–624.

Borden, J. H. (1985). Aggregation pheromones. *In* "Comprehensive Insect Physiology, Biochemistry and Pharmacology" (G. A. Kerkut and L. I. Gilbert, eds), Vol. 9, pp. 257–285. Pergamon Press, Oxford.

Brader, L. (1964). Étude de la relation entre le scolyte des rameaux du caféier *Xyleborus compactus* Eichh. (*X. morstatti* Hag.), et sa plante-hôte. *Meded. Landbouwhogesch. Wageningen* **64**(7), 1–109.

Brand, J. M., Bracke, J. W., Markovetz, A. J., Wood, D. L., and Browne, L. E. (1975). Production of verbenol pheromone by a bacterium isolated from bark beetles. *Nature (London)* **254**, 136–137.

Brand, J. M., Bracke, J. W., Britton, L. N., Markovetz, A. J., and Barras, S. J. (1976). Bark beetle pheromones: production of verbenone by a mycangial fungus of *Dendroctonus frontalis*. *J. Chem. Ecol.* **2**, 195–199.

Bridges, J. R. (1981). Nitrogen-fixing bacteria associated with bark beetles. *Microb. Ecol.* **7**, 131–137.

Bridges, J. R. (1983). Mycangial fungi of *Dendroctonus frontalis* (Col., Scolyt.) and their relationship to beetle population trends. *Environ. Entomol.* **12**, 858–861.

Bridges, J. R., and Norris, D. M. (1977). Inhibition of reproduction of *Xyleborus ferrugineus* by ascorbic acid and related chemicals. *J. Insect Physiol.* **23**, 497–501.

Bridges, J. R., Marler, J. E., and McSparrin, B. H. (1984). A quantitative study of the yeasts and bacteria associated with laboratory-reared *Dendroctonus frontalis* Zimm. (Coleopt. Scolytidae). *Z. Angew. Entomol.* **97**, 261–267.

Browne, F. G. (1961a). The biology of Malayan Scolytidae and Platypodidae. *Malay. For. Rec.* **22**, 1–255.

Browne, F. G. (1961b). Preliminary observations on *Doliopygus dubius* (Samps.) (Coleopt. Platypodidae). *Rep. W. Afr. Timb. Borer Res. Unit, Kumasi* **4**, 15–30.

Büchner, P. (1965). "Endosymbiosis of Animals with Plant Microorganisms." Wiley, New York.

Cachan, P. (1957). Les Scolytoidea mycétophages des forêts de Basse Côte-d'Ivoire. Problèmes biologiques et écologiques. *Rev. Pathol. Vég. Entomol. Agric. Fr.* **36**, 1–126.

Callaham, R. Z., and Shifrine, M. (1960). The yeasts associated with bark beetles. *For. Sci.* **6**, 146–154.

Castello, J. D., Shaw, C. G., and Furniss, M. M. (1976). Isolation of *Cryptoporus volvatus* and *Fomes pinicola* from *Dendroctonus pseudotsugae*. *Phytopathology* **66**, 1431–1434.

Chararas, C., Rivière, K., Ducauze, C., Delpui, G., Rutledge, D., and Cazelles, M.-T. (1980). Bioconversion d'un composé terpénique sous l'action d'une bactérie du tube digestif de *Phloeosinus armatus* (Coleoptera, Scolytidae). *C.R. Hebd. Séances Acad. Sci., Ser.D.* **291**, 299–302.

Conn, J. E., Borden, J. H., Hunt, D. W. A., Holman, J., Whitney, H. S., Spanier, O. J., Pierce, H. D., Jr., and Oehlschlager, A. C. (1984). Pheromone production by axenically reared *Dendroctonus ponderosae* and *Ips paraconfusus* (Coleoptera: Scolytidae). *J. Chem. Ecol.* **10**, 281–290.

Crowson, R. A. (1967). "The Natural Classification of the Families of Coleoptera". E. W. Classey, Hampton, Middlesex.

Crowson, R. A. (1981). "The Biology of Coleoptera". Academic Press, New York and London.

Doane, R. W., and Gilliland, O. J. (1929). Three Californian ambrosia beetles. *J. Econ. Entomol.* **22**, 915–921.

Dowding, P. (1969). The dispersal and survival of spores of fungi causing bluestain in pine. *Trans. Br. Mycol. Soc.* **52**, 125–137.

Dowding, P. (1979). The evolution of insect–fungus relationships in the primary invasion of forest timber. *In* "Invertebrate–Microbial Interactions" (J. M. Anderson, A. D. M. Rayner and D. W. H. Walton, eds), pp. 135–153. Cambridge Univ. Press, Cambridge.

Elliott, H. J., Madden, J. L., and Bashford, R. (1983). The association of ethanol in the attack behaviour of the mountain pinhole borer, *Platypus subgranosus* Schedl (Coleoptera, Curculionidae, Platypodinae). *J. Entomol. Soc. Aust.* **22**, 299–302.

Farris, S. H. (1963). Ambrosia fungus storage in two species of *Gnathotrichus* Eichhoff (Coleoptera: Scolytidae). *Can. Entomol.* **95**, 257–259.

Farris, S. H. (1965). Repositories of symbiotic fungus in ambrosia beetle *Monarthrum scutellare* Lec. (Coleoptera: Scolytidae). *Proc. Entomol. Soc. B. C.* **62**, 30–33.

Farris, S. H. (1969). Occurrence of mycangia in the bark beetle *Dryocoetes confusus* (Coleoptera: Scolytidae). *Can. Entomol.* **101**, 527–532.

Farris, S. H., and Funk, A. (1965). Repositories of symbiotic fungus in the ambrosia beetle, *Platypus wilsoni* Swaine (Coleoptera: Platypodidae). *Can. Entomol.* **97**, 527–532.

Fernando, E. F. W. (1960). Storage and transmission of ambrosia fungus in the adult *Xyleborus fornicatus* (Eich.) (Coleoptera: Scolytidae). *Ann. Mag. Nat. Hist.* Ser. 13, **2**, 475–480.

Finnegan, R. J. (1963). The storage of ambrosia fungus spores by the pitted ambrosia beetle, *Corthylus punctatissimus* Zimm. (Coleoptera: Scolytidae). *Can. Entomol.* **95**, 137–139.

Francke-Grosmann, H. (1952). Über die Ambrosiazucht der beiden Kiefernborkenkäfer *Myelophilus minor* Htg. und *Ips acuminatus* Gyll. *Medd. Statens Skogsforskningsinst. (Swed.)* **41**(6), 1–52.

Francke-Grosmann, H. (1956). Hautdrüsen als Träger der Pilzsymbiose bei ambrosiakäfern. *Z. Morphol. Oekol. Tiere* **45**, 275–308.

Francke-Grosmann, H. (1958). Über die Ambrosiazucht holzbrütender Ipiden im Hinblick auf das System. *Verh. Dtsch. Ges. Angew. Entomol.* **14**, 139–144.

Francke-Grosmann, H. (1963). Die Übertragung der Pilzflora bei dem Borkenkäfer *Ips acuminatus* Gyll. *Z. Angew. Entomol.* **52**, 355–361.

Francke-Grosmann, H. (1965). Ein Symbioseorgan bei dem Borkenkäfer *Dendroctonus frontalis* Zimm. (Coleoptera, Scolytidae). *Naturwissenschaften* **52**, 143.

Francke-Grosmann, G. (1966). Über Symbiosen von xylomycetophagen und phloeophagen Scolytoidea mit holzbewohnenden Pilzen. *Beih. Mater. Org.* **1**, 503–522.

Francke-Grosmann, H. (1967). Ectosymbiosis in wood-inhabiting insects. *In* "Symbiosis" (S. M. Henry, ed.), Vol. 2, pp. 141–205. Academic Press, New York and London.

Francke-Grosmann, H. (1975). Zur epizoischen und endozoischen Übertragung der symbiotischen Pilze des Ambrosiakäfers *Xyleborus saxeseni* (Coleoptera: Scolytidae). *Entomol. Ger.* **1**, 279–292.

Franklin, R. T. (1970). Observations on the blue stain-southern pine beetle relationship. *J. Georgia Entomol. Soc.* **5**, 53–57.

French, J. R. J., and Roeper, R. A. (1972). Observations on *Trypodendron rufitarsis* (Coleoptera: Scolytidae) and its primary symbiotic fungus, *Ambrosiella ferruginea*. *Ann. Entomol. Soc. Am.* **65**, 282.

French, J. R. J., and Roeper, R. A. (1973). Patterns of N-utilisation between the ambrosia beetle *Xyleborus dispar* and its ambrosia fungus. *J. Insect Physiol.* **19**, 593–605.

French, J. R. J., and Roeper, R. A. (1975). Studies on the biology of the ambrosia beetle *Xyleborus dispar* (F.) (Coleoptera: Scolytidae). *Z. Angew. Entomol.* **78**, 241–247.

Furniss, M. M., Woo, J. Y., Deyrup, M. A., and Atkinson, T. H. (1987). Prothoracic mycangium of pine-infesting *Pityoborus* spp. (Coleoptera: Scolytidae). *Ann. Entomol. Soc. Am.* **80**, 692–696.

Giese, R. L. (1967). The Columbian timber beetle, *Corthylus columbianus* (Coleoptera: Scolytidae). V. A description of the mycangia. *Can. Entomol.* **99**, 54–58.

Gouger, R. J. (1972). Interrelations of *Ips avulsus* (Eichh.) and associated fungi. *Diss. Abstr.* **32**, 6453-B.

Graham, K. (1967). Fungal–insect mutualism in trees and timber. *Ann. Rev. Entomol.* **12**, 105–126.

Gusteleva, L. A. (1982). The interaction of wood-decomposing insects with micro-organisms. *In* "Konsortivnye Svyazi Dereva i Dendrofil'nykh Nasekomykh" (A. S. Isaev, ed.), pp. 56–67. Novosibirsk, USSR.

Haack, R. A., and Slansky, F. (1987). Nutritional ecology of wood-feeding Coleoptera, Lepidoptera, and Hymenoptera. *In* "Nutritional Ecology of Insects, Mites and Spiders" (F. Slansky Jr. and J. G. Rodriguez, eds), pp. 449–486. Wiley, New York.

Haanstad, J. O., and Norris, D. M. (1985). Microbial symbionts of the ambrosia beetle, *Xyloterinus politus*. *Microb. Ecol.* **11**, 267–276.

Happ, G. M., Happ, C. M., and Barras, S. J. (1971). Fine structure of the prothoracic mycangium, a chamber for the culture of symbiotic fungi, in the Southern pine beetle. *Tissue Cell* **3**, 295–308.

Happ, G. M., Happ, C. M., and Barras, S. J. (1975). Bark beetle–fungal symbiosis. III. Ultrastructure of conidiogenesis in a *Sporothrix* ectosymbiont of the Southern pine beetle. *Can. J. Bot.* **53**, 2702–2711.

Happ, G. M., Happ, C. M., and Barras, S. J. (1976a). Bark beetle–fungal symbiosis. II. Fine structure of a basidiomycetous ectosymbiont of the Southern pine beetle. *Can. J. Bot.* **54**, 1049–1062.

Happ, G. M., Happ, C. M., and French, J. R. J. (1976b). Ultrastructure of the mesonotal mycangium of an ambrosia beetle, *Xyleborus dispar* (F.) (Coleoptera: Scolytidae). *Int. J. Insect Morphol. Embryol.* **5**, 381–391.

Hara, A. H., and Beardsley, J. W. (1979). The biology of the black twig borer, *Xylosandrus compactus* (Eichhoff), in Hawaii. *Proc. Hawaii. Entomol. Soc.* **23**, 55–70.

Harris, J. A., Campbell, K. G., and Wright, G. M. (1976). Ecological studies on the horizontal borer *Austroplatypus incompertus* (Schedl) (Coleoptera: Platypodidae). *J. Entomol. Soc. Aust.* **9**, 11–21.

Hosking, G. P. (1973). *Xyleborus saxeseni*, its life history and flight behaviour in New Zealand. *N.Z. J. For. Sci.* **3**, 37–53.

Janzen, D. H. (1985). The natural history of mutualisms. *In* "The Biology of Mutualism. Ecology and Evolution". (D. H. Boucher, ed.), pp. 40–99. Oxford Univ. Press, Oxford.

Kabir, A. K. M. F., and Giese, R. L. (1966a). The Columbian timber beetle, *Corthylus columbianus* (Coleoptera: Scolytidae). I. Biology of the beetle. *Ann. Entomol. Soc. Am.* **59**, 883–894.

Kabir, A. K. M. F., and Giese, R. L. (1966b). The Columbian timber beetle, *Corthylus columbianus* (Coleoptera: Scolytidae). II. Fungi and staining associated with the beetle in soft maple. *Ann. Entomol. Soc. Am.* **59**, 894–902.

Kalshoven, L. G. E. (1960). A form of commensalism occurring in *Xyleborus* species? *Entomol. Ber.* **20**, 118–120.

Kaneko, T. (1965). Biology of some scolytid ambrosia beetles attacking tea plants. I. Growth and development of two species of scolytid beetles reared on sterilised tea plants. *Jpn. J. Appl. Entomol. Zool.* **9**, 211–216.

Kingsolver, J. G., and Norris, D. M. (1977). The interaction of *Xyleborus ferrugineus* (Coleoptera: Scolytidae) behavior and initial reproduction in relation to its symbiotic fungi. *Ann. Entomol. Soc. Am.* **70**, 1–4.

Kirkendall, L. R. (1983). The evolution of mating systems in bark and ambrosia beetles (Coleoptera: Scolytidae and Platypodidae). *J. Linn. Soc. London, Zool.* **77**, 293–352.

Kok, L. T. (1979). Lipids of ambrosia fungi in the life of mutualistic beetles. *In* "Insect–Fungus symbiosis". (L. R. Batra, ed.), pp. 33–52. Halsted Press, Chichester, Sussex.

Kok, L. T., Norris, D. M., and Chu, H. M. (1970). Sterol metabolism as a basis for mutualistic symbiosis. *Nature (London)* **225**, 661–662.

Leufvén, A., Bergström, G., and Falsen, E. (1984). Interconversion of verbenols and verbenone by identified yeasts isolated from the spruce bark beetle, *Ips typographus. J. Chem. Ecol.* **10**, 1349–1361.

Lhoste, J., and Roche, A. (1961). Anatomie comparée des organes transporteurs de champignons chez quelques Scolytoidea. *Proc. Int. Congr. Entomol., 11th, 1960* Vol. 1, pp. 385–387.

Livingston, R. L., and Berryman, A. A. (1972). Fungus transport structures in the fir engraver, *Scolytus ventralis* (Coleoptera: Scolytidae). *Can. Entomol.* **104**, 1793–1800.

Lowe, R. E., Giese, R. L., and McManus, M. L. (1967). Mycetangia of the ambrosia beetle, *Monarthrum fasciatum. J. Invertebr. Pathol.* **9**, 451–458.

Martin, M. M. (1979). Biochemical implications of insect mycophagy. *Biol. Rev. Cambridge Philos. Soc.* **54**, 1–21.

Moeck, H. A. (1981). Ethanol induces attack on trees by spruce beetles, *Dendroctonus rufipennis* (Coleoptera: Scolytidae). *Can. Entomol.* **113**, 939–942.

Moore, G. E. (1971). Mortality factors caused by pathogenic bacteria and fungi of the Southern pine beetle in North Carolina. *J. Invertebr. Pathol.* **17**, 28–37.

Nakashima, T. (1971). Notes on the associated fungi and the mycetangia of the ambrosia beetle, *Crossotarsus niponicus* Blandford (Coleoptera: Platypodidae). *Appl. Entomol. Zool.* **6**, 131–137.

Nakashima, T. (1972). Notes on the mycangia of the ambrosia beetles, *Platypus severini* Blandford and *P. calamus* Blandford (Coleoptera: Platypodidae). *Appl. Entomol. Zool.* **7**, 217–225.

Nakashima, T. (1975). Several types of the mycetangia found in platypodid ambrosia beetles (Coleoptera: Platypodidae). *Insecta Matsumurana* (N.S) **7**, 1–69.

Nakashima, T. (1982). Function and location of mycetangia in ambrosia beetles. *In* "The Ultrastructure and Functioning of Insect Cells" (H. Akai *et al.*, eds), pp. 87–90. Society for Insect Cells, Japan.

Nakashima, T., Iizuka, T., Ogura, K., Maeda, M., and Tanaka, T. (1982). Isolation of some microorganisms associated with five species of ambrosia beetles and two kinds of antibiotics produced by Xv-3 strain in these isolates. *J. Fac. Agric., Hokkaido Univ.* **61**, 60–72.

Neger, F. W. (1911). Zur Übertragung des Ambrosiapilzes von *Xyleborus dispar. Naturwiss. Z. Forst Landwirtsch.* **9**, 223–225.

Nord, J. C. (1972). Biology of the Columbian timber beetle, *Corthylus columbianus* (Coleoptera: Scolytidae), in Georgia. *Ann. Entomol. Soc. Am.* **65**, 350–358.

Norris, D. M. (1976). Chemical interdependencies among *Xyleborus* spp. ambrosia beetles and their symbiotic microbes. *Beih. Mater. Org.* **3**, 479–488.

Norris, D. M. (1979). The mutualistic fungi of xyleborine beetles. *In* "Insect–Fungus Symbiosis" (L. R. Batra, ed.), pp. 53–63. Halsted Press, Chichester, Sussex.

Nunberg, M. (1951). Contribution to the knowledge of prothoracic glands of Scolytidae and Platypodidae (Coleoptera). *Ann. Mus. Zool. Pol.* **14**, 261–265.

Paine, T. D., and Birch, M. C. (1983). Acquisition and maintenance of mycangial fungi by *Dendroctonus brevicomis* LeConte (Coleoptera: Scolytidae). *Environ. Entomol.* **12**, 1384–1386.

Peleg, B., and Norris, D. M. (1972). Bacterial symbiote activation of insect parthenogenetic reproduction. *Nature (London) New Biol.* **236**, 111–112.

Roberts, H. (1961). The adult anatomy of *Trachyostus ghanaensis* Schedl (Platypodidae), a W. African beetle, and its relationship to changes in adult behaviour. *Rep. W. Afr. Timb. Borer Res. Unit, Kumasi* **4**, 31–38.

Roberts, H. (1968). Notes on the biology of ambrosia beetles of the genus *Trachyostus* Schedl (Coleoptera: Platypodidae) in West Africa. *Bull. Entomol. Res.* **58**, 325–352.

Roeper, R. A., and French, J. R. J. (1978). Observations on *Monarthrum dentiger* (Coleoptera: Scolytidae) and its primary symbiotic fungus *Ambrosiella brunnea* (Fungi Imperfecti) in California. *Pan-Pac. Entomol.* **54**, 68–69.

Schedl, K. E. (1972). "Monographie der Familie Platypodidae, Coleoptera". Junk, Den Haag.

Schedl, W. (1962). Ein Beitrag zur Kenntnis der Pilzübertragungsweise bei xylomycetophagen Scolytiden (Coleoptera). *Sitzungsber. Oesterr. Akad. Wiss.* Abt. I. **171**, 363–387.

Schneider, I. (1976). Untersuchungen über die biologische Bedeutung der Mycetangien bei einigen Ambrosiakäfern. *Beih. Mater. Org.* **3**, 489–497.

Schneider, I., and Rudinsky, J. A. (1969). Mycetangial glands and their seasonal changes in *Gnathotrichus retusus* and *G. sulcatus*. *Ann. Entomol. Soc. Am.* **62**, 39–43.

Stevens, G. C. (1986). Dissection of the species–area relationship among wood-boring insects and their host plants. *Am. Nat.* **128**, 35–46.

Swift, M. J., and Boddy, L. (1984). Animal–microbial interactions in wood decomposition. *In* "Invertebrate–Microbial Interactions" (J. M. Anderson, A. D. M. Rayner and D. W. H. Walton, eds), pp. 89–131. Cambridge Univ. Press, Cambridge.

Takagi, K. (1967). The storage organ of symbiotic fungus in the ambrosia beetle *Xyleborus rubricollis* Eichhoff (Coleoptera: Scolytidae). *Appl. Entomol. Zool.* **2**, 168–170.

Takagi, K., and Kaneko, T. (1965). Biology of some scolytid ambrosia beetles attacking tea plants. II. Spore storage organ of tea root borer, *Xyleborus germanus* Blandford. *Jpn. J. Appl. Entomol. Zool.* **9**, 247–248.

Weber, B. C., and McPherson, J. E. (1984). The ambrosia fungus of *Xylosandrus germanus* (Coleoptera: Scolytidae). *Can. Entomol.* **116**, 281–283.

Whitney, H. S. (1971). Association of *Dendroctonus ponderosae* (Coleoptera: Scolytidae) with blue stain fungi and yeasts during brood development in lodgepole pine. *Can. Entomol.* **103**, 1495–1503.

Whitney, H. S. (1982). Relationships between bark beetles and symbiotic organisms. *In* "Bark Beetles in North American Conifers" (J. B. Mitton and K. B. Sturgeon, eds), pp. 183–211. Univ. of Texas Press, Austin.

Whitney, H. S., and Farris, S. H. (1970). Maxillary mycangium in the mountain pine beetle. *Science* **167**, 54–55.

Wood, S. L. (1973). On the taxonomic status of Platypodidae and Scolytidae. *Great Basin Nat.* **33**, 77–90.

Wood, S. L. (1986). A reclassification of the genera of Scolytidae. *Great Basin Nat. Mem.* **10**, 1–126.

Wood, S. L. (1982). Bark and ambrosia beetles of North and Central America (Coleoptera: Scolytidae), a taxonomic monograph. *Great Basin Nat. Mem.* **6**, 1–1359.

Zimmermann, G., and Butin, H. (1973). Untersuchungen über die Hitze- und Trockenresistenz holzbewohnender Pilze. *Flora (Jena)* **162**, 393–419.

6

Adaptive Pathways in Scolytid–Fungus Associations

ALAN A. BERRYMAN

I. INTRODUCTION

Many scolytids (H25.63) have evolved extremely interesting and complex relationships with fungi, some using fungi as food and others using them as weapons to kill trees. But how these symbiotic relationships arose, and to what extent they may still be evolving has not been addressed in detail. In this paper some speculations are presented on the co-adaptive pathways that could have given rise to present-day associations between fungi and scolytid beetles. These are the speculations of an entomologist and ecologist, not a mycologist nor evolutionary biologist, and therefore lean towards the insect more than the fungus. However, it is essential to examine the role of fungi in the physiological aspects of host tree defence, for without the knowledge of conifer reactions to fungal invasion, it would be difficult or impossible to unravel the mystery of bark beetle population ecology. In view of this, perhaps the paper would have been better entitled "Adaptive pathways in *conifer*–scolytid–fungus associations."

In preparing this paper a question that other evolutionary ecologists have encountered when thinking about mutualism was squarely faced; namely, what are the evolutionary roots of mutually beneficial interactions? For instance, are the relationships between scolytid beetles and fungi truly mutualistic, with both parties gaining fitness advantages that exceed their costs, or are they more of a one way association—exploitation or commensalism? It seems to the author that the common assumption of mutualism is tenuous in certain scolytid–fungus relationships. Therefore, this paper will end by exploring this question.

II. ADAPTIVE STRATEGIES OF SCOLYTIDAE

A. Feeding Strategies

Scolytid beetles can be divided into three major groups according to their feeding strategies:

1. Saprophages

These feed almost exclusively on the bark of dead trees or tree parts. Although their diet may be supplemented by microorganisms, strict associations with specific fungi are unlikely to occur and specialized fungus-containing organs (mycangia) are not usually found. In these insects, the association with fungi is more coincidental than symbiotic (here symbiosis is used to mean an adapted association, with one or both members having adaptations that further the relationship). Examples can be found in the sour sap beetles of the genus *Hylurgops*, some species of which produce several generations in old disintegrating logs or stumps (Bright and Stark, 1973). Many other scolytids are probably saprophagous, but the literature dealing with these species is scant because they cause no economic damage.

2. Phytophages

This group, commonly known as bark beetles, feed on the living tissue of trees and often depend on fungal pathogens to kill and condition their host. Most of them have both specific and coincidental relationships with pathogenic fungi, many of which belong to the genera *Ophiostoma* (F4.24) and *Ceratocystis* (F4.24), as well as with yeasts and, in some cases, basidiomycetes (Graham, 1967; Barras and Perry, 1972; Whitney and Cobb, 1972).

The spores of fungal pathogens are often seen adhering to the external body parts of bark beetles and may also be cultured from mycangia (Barras and Perry, 1972; Whitney and Cobb, 1972). Within the attacked tree, pathogenic fungi can be isolated from phloem tissues in advance of beetle borings (Wong and Berryman, 1977) and also deep within the sapwood, quite remote from beetle activity (Whitney and Cobb, 1972). It is this sapwood penetration and rupturing of the water-conducting system which is believed to cause the rapid death of the tree (Mathre, 1964; Horntvedt *et al.*, 1983; Raffa and Berryman, 1983).

Yeasts can be isolated from phloem tissue near beetle galleries, from mycangia, and from the intestinal tracts of larvae and adults and, therefore, are probably important nutritionally (Barras and Perry, 1972; Callaham and Shifrine, 1960). However, the primary diet of phytophagous scolytids is bark tissue. Reproduction and survival are not apparently dependent on *specific* fungal foods, as bark beetles can survive and reproduce *in vitro* on diets consisting of phloem tissue supplemented with brewer's yeast (Bedard, 1966), and can also be reared in sterilized phloem alone (Strongman, 1987). However, the yeast additives in these diets usually enhance bark beetle development and survival (Whitney, 1982).

Phytophagous scolytids that are closely associated with fungi often possess mycangia from which the specific fungi can be isolated (Barras and Perry, 1971, 1972). They also orientate during attack to the detoxification products of host defensive chemicals (oxidized and hydroxylated monoterpenes) which may be synthesized by bacteria and fungi (Brand *et al.*, 1975, 1976; Byers and Wood, 1981). It is this pheromonal communication that synchronizes the mass attack, resulting in multiple fungal infections that overwhelm the defences of the tree and lead to its eventual death (Mathre, 1964; Raffa and Berryman, 1983, 1987; Horntvedt *et al.*, 1983). Thus it is not surprising that bark beetles are our most dangerous scolytids, for their associations with pathogens and mass attack behaviour give rise to dynamic instabilities at the population level that can result in the widespread destruction of forests and shade trees (Berryman *et al.*, 1984; Brasier, 1987). For this reason, the tree-killing bark beetles are our most intensively studied scolytids.

Within the phytophagous group, a few bark beetles seem to be relatively independent of fungal symbionts. These phytophages, most notably the *Dendroctonus* species, *D. micans, D. valens, D. punctatus* and *D. terebrans*, feed on living trees more as parasites than as predators, and consequently do not normally kill their host. Adult beetles do not have aggregation pheromones and so the insects are usually solitary, except during outbreaks. However, larvae of *D. micans* do produce aggregation pheromones which apparently induce the gregarious larvae to feed in organized clusters within the common brood chambers (Gregoire, 1985). It is not known if the larvae of other solitary scolytids produce similar aggregation pheromones.

The solitary dendroctonids may suppress fungal growth within their brood chambers by manipulating the flow of resin from their living hosts. They do this by boring downwards to increase resin flow into their galleries and upwards when resin flow is excessive, the resin ducts being drained above their galleries (Wood, 1963). As conifer resin is known to suppress fungal growth (Raffa and Berryman, 1987), and as fungi never colonize resin-saturated wood (pers. obs.), this behaviour in the solitary scolytids is probably intended to suppress fungal growth.

3. Mycetophages

These so-called "ambrosia beetles" feed almost exclusively on fungi (ambrosial fungi) which are cultivated in galleries bored into the sapwood of dead trees or timber. Ambrosia beetles have close, and very specific relationships with several genera of fungi (Kok, 1979; Norris, 1979). Like the phytophages, fungus-feeding scolytids usually possess specialized mycangia and orient to aggregation pheromones. They all use fungi primarily for food, and their reproduction and survival are entirely dependent on this diet (Norris, 1979). Thus, although ambrosia beetle pheromones provide them with the potential to mass attack and kill living trees, this rarely if ever occurs. This suggests that ambrosia beetles have co-adapted with the nutritionally important yeast-like fungi rather than with the pathogens. Perhaps the elimination of pathogenic forms, which would be the first to invade plant tissues, is a necessary condition for the evolution of the mycetophagous strategy.

It is rather remarkable that the phytophagous and mycetophagous scolytids are so similar in their behaviour and morphological adaptations, both having mycangial structures and pheromone communication. This might lead one to suspect a common ancestor possessing both attributes. We will see later, however, that this is unlikely.

It seems probable that both adaptive strategies, phytophagy and mycetophagy, arose from a common saprophagous ancestor and, therefore, that the saprophagous strategy is primitive. If this is true, then clues to the evolutionary pathways leading to the two major strategies may be found in mycangial morphology and pheromone chemistry, both of which seem to have evolved separately in the two groups.

B. Scolytid Mycangia

The term mycangia was coined by Batra (1963) to describe specialized structures that contain fungi. Scolytid mycangia occur in many places on the insect body but all are invaginations of the cuticle forming pits, sacs, pouches or tubes within which the fungal hyphae or spores are contained.

Most mycangia are also associated with secretory glands and frequently contain waxy substances. The secretions of some ambrosia beetle mycangia contain fatty acids, phospholipids, sterols and large amounts of amino acids, the latter seeming to control the growth and form of the mycangial fungi (Norris, 1979).

Scolytid mycangia can be conveniently classified according to the part of the insect's body on which they are found. On the head, mycangia occur most frequently in the oral cavity, either as membranous pouches at the base of the mandibles or in the cardines of the maxillae (Francke-Grosmann, 1963; Farris, 1969; Whitney and Farris, 1970; Norris, 1979) (Fig. 1A). However, they also occur as numerous small pits on the top of the cranium (Livingston and Berryman, 1972) (Figs 1B,C). On the thorax, mycangia range from sac-like structures on each side of the prothorax to pleural pits, intersegmental pouches, subcoxal cavities and pronotal collars (Fig. 1D) or pubescent areas (Fig. 1E, F) (Francke-Grosmann, 1963; Farris, 1965; Barras, 1967; Barras and Perry, 1971; Furniss *et al.*, 1987), while on the elytra, mycangia are found as sclerotized pouches or cup-like pits.

Clearly, scolytid mycangia are extraordinary for their variability and apparent disorder within and between taxonomic groups. For example, mandibular pouches are found, not only in the ambrosia beetle genera *Xyleborus* and *Gnathotrichus*, but also in the bark beetles *Ips acuminatus* and *Dryocoetes confusus* (Francke-Grosmann, 1963; Farris, 1969). On the other hand, most members of the genus *Dendroctonus* have pronotal collar mycangia (Fig. 1D), with the exception of *Dendroctonus ponderosae* which has a mycangium in the maxillary cardines (Fig. 1A) (Whitney and Farris, 1970). The ambrosia beetle genus *Xyleborus*, one of the largest insect genera, has the greatest diversity of mycangia, some species having mandibular pouches, others intersegmental promesonotal or prothoracic pouches, and yet others elytral pouches.

The absence of clear phylogenic lines amongst scolytid mycangia suggests that they are fairly recent adaptations which have evolved simultaneously and independently within and between many of the modern genera; that is, as a result of parallel rather than divergent evolution. This observation seems to support the supposition that primitive scolytids lived in close but coincidental association with the fungi that would later become such an important part of their life-histories, and that would provide the selective pressure for the evolution of mycangial structures.

C. Scolytid Pheromones

During attack on living or even dead trees, scolytids come into contact with plant defensive chemicals such as terpenes, phenols and resin acids. These

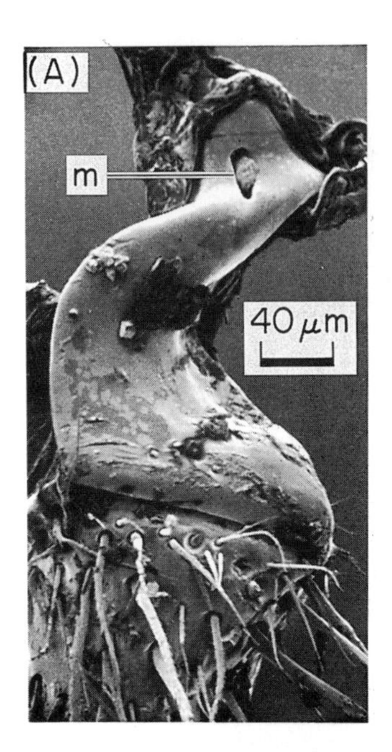

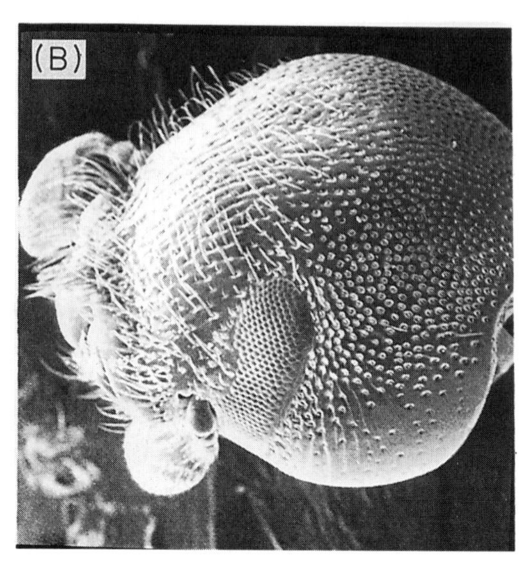

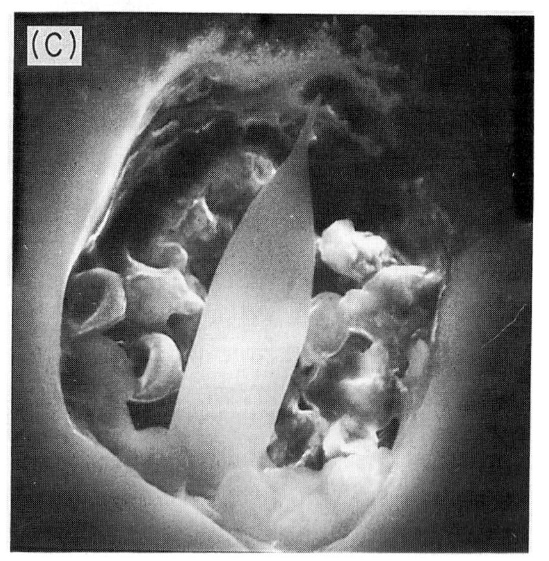

Fig. 1. (A) The maxillary mycangium of the mountain pine beetle. (From Canadian Forestry Service, Whitney and Farris, 1970, copyright 1970 by the American Association for the Advancement of Science.) (B, C,) Mycangial pits on the head of the fir engraver beetle. (From Livingston and Berryman, 1972, with the permission of the Entomological Society of Canada.) (D) Pronotal collar mycangium (M) of the southern pine beetle. (From Barras, 1975, with the permission of Verlag Paul Pary, Hamburg.) (E) Pubescent prothoracic mycangium of *Pityoborus* spp. (From Furniss *et al.*, 1987, reprinted from the Annals of the Entomological Society of America.) (F) Spores and setae in *Pityoborus* sp. mycangium. (From Furniss *et al.* 1987, reprinted from the *Annals of the Entomological Society of America.*)

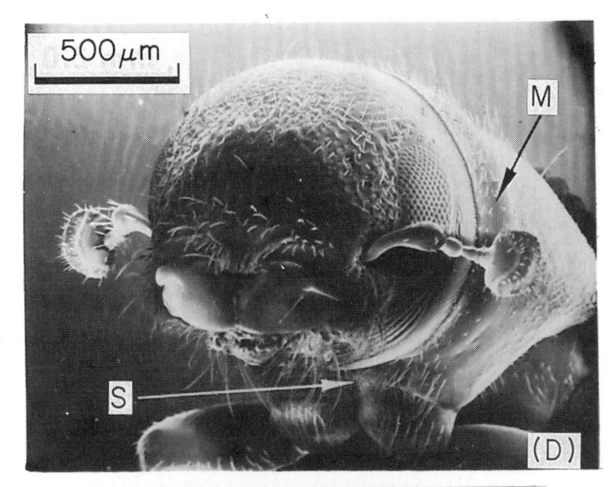
500 μm
M
S
(D)

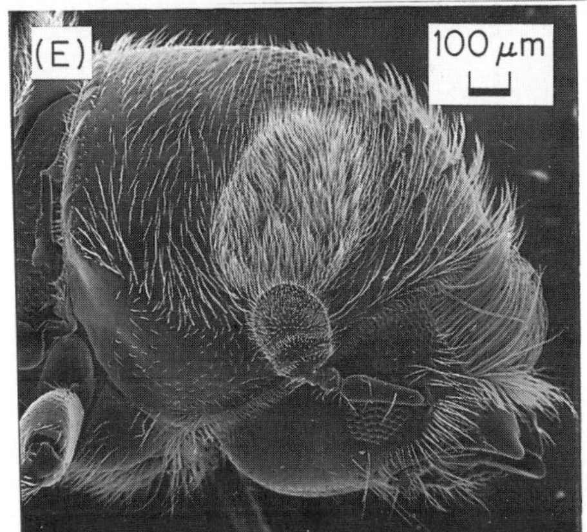
(E)
100 μm

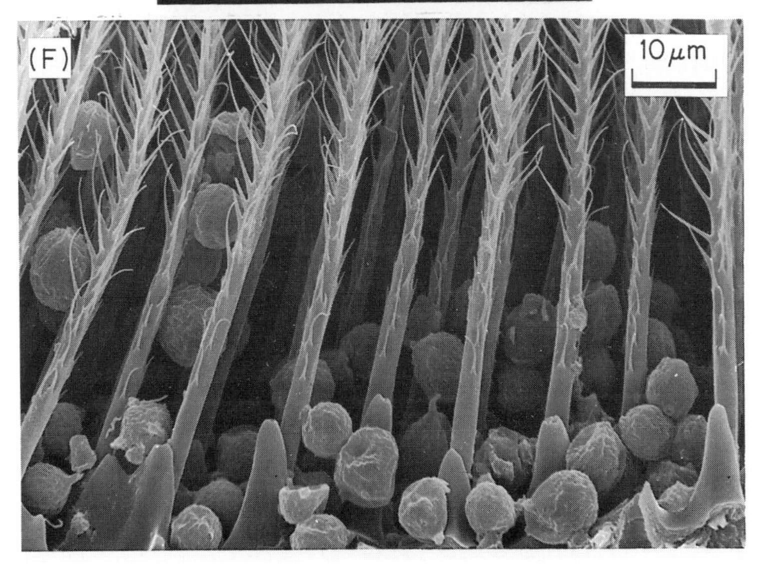
(F)
10 μm

compounds can be toxic, repellent or inhibitory to the beetle and/or associated fungi and must be detoxified before or during digestion (Smith, 1963; Reid and Gates, 1970; Coyne and Lott, 1976, Bordasch and Berryman, 1977; Raffa *et al.*, 1985; Raffa and Berryman, 1987). Detoxification involves the oxidation of terpenoid compounds to terpene alcohols and ketones either by the beetle, its associated microorganisms, or both (Vité *et al.*, 1972; Hughes and Renwick, 1977; Brand *et al.*, 1975, 1976; Byers and Wood, 1981; Conn *et al.*, 1984; Leufven *et al.*, 1984). These detoxification products are thought to be the chemicals that came to be used as pheromones by scolytid beetles.

Scolytid pheromones are generally multifunctional, and can be attractants, acting as sex and/or aggregation pheromones, or repellents, acting as anti-aggregation or overpopulation pheromones. Most compounds acting as attractants are alcohols while repellents are usually, but not always, ketones (Borden, 1982).

Microorganisms may also be involved in pheromone production. For example, Brand *et al.* (1975) showed that *Bacillus cereus*, isolated from the gut of *Ips paraconfusus*, produced the pheromone verbenol when the monoterpene α-pinene was added to the growth medium. Furthermore, mycangial fungi from *Dendroctonus frontalis* can transform the aggregation pheromone verbenol into the antiaggregation pheromone verbenone as well as synthesize sulcatol, a pheromone of the ambrosia beetles *Gnathotrichus sulcatus* and *Platypus flavicornis* (Brand *et al.*, 1976; Brand and Barras, 1977). As terpenes are toxic, or at least inhibitory, to many bark beetles, their transformation by microbes into attractive, non-toxic chemicals represents a significant adaptive trait to the beetle receiving the stimulus, for it signals a tree that is being successfully detoxified by microorganisms (Raffa and Berryman, 1980, 1987). Likewise, the evolution of avoidance behaviour to compounds produced by mycangial fungi could also be adaptive to the recipient beetle, indicating that the tree is fully occupied and, therefore, that insufficient food remains to rear its offspring.

Unlike the mycangia, pheromones do show some degree of phylogenic consistency amongst scolytid groups (Borden, 1982; Wood, 1982). For example, the pheromones brevicomin and frontalin are used by many members of the genus *Dendroctonus*, ipsenol and ipsdienol by the genera *Ips* and the closely related *Pityokteines*, and sulcatol is the major aggregation pheromone in the ambrosia beetle genus *Gnathotrichus*. On the other hand, some compounds are much more ubiquitous. For example, the products of α-pinene, verbenol and verbenone, have pheromonal activity in the genera *Dendroctonus*, *Ips*, and *Blastophagus* (Borden, 1982). Yet other materials, such as myrtenol in *D. frontalis*, heptanol in *D. jeffreyi*, seudenol in *D. pseudotsugae* and *D. rufipennis*, and methylbutanol in *Ips typographus* and *I. cembrae* seem to be much more specific (it is of some interest that methylbutanol only appears as a pheromone in the old world *Ips* [Wood, 1982]). These more consistent phylogenic relationships suggest that behavioural responses to microbial or beetle produced detoxification products have much deeper evolutionary roots, and that the scolytids may have followed the evolutionary pathway outlined in Fig. 2. Alternative pathways can also be visualized in this figure, of course.

According to this scheme, primitive scolytids probably sprang from weevil-like ancestors that lived a non-colonial life in dead and decaying phloem and wood (Schedl, 1958) along with many other arthropods, fungi, bacteria and other saprophagous organisms. Competition for resources must have been extreme in this environment, and selection would favour genotypes that could detect the presence of dead trees at the earliest possible moment. Ethanol, one of the early fermentation products of microbial activity in dead wood, is known to be a primary attractant to certain ambrosia and bark beetles (Borden, 1982). Ethanol, therefore, could well have been the first attractant, and adaptation to orientate to this alcohol would have enabled the ancestral sceltyids to become pioneers in the

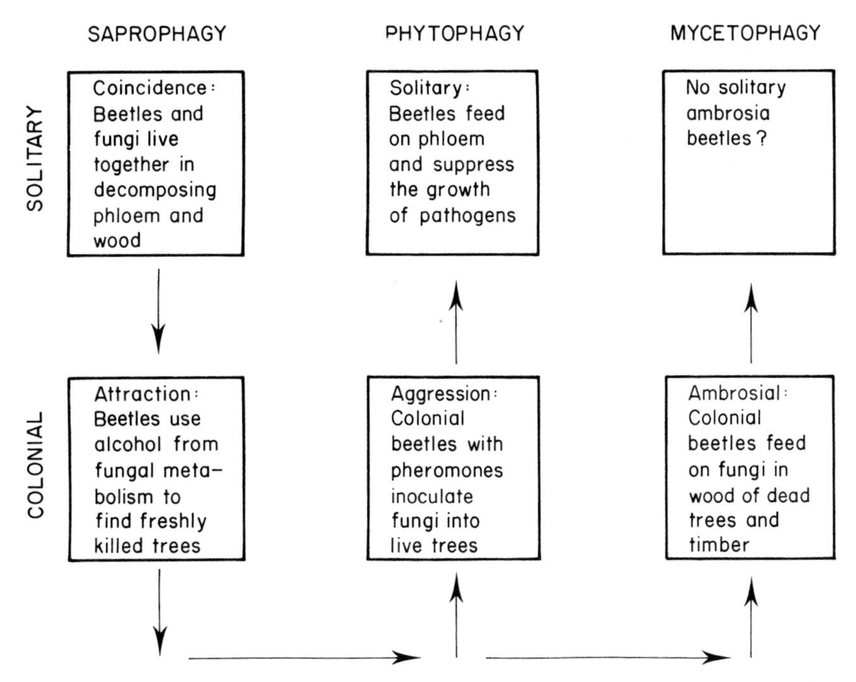

Fig. 2. Possible adaptive pathways in the Scolytidae: from saprophagy to colonial or solitary phytophagy and mycetophagy.

utilization of newly fallen trees. Responding to attractive host compounds would lead, automatically, to beetle aggregations on suitable host material and the beginning of the colonial life-style so characteristic of the Scolytidae. The stage would then be set for co-adaptation of the insects with primary fungi, including the yeasts, well known for their production of alcohols, and pathogens, which by their very nature would often have been associated with dead trees. At this point it is easy to visualize the evolution of the two major pathways towards phytophagy and mycetophagy. In the former, scolytids could have evolved to orientate to other alcohols, which were being metabolized by pathogenic fungi from the defensive secretions of living trees; e.g. the synthesis of verbenol from α-pinene. This adaptation would bring them into contact with fungal pathogens which they could then transport, at first accidentally, to healthy hosts, thereby setting up a future food supply, as for example occurs in Dutch elm disease (Webber and Gibbs, this volume). The natural progression of this strategy would be, on the one hand, towards more aggressive bark beetles and closer ties with the pathogenic fungi, including the appearance of mycangia, and on the other, a parasitic strategy, where the beetles lived a solitary existence independently of fungi. The adaptive value of the latter could be three-fold. First, the solitary life-style would conserve food for their siblings, because the tree is neither killed nor shared with conspecifics and, therefore, potentially could be utilized by several generations. Second, limiting fungal growth would reduce the negative impact of fungal contaminants, such as entomopathogens and competitors for host resources, on beetle survival (Barras, 1970; Whitney, 1982). Third, disassociation with fungi would reduce the amount of secondary resinosis in the phloem, where the larvae feed. It is, after all, the fungi that induce massive hypersensitive reactions in the parenchyma cells of healthy conifers (Berryman, 1972; Wong and Berryman, 1977; Christiansen and Horntvedt, 1985). By suppressing fungal growth, the beetles would prevent the tree from saturating its phloem with indigestible resinous compounds. A stealthy strategy such as this would be essential for an insect living within the tissues of a living conifer (Rhoades, 1985). The habit of sibling mating and the tendency for broods to attack successively the same tree, as epitomized by the great spruce bark beetle *D. micans*, might also evolve under these circumstances. Thus, solitary parasitism seems to be the culmination of the phytophagous scolytid line.

The mycetophagous pathway involves the utilization of a different plant tissue, the xylem, by taking advantage of the ability of fungi to mobilize scarce and diffuse food reserves. In so doing the ancestral ambrosia beetles would avoid competition with their phloem-feeding relatives by opening up the food reserves of the xylem. The natural progression along this pathway would lead to a greater dependency on the fungi as food and the parallel evolution of various kinds of mycangia. However, there seems to have been no evolutionary path leading towards solitary mycetophagy. Likewise, the ambrosia beetles do not seem to use pheromones as weapons to kill healthy trees. The apparent absence of aggressive ambrosia beetles, even though they possess one of the tools of aggression, an aggregation pheromone, suggests that the other tool, a fungal pathogen, is missing. This in turn suggests that specialization for a strict fungal diet *and* association with fungal pathogens is not an evolutionary stable strategy. It may well be that the pathogens, particularly *Ceratocystis* (F4.11) which quickly colonize the sapwood, tie up the nutrients in a form unavailable to the beetles.

The absence of solitary ambrosia beetles is probably due to the fact that dead host plants cannot be utilized by successive generations of beetles. Thus, with no advantage in saving the tree for their offspring, and with benefits in obtaining mates and fungus nuclei, there would be little selective pressure for a solitary life-style.

III. MUTUALISM OR EXPLOITATION?

Mutualism has always been a difficult subject to discuss in evolutionary terms. Although one can see the benefits of the association for the two organisms, it does not mean that the situation arose through a series of reciprocal adaptations that were beneficial to both participants (Thompson, 1986). The problem is seen in the relationship between humans and, for example, wheat. A Martian landing on Earth might see a mutually beneficial relationship, with wheat plants providing humans with food and humans husbanding and protecting the plant. They might think that the evolution of large grains was an adaptation of the wheat to ensure human care and protection. Yet the truth is that genetic adaptations in the plant was the result of human breeding experiments and, from this viewpoint, the relationship appears more exploitative than mutualistic. On the other hand, this highly modified grass has been spread to all corners of the Earth and there are undoubtably more individual plants now than ever before. From this standpoint we may well have improved the fitness of this species, at least while it exists within the carefully husbanded environment created by *Homo sapiens*. The price of such “mutualism” is dependency, for without the partner one or both species may be placed in a very hazardous position, unable to compete with or defend against more independent species.

We have seen that members of the Scolytidae undoubtedly benefit from their association with fungi but that others succeed without them. What the interaction has done is to enable the insects to radiate into new habitats and to open up new ecological niches. This is particularly apparent in the beetles which use fungi as food and as weapons against living trees. Whether the fungi have benefitted equally is more difficult to tell. Is the "secure trip to a new suitable woody substrate, and tender devoted care, once there" (Norris, 1979) truly worth being used as food? Is the ambrosial growth habit a fungal adaptation to ensure beetle co-operation, or are the scolytids merely controlling the growth form of the fungus? (Norris, 1979). As Whistler (1979) notes, ambrosia beetles are "fastidious farmers". But so is *Homo sapiens*! The question is: Is farming a mutualistic or an exploitative process? In the end it boils down to whether the fitness benefits of the association exceed its costs for both species. This is not necessarily an easy question to answer, but let us hope it provokes discussion.

Acknowledgements

I would like to thank Drs K. F. Raffa, D. W. A. Hunt and J. F. Webber for their comments on this manuscript, which is not to imply that they agree with my interpretations. Mal Furniss and Stu Whitney provided useful information and photographs. I would also like to thank Dr Joan Webber, the Royal Entomological Society, and the British Mycological Society for making it possible for me to contribute to this symposium. The American Association for the Advancement of Science and the Entomological Societies of America and Canada are acknowledged for granting permission to reprint photographs of scolytid mycangia.

REFERENCES

Barras, S. J. (1967). Thoracic mycangium of *Dendroctonus frontalis* (Coleoptera: Scolytidae) is synonymous with a secondary female character. *Ann. Entomol. Soc. Am.* **60**, 486–487.

Barras, S. J. (1970). Antagonism between *Dendroctonus frontalis* and the fungus *Ceratocystis minor*. *Ann. Entomol. Soc. Am.* **63**, 1187–1190.

Barras, S. J. (1975). Release of fungi from mycangia of southern pine beetles observed under a scanning electron microsope. *Z. Angew. Entomol.* **79**, 173–176.

Barras, S. J., and Perry, T. (1971). Gland cells and fungi associated with prothoracic mycangium of *Dendroctonus adjunctus* (Coleoptera: Scolytidae). *Ann. Entomol. Soc. Am.* **64**, 123–126.

Barras, S. J., and Perry, T. (1972). Fungal symbionts in the prothoracic mycangium of *Dendroctonus frontalis*: (Coleoptera: Scolytidae). *Z. Angew. Entomol.* **71**, 95–104.

Batra, L. R. (1963). Ecology of ambrosia fungi and their dissemination by beetles. *Trans. Kans. Acad. Sci.* **66**, 213–236.

Bedard, W. D. (1966). A ground phloem medium for rearing immature bark beetles (Scolytidae). *Ann. Entomol. Soc. Am.* **59**, 931–938.

Berryman, A. A. (1972). Resistance of conifers to invasion by bark beetle–fungus associations. *BioScience* **22**, 598–602.

Berryman, A. A., Stenseth, N. C., and Wollkind, D. J. (1984). Metastability in forest ecosystems infested by bark beetles. *Res. Popul. Ecol.* **26**, 13–29.

Bordasch, R. P., and Berryman, A. A. (1977). Host resistance to the fir engraver beetle, *Scolytus ventralis* (Coleoptera: Scolytidae). 2. Repellency of *Abies grandis* resin and some monoterpenes. *Can. Entomol.* **109**, 95–100.

Borden, J. H. (1982). Aggregating pheromones. *In* "Bark Beetles in North American Conifers" (J. B. Mitton and K. B. Sturgeon, eds), pp. 74–139. Univ. Texas Press, Austin.

Brand, J. M., and Barras, S. J. (1977). The major volatile constituents of a basidiomycete associated with the southern pine beetle. *Lloydia* **40**, 319–399.

Brand, J. M., Bracke, J. W., Markovetz, A. J., Wood, D. L., and Browne, L. E. (1975). Production of verbenol pheromone by a bacterium isolated from bark beetles. *Nature (London)* **254**, 136–137.

Brand, J. M., Bracke, J. W., Britton, L. N., Markovetz A. J., and Barras, S. J. (1976). Bark beetle pheromones: Production of verbenone by a mycangial fungus of D. frontalis. *J. Chem. Ecol.* **2**, 195–199.

Brasier, C. M. (1987). Recent genetic changes in *Ophiostoma ulmi* populations: the threat to the future of the elm. *In* "Populations of Plant Pathogens: their Dynamics and Genetics" (M. S. Wolfe and C. E. Caten, eds), pp. 213–226. Blackwell Scientific, Oxford.

Bright, D. E., and Stark, R. W. (1973). The bark and ambrosia beetles of California. Coleoptera: Scolytidae and Platpodidae. *Bull. Calif. Insect Surv.* **16**, 1–169.

Byers, J. A., and Wood, D. L. (1981). Antibiotic-induced inhibition of pheromone synthesis in a bark beetle. *Science* **213**, 763–764.

Callaham, R. I., and Shifrine, M. (1960). The yeasts associated with bark beetles. *For. Sci.* **6**, 146–154.

Christiansen, E., and Horntvedt, R. (1985). Combined *Ips Ceratocystis* attack on Norway spruce, and defensive mechanisms of the trees. *Z. Angew. Entomol.* **96**, 110–118.

Conn, J. E., Borden, J. H., Hunt, W. A., Holman, J., Whitney, H. S., Spanier, O. J., Pierce, H. D., and Oehlschlager, A. C. (1984). Pheromone production by axenically-reared *Dendroctonus ponderosae* and *Ips paraconfusus* (Coleoptera: Scolytidae). *J. Chem. Ecol.* **10**, 281–290.

Coyne, J. F., and Lott, L. H. (1976). Toxicity of substances in pine resin to southern pine beetle. *J. Georgia Entomol. Soc.* **11**, 301–305.

Farris, S. H. (1965). Repositories of symbiotic fungus in the ambrosia beetle *Monarthrum scutellare* LeC (Coleoptera: Scolytidae). *Proc. Entomol. Soc. B. C.* **62**, 30–33.

Farris, S. H. (1969). Occurrence of mycangia in the bark beetle *Dryocoetes confusus* (Coleoptera: Scolytidae). *Can. Entomol.* **101**, 527–532.

Francke-Grosmann, H. (1963). Some new aspects in forest entomology. *Annu. Rev. Entomol.* **8**, 415–438.

Furniss, M. M., Woo, J. Y., Deyrup, M. D., and Atkinson, T. H. (1988). Prothoracic mycangium on pine-infesting *Pityoborus* spp. (Coleoptera: Scolytidae). *Ann. Entomol. Soc. Am.* **80**, 692–696.

Graham, K. (1967). Fungal–insect mutualism in trees and timber. *Annu. Rev. Entomol.* **12**, 105–126.

Gregoire, J. C. (1985). Host colonization strategies in *Dendroctonus*: Larval gregariousness or mass attack by adults? *In* "Proceedings of IUFRO Conference on The Role of the Host in the Population Dynamics of Forest Insects" (L. Safranyik, ed.), pp. 147–154. Pacific Forest Research Centre, Victoria, BC.

Horntvedt, R., Christiansen, E., Sodheim, H., and Wang, S. (1983). Artificial inoculation with *Ips typographus*—associated blue stain fungi can kill healthy Norway spruce trees. *Medd. Nor. Inst. Skogforsk.* **38**, 1–20.

Hughes, P. R., and Renwick, J. A. A. (1977). Neural and hormonal control of pheromone biosynthesis in the bark beetle *Ips paraconfusus. Physiol. Entomol.* **2**, 117–123.

Kok, L. T. (1979). Lipids of ambrosia fungi and the life of mutualistic beetles. *In* "Insect–Fungus Symbiosis: Nutrition, Mutualism and Commensalism" (L. R. Batra, ed.), pp. 33–52. John Wiley and Sons, New York.

Leufven, A., Bergstrom, G., and Folsen, E. (1984). Interconversion of verbenols and verbenone by identified yeasts isolated from the spruce bark beetle *Ips typographus*. *J. Chem. Ecol.* **10**, 1349–1361.

Livingston, R. L., and Berryman, A. A. (1972). Fungus transport structures in the fir engraver, *Scolytus ventralis* (Coleoptera: Scolytidae). *Can. Entomol.* **104**, 1793–1800.

Mathre, D. E. (1964). Studies on the pathogenicity to ponderosa pine of *Ceratocystis* spp. associated with bark beetles in California. PhD Thesis, Univ. California, Davis.

Norris, D. M. (1979). The mutualistic fungi of Xyleborini beetles. *In* "Insect–Fungus Symbiosis: Nutrition, Mutualism and Commensalism" (L. R. Batra, ed.), pp. 53–64. John Wiley and Sons, New York.

Raffa, K. F., and Berryman, A. A. (1980). Flight responses and host selection by bark beetles. *In* "Proceedings of the Second IUFRO Conference on Dispersal of Forest Insects: Evaluation, Theory and Management Implications" (A. A. Berryman and L. Safranyik, eds), pp. 213–233. Washington State. Univ. Pullman.

Raffa, K. F., and Berryman, A. A. (1983). The role of host plant resistance in the colonization behavior and ecology of bark beetles. *Ecol. Monogr.* **53**, 27–49.

Raffa, K. F., and Berryman, A. A. (1987). Interacting selective pressures in conifer–bark beetle systems: A basis for reciprocal adaptations? *Am. Nat.* **129**, 234–262.

Raffa, K. F., Berryman, A. A., Simasko, J., Teal, W., and Wong, B. L. (1985). Effects of grand fir monoterpenes on the fir engraver beetle (Coleoptera: Scolytidae) and its symbiotic fungi. *Environ. Entomol.* **14**, 552–556.

Reid, R. W., and Gates, H. (1970). Effects of temperature and resin on hatch of eggs of the mountain pine beetle. *Can. Entomol.* **102**, 617–622.

Rhoades, D. F. (1985). Offensive–defensive interactions between herbivores and plants: Their relevance in herbivore population dynamics and ecological theory. *Am. Nat.* **125**, 205–238.

Schedl, K. E. (1958). Breeding habits of arboricole insects in Africa. *Proc. Int. Congr. Entomol., 10th, 1957* **1**, 183–197.

Smith, R. H. (1963). Toxicity of pine resin vapors to three species of *Dendroctonus* bark beetles. *J. Econ. Entomol.* **56**, 327–831.

Strongman, D. B. (1987). A method for rearing *Dendroctonus ponderosae* Hopk. (Coleoptera: Scolytidae) from eggs to pupae on host tissue with or without a fungal complement. *Can. Entomol.* **119**, 207–208.

Thompson, J. N. (1986). Constraints on arms races in coevolution. *Trends in Ecol. Evol.* **1**, 105–107.

Vité, J. P., Bakke, A., and Renwick, J. A. A. (1972). Pheromones in *Ips* (Coleoptera: Scolytidae): Occurrence and production. *Can. Entomol.* **104**, 1967–1975.

Whistler, H. C. (1979). The fungi versus the arthropods. *In* "Insect–Fungus Symbiosis: Nutrition, Mutualism and Commensalism" (L. R. Batra, ed.), pp. 1–32. John Wiley and Sons, New York.

Whitney, H. S. (1982). Relationships between bark beetles and symbiotic organisms. *In* "Bark Beetles in North American Conifers" (J. B. Mitton and K. B. Sturgeon, eds), pp. 183–211. Univ. Texas Press, Austin.

Whitney, H. S., and Cobb, F. W. (1972). Non-staining fungi associated with the bark beetle *Dendroctonus brevicomis* (Coleoptera: Scolytidae) on *Pinus ponderosa*. *Can. J. Bot.* **50**, 1943–1945.

Whitney, H. S., and Farris, S. H. (1970). Maxillary mycangium in the mountain pine beetle. *Science* **167**, 54–55.

Wong, B. L., and Berryman, A. A. (1977). Host resistance to the fir engraver beetle. 3. Lesion

development and containment of infection by resistant *Abies grandis* inoculated with *Trichosporium symbioticum*. *Can. J. Bot.* **55**, 2358–2365.
Wood, D. L. (1982). The role of pheromones, kairomones, and allomones in the host selection and colonization behavior of bark beetles. *Annu. Rev. Entomol.* **27**, 411–446.
Wood, S. L. (1963). A revision of the bark beetle genus *Dendroctonus* Erickson (Coleoptera: Scolytidae). *Great Basin Nat.* **23**, 1–117.

7

Insect Dissemination of Fungal Pathogens of Trees

J. F. WEBBER AND J. N. GIBBS

I. INTRODUCTION

Among plant pathologists there is often a generalized idea that insect dissemination is the exception rather than the rule. This is reflected in the remarks of Ingold (1978) who, in reviewing the dispersal of micro-organisms, suggested that "comparatively few fungal pathogens are spread by insects". If we consider herbaceous plants this does indeed appear to be so. The classical case of insect dispersal of conidia of *Claviceps purpurpeum* (F4.4), the cause of ergot of rye, is well known (see Atanasoff, 1920) but a search of the literature reveals very few studies dealing with insect dispersed pathogens. Moreover, no one particular taxonomic group of insects seems to be strongly associated with the vectoring of herbaceous plant diseases. Instead the known examples range from transmission of anther smuts (*Ustilago* spp. (F5.23)) in the Caryophyllaceae by butterflies, bees and moths (Jennersten, 1983; and see Fig. 1), to the spread of *Monilinia* spp. (F4.17) to wild and cultivated blueberries by at least 22

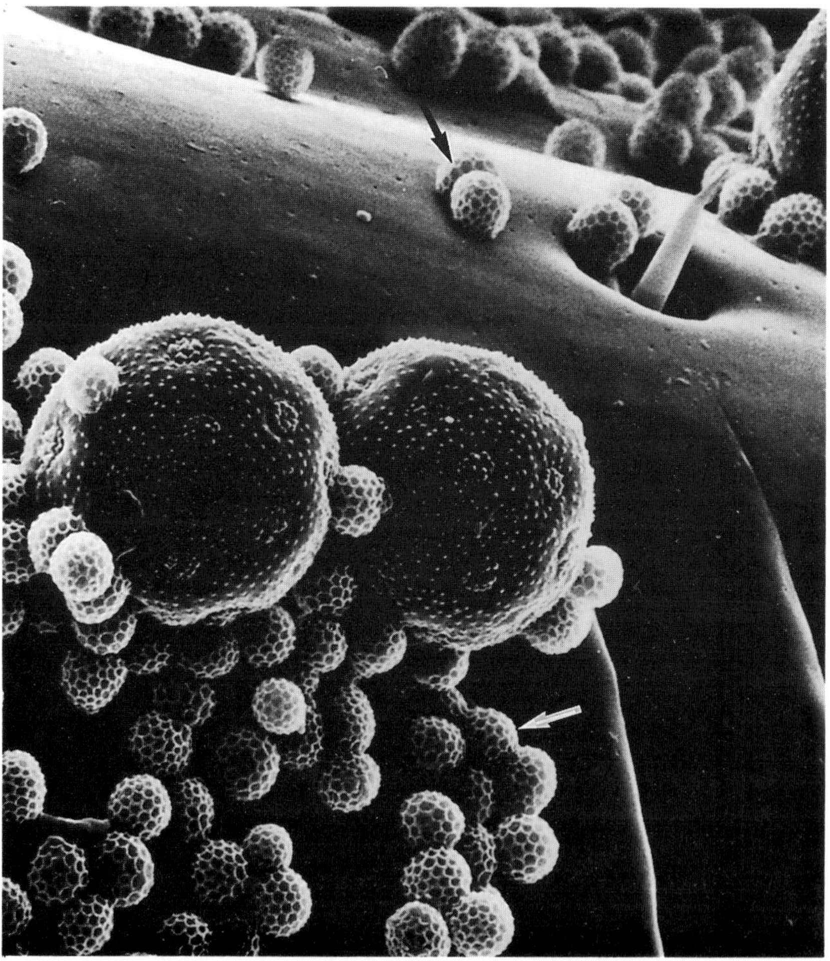

Fig. 1. Scanning electron micrograph of *Ustilago violacea* (F5.23) spores (arrowed) and *Viscaria* pollen grains on the proboscis of the butterfly species *Haemorrhagia tityus*. (Reproduced with kind permission from Jennersten, 1983.)

species of insects (Batra and Batra, 1985). In contrast, some of the most damaging pathogens of woody plants are transmitted by insects, and in these cases most of the fungi are concentrated in only a few genera while the vectors are typically bark beetles or weevils.

A. Dissemination for Fertilization

Although attention in this paper is concentrated on the dissemination of fungi for disease transmission, dissemination of the spores of fungal pathogens may also act in the cause of fertilization. Many fungi exist as two (or more) mating types and these have to be brought together for the life-cycle to be completed. In the rusts (F5.22), one of the most economically important groups of plant pathogens, insects play a major role in this process (Leach, 1940). Rust fungi have several different kinds of fruiting structures, one of which—the "pycnium"—exudes drops of liquid containing spores which will not germinate under normal conditions. Their function was a subject of some speculation until Craigie (1931), working with the sunflower rust *Puccinia helianthi*, discovered that flies attracted by the sweetness of the liquid would transfer spores of one mating type to another. As a result, cross-fertilization often occurred and aecidiospores, the next spore stage in the life-cycle, were produced.

Most texts on the rusts highlight the part played by insects in fertilization, but remarkably few detailed studies on the process have been carried out. It could be a productive field for research. Savile (1976) has drawn attention to the fact that the nectar produced by "primitive" rusts, such as *Chrysomyxa* on spruce, has a carrion-like smell while the more "modern" rusts, like the *Puccinia* species, have a fragrant flower-like scent. He suggests that more detailed analysis of the nectar of different rusts might provide an indication of the insects present when these rust species originated.

With some fungal pathogens there may also be a link between dissemination for fertilization and dissemination for transmission. For example, in *Ceratocystis fagacearum* (F4.24), the cause of oak wilt (Section III, B), the switch from asexual spore to sexual spore production following fertilization prolongs the period over which inoculum is available for disease transmission.

B. Disease Transmission

Table I gives a list of tree pathogens, their insect vectors and their host plants. They are divided into three main taxonomic groups—the Oomycotina (F2.3), the Basidiomycotina (F5), and the Ascomycotina (F4) plus closely allied Fungi Imperfecti (F6).

Although the Basidiomycotina include many important fungal pathogens

TABLE I. Insect-disseminated fungal pathogens of trees.

	Pathogen	Vector	Host
Basidiomycotina	*Amylostereum areolatum* (F5.8)	Woodwasps (H33.1) (*Sirex* spp.)	Pine
	A. chailettii	Woodwasps (*Sirex* and *Urocerus* spp.)	Pine
	Heterobasidion annosum (F5.8)	*Flies (H28)	Spruce, larch and fir
Oomycotina	*Phytophthora palmivora* (F2.3)	Ants (H33.9)	Cocoa
	P. megakarya	Ants	Cocoa
Ascomycotina	*Ceratocystis clavigera* (F4.24)	Bark beetles (H25.63) (*Dendroctonus ponderosae*)	Pine
	C. fagacearum	Nitidulid beetles (H25.30)	Oak
		Bark beetles (H25.63) (*Pseudopityophthorus* spp.)	
	C. fimbriata	Nitidulid beetles (H25.30)	Almond, apricot and peach
	C. fimbriata	*Ambrosia beetles (H25.63) (*Xyloborus ferrugineus*)	Cocoa
	C. laricicola	Bark beetles (H25.63) (*Ips cembrae*)	Larch
	Cryphonectria (*Endothia*) *parasitica* (F4.7)	*Coleoptera (H25)	Chestnut
	Ophiostoma ips (F4.24)	Bark beetles (H25.63)(*Ips* spp.)	Pine
	O. minor	Bark beetles (*Dendroctonus* spp.)	
	O. montia	Bark beetles (*Dendroctonus* spp.)	Pine
	O. polonica	Bark beetles (*Ips typographus*)	Spruce
	O. piceae	Bark beetles (*D. rufipennis*)	Spruce
	O. pilifera	Bark beetles	Pine
	O. ulmi	Bark beetles (*Scolytus* spp. and *Hylurgopinus rufipes*)	Elm and Zelkova
Fungi Imperfecti	*Fusarium solani* (F6.2)	Bark beetles	Cocoa
	Leptographium terebrantis (F6.2)	*Bark beetles (*Dendroctonus* spp.)	Pine
	Trichosporium symbioticum (F6.2)	Bark beetles (*S. ventralis*)	Fir
	Verticicladiella serpens (F6.2)	*Bark beetles (*Hylastes* spp.)	Pine
	V. wageneri	Bark beetles and weevils (H.25.63) (*Hylastes, Steremnius* and *Pissodes* spp.)	Pine and Douglas fir

The decision was taken to restrict the list of fungi in this table to those for which good evidence of pathogenicity was available. In some cases evidence for insect transmission is circumstantial and these are indicated thus *.

of trees, the only indubitable case of insect transmission concerns the *Amylostereum* species that are transmitted by woodwasps (H33.1) of the genera *Sirex* (Fig. 2A) and *Urocerus* (Madden and Coutts, 1977; Talbot, 1977). In most parts of the world these organisms do little damage but in Australia and New Zealand, *Sirex noctilio* (H33.1) and *Amylostereum areolatum* (F5.8) can cause significant losses in *Pinus radiata* plantations, especially during periods of drought.

The fungus is carried by the adult siricids in a pair of small invaginated intersegmental sacs protruding into the body. These are connected by ducts to the anterior end of the ovipositor. During oviposition, spores of the fungus are "injected" into the sapwood of trees and developing mycelium then invades the wood around the oviposition hole and larval tunnels. As is commonly the case with xylem pathogens, a zone of reduced moisture content develops around the tissue occupied by *Amylostereum* and this ensures that the *Sirex* eggs hatch and the larvae develop in relatively dry wood. In addition, host resinosis is reduced in colonized tissue, and this also favours larval development. Female larvae from the second instar onwards, carry the fungus in deep skin folds called hypopleural organs sited in the abdomen. The young adult then acquires the fungus during eclosion when, by reflex actions, it breaks up the hypopleural organs in the cast-off larval skin. Interestingly, inoculation with the fungus alone has relatively little effect and the insect's mucus secretions are believed to play an important part in "conditioning" the tree for invasion by *A. areolatum* (Coutts, 1969).

In contrast to the intimacy of this relationship, a very casual association may exist between the serious root rot pathogen *Heterobasidion annosum* (*Fomes annosus* (F5.8)) and a variety of insect species. As the pathogen often produces its conidial *Oedocephalum* (F6.2) stage on freshly exposed wood surfaces, it has been suggested that the spores are transmitted to healthy trees by flies and other animals. However, no critical evidence to support this hypothesis has yet been produced (Hunt and Cobb, 1982).

The only Oomycetes (F2.3) represented in Table I are *Phytophthora palmivora* and *Phytophthora megakarya*. They cause black pod disease of cocoa and can also cause stem cankers and wilt of flower cushions. Although these fungi can be disseminated via rain splash and contact with plant litter, several species of ants can also be instrumental in disease transmission. This was first demonstrated in Ghana by Evans (1971) and subsequently in Nigeria by Okaisabor (1974) and Taylor and Griffin (1981). Transmission occurs when ants collect particles of soil contaminated with *Phytophthora* and build them into tents on the cocoa plants, including on the pods (Fig. 2B). These can then become infected with the pathogen, which appears to be capable of invading both wounded and uninjured pods.

The majority of fungi in Table I are either Ascomycotina or Fungi

Fig. 2. (A) *Sirex* woodwasp, vector of basidiomycete *Amylostereum*, ovipositing on pine. (Forestry Commission copyright.) (B) A soil tent (arrowed) produced by ants, causing *Phytophthora megakarya* infection of cocoa pods. (Photograph kindly supplied by C. M. Brasier.)

Imperfecti, and almost all the ascomycetes fall within two genera of the family Ophiostomataceae (F4.24)—*Ceratocystis* and *Ophiostoma*. In addition, most of the imperfect fungi are members of *Leptographium* or *Verticicladiella*, genera that would be expected to have an *Ophiostoma* perfect state. Of the few exceptions, one—*Trichosporium symbioticum* (F6.2), which causes brown bark and stem lesions in *Abies* (Wright, 1935)—has been ascribed to what is now regarded as an invalid genus. It now seems possible that this fungus is also a member of the Ophiostomataceae (F4.24). Another exception, *Fusarium solani*, is known to be carried by elm bark beetles but there is little indication of it causing disease in elm (Webber and Hedger, 1986). However, Prior (1986) has recently shown that *F. solani* can grow extensively in healthy wood of cocoa when introduced to the tree by bark beetles invading *Phytophthora palmivora* bark cankers.

The remaining fungus in Table I *Cryphonectria parasitica* (F4.7), is well known for having caused the almost total destruction of the American chestnut. At one time or another a wide range of insect species and birds, mammals and even molluscs have been implicated in transmission, but probably only wound-making insects can act as significant vectors (Anagnostakis, 1987). Recently, however, interest in insect dissemination has increased with the discovery of hypovirulent forms of the pathogen. These have largely lost their pathogenic properties following virus infection and can pass on the virus to healthy forms of the fungus through contact followed by hyphal fusion. It now appears that insects may play an important part in this process by moving fragments of hypovirulent *Cr. parasitica* from "cured" to active cankers and thereby effect the conversion of the latter to hypovirulence (Russin *et al.*, 1984).

The economic significance of the Ophiostomataceae (F4.24) has ensured they have been the subject of a great deal of research. Evaluation of this research can give many useful insights into the process of insect transmission. Therefore, using this group of fungi as examples, the subject will be approached in two ways. Firstly the various stages in the transmission process will be identified, and secondly, a number of case studies will be examined and used to explore different aspects of disease transmission.

II. THE STAGES OF TRANSMISSION

A. Establishment of the Association between Insect and Fungus in the Diseased Tree

For a vector relationship to be established, the insect and the fungus must coincide in space and time within the infected host. A number of factors can be identified which may influence this process.

1. The Distribution of the Fungus in the Host During Pathogenesis

Among diseases caused by the Ophiostomataceae quite contrasting patterns of host invasion can be found. Some of them are termed "vascular wilts", indicating that they are diseases in which the fungus typically shows extensive vertical distribution in the xylem elements (vessels or tracheids), but in which it is unable to colonize xylem parenchyma, the medullary rays, the cambium and inner bark (phloem tissue) until the host becomes moribund. Examples include Dutch elm disease caused by *Ophiostoma ulmi* (F4.24), oak wilt (*Ceratocystis fagacearum* (F4.24)), and black stain root disease (*Verticicladiella wageneri* (F6.2)) which affects pines and Douglas fir. By contrast there are the "vascular stain diseases", in which growth of the fungus in the rays and in the ray parenchyma is a major feature; movement through the vessels does occur but this is more restricted than is the case with the vascular wilts. The fungus can usually also invade the inner bark. Insect-vectored examples include *Ceratocystis* canker of deciduous fruit trees caused by *Ceratocystis fimbriata* and the "pathogenic blue stains" of conifers, caused by various fungi including *Ceratocystis clavigera, Ceratocystis laricicola, Ophiostoma polonica* and *Verticicladiella serpens*. These two different patterns of distribution within the diseased tree have a significant influence on the opportunities that exist for the establishment of a link with potential insect vectors.

2. Synchrony between Insect and Fungus

Specialized plant pathogens of the types discussed above (Section II, A, 1) are characterized by an expanding phase of parasitic growth in the living host and a declining phase of saprophytic growth in the dead host (Garrett, 1956). This is because qualities which fit the fungus for parasitism may put it at a disadvantage when it comes to prolonged saprophytic survival in competition with other microorganisms. Garrett expressed this as "low competitive saprophytic ability". Therefore, if an association between insect and tree pathogen is to result in the disease being vectored and perhaps to evolve into a close symbiosis, any potential vector must have a life-cycle which not only coincides with the pathogen, but is completed before the decline of the fungus during its saprophytic phase. Clearly then, an insect with a long period of development in the tree is unlikely to have the potential to become the vector of an ephemeral fungal pathogen.

In this context, Kile and Hall (1988) have recently shown that very large numbers of the mountain pin hole borer, *Platypus subgranosus* (H25.63), breed in *Nothofagus* trees killed by *Chalara australis* (F6.2). This fungus is a vascular stain pathogen, and its life-style and intimate contact with an insect immediately suggest an insect-disseminated disease. However, the *Platypus* has a 2–2½ year life-cycle, whereas the fungus survives in the stems of dead trees for only 12–18 months. Hence, the lack of synchrony between the two species is a major factor in preventing *P. subgranosus* from vectoring the pathogen.

B. Transport from Diseased to Healthy Trees

By definition a vector must move from a diseased to a healthy tree, or at least from a diseased part to a healthy part of the same tree. Nonetheless, insufficient attention has sometimes been paid to this aspect of an insect's behaviour during the evaluation of its role as a vector. For example, it is a critical issue in the much reviewed but little researched "*Xyleborus–Ceratocystis* complex" which causes serious damage to cocoa in much of South America. The causal agent, *Ceratocystis fimbriata* (F4.24), produces a vascular stain disease and various scolytid beetles (H25.63), mostly *Xyleborus*, have been found to breed in the diseased trees. However, although frequently cited as vectors, their importance is unresolved, largely because it is often not known whether they attack healthy trees or merely colonize those which are already diseased (Saunders, 1965). One species (*Xyleborus ferrugineus*) does seem to be a primary invader, but even so, in Trinidad at least, dissemination of *C. fimbriata* probably results principally from the beetles boring into dying trees and releasing a mixture of wood frass and fungal spores. The wind-blown frass may then infect healthy trees, providing that suitable wounds are available (Iton, 1961; Iton and Conway, 1961).

Once a potential vector has acquired an inoculum of pathogen spores, the way in which the spores are carried, the behaviour of the vector and the environmental conditions during dissemination can all be instrumental in influencing the chances of successful disease transmission. With many vectors, spores are simply carried externally on the insect's exoskeleton and so are directly exposed to desiccation and ultra-violet radiation during flight. Under these circumstances the period of flight and the time taken by a vector to locate suitable host material for feeding or breeding may be critical if the fungus is to survive. By contrast, fungi are usually well protected from these rigours if they are contained within the specialized carrying organs, mycangia, possessed by many bark and wood-boring beetles (H25) (Batra, 1963; Francke-Grosmann, 1967). However, very few of the fungi transported in mycangia are actually pathogenic, although *Dendroctonus ponderosae* (H25.63) is a good example of an insect which does transport pathogenic fungi in this way.

Temperature may also play an important part in disease transmission by regulating insect behaviour. Although most vectors are commonly assumed to fly, the threshold temperatures for flight can often be quite high, at least 20°C in the case of the Dutch elm disease scolytid vectors (Fransen, 1939).

Indeed, the very temperatures which encourage insect activity can often be detrimental to the fungi they transport. Thus crawling may be more important than is generally recognized, and is crucial in the dissemination of *Verticicladiella wageneri* (F6.2) by various root beetles and weevils, which actively seek out stressed and diseased roots for oviposition (Section III, C).

C. Transfer and the Establishment of Infection

Once the vector has reached a potential host tree, successful transfer of the fungus to susceptible plant tissue and the initiation of sustained disease again depend on several factors. These include the nature of the infection court, and the quantity of inoculum that is introduced to it.

Wounds are required in all known cases of infection by the Ophiostomataceae (F4.24) involving insect transmission, and, where bark beetles and weevils are involved, the insects make the wounds themselves. However, with some other kinds of insects, such as the sap-feeding nitidulids (H25.30), there is a dependence on wounds made by external agencies. The depth of the wound may also be important. With the vascular wilt diseases like Dutch elm disease and oak wilt, the wound must reach the wood if the pathogen is successfully to penetrate and invade the xylem elements (Gibbs and French, 1980), whereas with a vascular stain disease like the *Ceratocystis* (F4.24) canker of fruit trees, a superficial wound exposing only live inner bark is all that is required (Devay *et al.*, 1968).

Furthermore, the arrival of viable fungal spores in a suitable wound does not necessarily lead to disease. Infection depends on the number and physiological condition of the spores, and on the effects of competing microorganisms (Sections III, A and B). Simultaneous host invasion from many infection points may also be necessary to overwhelm the resistance of a host which might otherwise be able to check the development of a single infection. This has been shown to be important with the vascular stain pathogen *Ophiostoma polonica* (F4.24) on Norway spruce, where doubling the number of inoculation points produced an almost eight-fold increase in the amount of wood which succumbed to the pathogen (Christiansen, 1985).

III. CASE STUDIES ON INSECT-DISSEMINATED FUNGAL PATHOGENS

The following case studies broadly represent two categories of insect–fungus association. Firstly, those in which the emphasis is on an aggressively pathogenic fungus and where dissemination is usually effected by several, albeit often closely related, species or genera of insects. Examples to be discussed here include *Ophiostoma ulmi* (F4.24), *Ceratocystis fagacearum* (F4.24) and *Verticicladiella wageneri* (F6.2), all vascular wilt organisms.

Secondly, there are those in which the emphasis is on a particular species of insect (invariably a bark beetle (H25.63)) which engages in mass attacks on its host and which often carries several species of the vascular stain fungi. The examples presented here include *Dendroctonus ponderosae* and its associates *Ceratocystis clavigera* and *Ophiostoma montia*, plus the intriguing relationship between *Dendroctonus frontalis* and *Ophiostoma minor*.

A. Dutch Elm Disease

Many people are now familiar with the ravages caused by Dutch elm disease. Two "new" aggressive forms of *Ophiostoma ulmi* (F4.24) (the EAN and NAN races) have recently spread throughout Europe, central Asia and North America (Brasier, 1987), disseminated by various elm bark beetles (H25.63) mainly of the genus *Scolytus* (Lanier and Peacock, 1981). In much of Europe the dominant vector is *S. scolytus* (Fig. 3A), but *S. multistriatus, S. kirschi* and *S. laevis* are all considered as potential vectors (Lekander *et al.*, 1977; Maslow, 1970). In North America both the native *Hylurgopinus rufipes* and introduced *S. multistriatus* are involved (Readio, 1935).

The outline of the disease cycle has been known since the 1930s and "starts" when a new generation of scolytids emerge from elm bark. They carry on their bodies spores of the pathogen, and while seeking out bark for fresh breeding material often feed in the twig crotches of healthy elms. Although this is not an obligatory part of the beetle life-cycle (Fisher, 1937; Choudhury, 1979), it is the means by which *O. ulmi* infects elms, the wounds made by the beetles (feeding grooves) serving as infection courts for the pathogen (Fig. 3B).

External symptoms of wilting and foliage-yellowing then follow, with concurrent development of internal wood discoloration as *O. ulmi* spreads through the outermost ring of xylem vessels. An infected tree may succumb to the disease within a single season or over several years, but eventually the bark of a dying elm can be colonized by breeding scolytids. As they enter to breed, the beetles may again carry in spores of *O. ulmi*, with the end result that the fungus colonizes the dying elm bark in association with the developing scolytid broods. This period of bark colonization by insect and fungus generally occurs during the winter, although it can be compressed into a few weeks in a summer generation, and ensures that pathogen and

Fig. 3. (A) Larger European elm bark beetle, *Scolytus scolytus*, vector of Dutch elm disease. (Forestry Commission copyright.) (B) Feeding grooves (arrowed) produced by *S. scolytus* on *Ulmus procera*.

vector are reunited by the time of beetle emergence in the early summer.

Put in these terms the basic disease cycle appears simple, but the association between fungus and insect is by no means a straightforward one. An illustration of this is provided during early stages of larval growth, when the developing gallery consists of a central maternal tunnel with outwardly fanning larval galleries. Often there is also an expanding lesion produced by *O. ulmi* (or occasionally by other bark fungi) centred on the maternal gallery, but it usually lies behind the line of feeding larvae which mine outwards into bark free of any fungal colonization (Fig. 4A and Webber *et al.*, 1988). Hence there is usually no contact between insect and pathogen during the initial period of colonization and, if it should occur, evidence suggests it may be inimical to larval development (Webber, 1981). Moreover, if larvae encounter discrete areas of colonized bark they often appear to avoid it and their galleries can be seen turning away (Fig. 4B).

The behaviour of the larvae prior to pupation can also have an important influence on the degree of association between vector and pathogen. Although *O. ulmi* may eventually colonize the whole gallery system, the critical factor is the extent to which it colonizes the pupal chambers where new adult beetles acquire their inoculum of spores. Pupation, however, can take place in a variety of locations—occasionally in the sapwood but more frequently in the inner and outer layers of bark (Kirby and Fairhurst, 1983). Furthermore, over two-thirds of pupal chambers lying in the inner bark, but usually less than a third of those present in the outer bark, contain sporulating *O. ulmi* (Webber and Brasier, 1984). Moreover, sporulation tends to be much sparser in chambers located in outer bark than in those in the inner bark. As a result the amount of inoculum picked up by beetles prior to emergence can vary enormously, some beetles carrying no *O. ulmi* and others carrying as many as 300,000 spores (Table II). In addition, the smaller vector species, such as *S. multistriatus* and *S. kirschi*, often carry significantly fewer spores than the larger species such as *S. scolytus* (Webber, 1988; and Table II). This may simply be a function of body size (Webber and Brasier, 1984), but seems more likely to result from pupation behaviour; smaller beetles pupating more frequently in outer bark, or in thinner bark that dries out quickly and as a consequence is less favourable for *O. ulmi* sporulation.

Since the elm scolytids have no specialized organs for fungal transport the process of transportation may pose further problems for *O. ulmi*. Although some of the spores are lodged in the crevices and pits that are distributed over the insect's surface (Fig. 5A, B), many are probably quite unprotected and consequently exposed to extreme environmental fluctuations during beetle flight. Laboratory studies have suggested that *O. ulmi* spores are very susceptible to the desiccation caused by relative humidities of $< 80\%$ and to

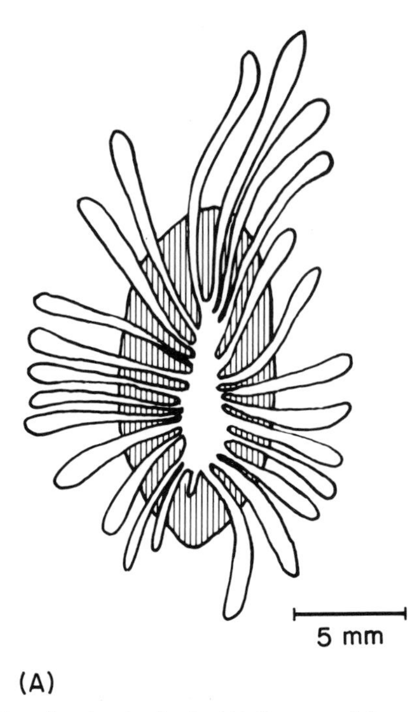

Fig. 4. Scolytid beetle breeding in elm bark. (A) Pattern of fungal colonization (hatched) associated with developing *Scolytus* broods. (B) Larval galleries turning away from a region of fungus colonized bark. Furthest extent of fungal lesion indicated by arrows.

the wavelengths of ultra-violet light encountered in daylight. Therefore, it is probably not surprising to find that the proportion of *S. scolytus* individuals carrying viable spores declines dramatically after a period of flight (Table III).

In the final stage of disease transmission, the transfer of the pathogen from vector to infection court appears to be generally straightforward, with most feeding beetles passing on their inoculum to the feeding groove (Webber and Brasier, 1984). However, infection of the underlying xylem occurs in only about 15% of the feeding grooves that acquire an inoculum of *O. ulmi* (Webber and Brasier, 1984). The important factor here appears to be the number of viable spores available for infection. Experiments with artificial feeding grooves have indicated that with *Ulmus procera* (English elm) at least 1000 spores are necessary to produce a single infection (Fig. 6 and Webber, 1987) and relatively few beetles carry this number of spores (Table II). Even an inoculum of 50,000 spores does not always guarantee successful infection although the higher the spore density, the more likely

(B)

TABLE II. Comparison of three scolytid species as vectors of *Ophiostoma ulmi*.

Location of sample material, and date	Beetle species	Mean body length (mm)	Sample size	Proportion of beetles carrying *O. ulmi*	Range of beetle spore loads	Median spore number
Crediton, England 1982	*S. multistriatus*	3.2	16	100%	5–150	19
	S. scolytus	5.5	48	90%	0–40,000	475
Guadalajara, Spain 1984	*S. kirschi*	2.5	51	8%	0–300	13
	S. multistriatus	3.1	51	35%	0–57,000	210
Rubena, Spain 1984	*S. kirschi*	–	50	6%	–	–
	S. multistriatus	3.4	48	63%	0–30,000	7
	S. scolytus	5.4	52	98%	0–343,000	1200

After Webber and Brasier, 1984; Webber, 1988.

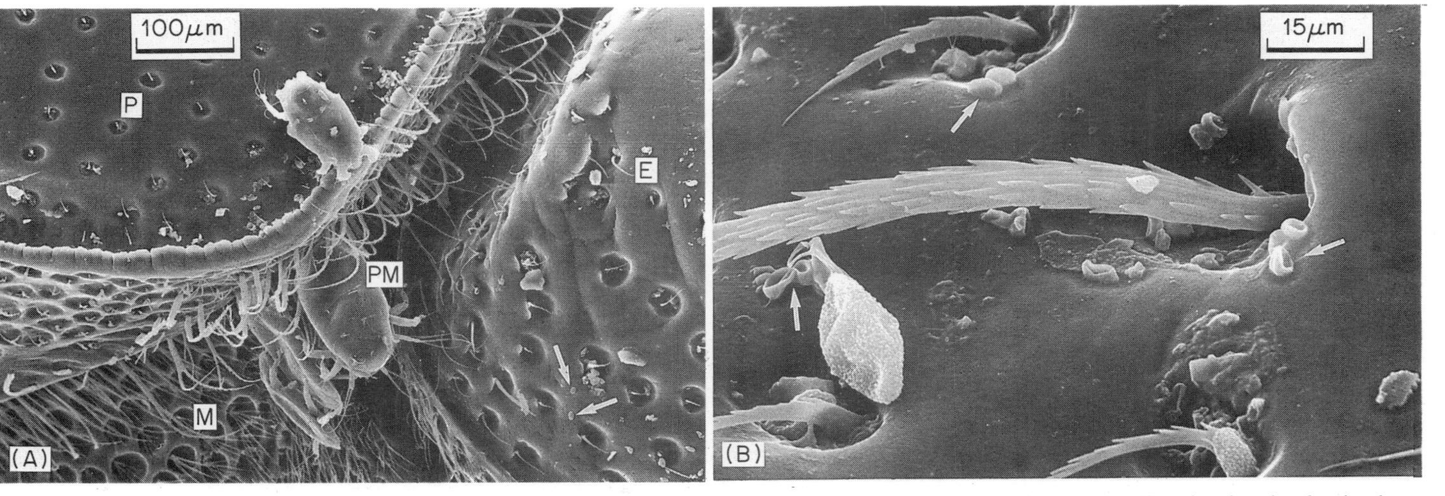

Fig. 5 Scanning electron micrographs of the cuticle of *Scolytus scolytus*. (A) Dorsal view of elytron (E) and pronotum (P), showing deeply pitted mesothorax surface (M) and attached phoretic mites (PM). Fungal spores corresponding to the shape and size of *Ophiostoma ulmi* conidia are arrowed. (B) Conidia of *O. ulmi* (arrowed) lodged in setal pits located on the elytra.

TABLE III. Percentage of beetles carrying *Ophiostoma ulmi* before and after emergence.

Year	Species	% of beetles carrying *O. ulmi*	
		Taken from pupal chambers	After emergence and flight
1938 (from Parker, 1939)	*S. scolytus*	56.5 (329)[a]	25.7 (412)
1980 (from Webber, 1981 and Brasier, 1984)	*S. scolytus*	80.4 (46)	23.0 (74)
1981 (from Webber, 1981 and Brasier, 1984)	*S. scolytus*	97.9 (48)	10.2 (69)

[a] Numbers in parentheses are those used to calculate percentages.

infection is to result. Infection rates may also be enhanced if humidity levels are high (Fig. 6).

B. Oak Wilt

Oak wilt has been known as a serious problem in the mid-western States of the USA since the 1940s (Henry *et al.*, 1944). The upsurge in this apparently indigenous disease seems, in part, to have been caused by the regeneration of highly susceptible populations of *Quercus ellipsoidalis* following fire damage to previously diverse woodland communities, and also by man's actions in opening up forests for logging and construction, with consequent tree damage (Gibbs and French, 1980).

Insect transmission occurs when nitidulid beetles (H25.30), attracted by exuding sap, carry *Ceratocystis fagacearum* (F4.24) to wounded oaks. These wounds then serve as infection courts, although only wounds made over a relatively short season from late April to early June are likely to result in infection (Gibbs and French, 1980; Juzwik and French, 1986). In addition, the wound must be fresh, as a lapse of only 24 hours from the time of wounding to the first visit by a nitidulid vector can significantly reduce the likelihood of infection (Kuntz and Drake, 1957). This is possibly due to the accumulation of antifungal host metabolites such as phenolics (Cobb *et al.*, 1965), but is also because the wounds are rapidly colonized by non-pathogenic fungi common to oak, which outcompete and thus exclude *C. fagacearum* (Gibbs, 1980a).

If the pathogen does successfully infect, it invades the xylem causing internal and external symptoms similar to those of Dutch elm disease. Initially the fungus is restricted to the xylem, but on the death of the tree it

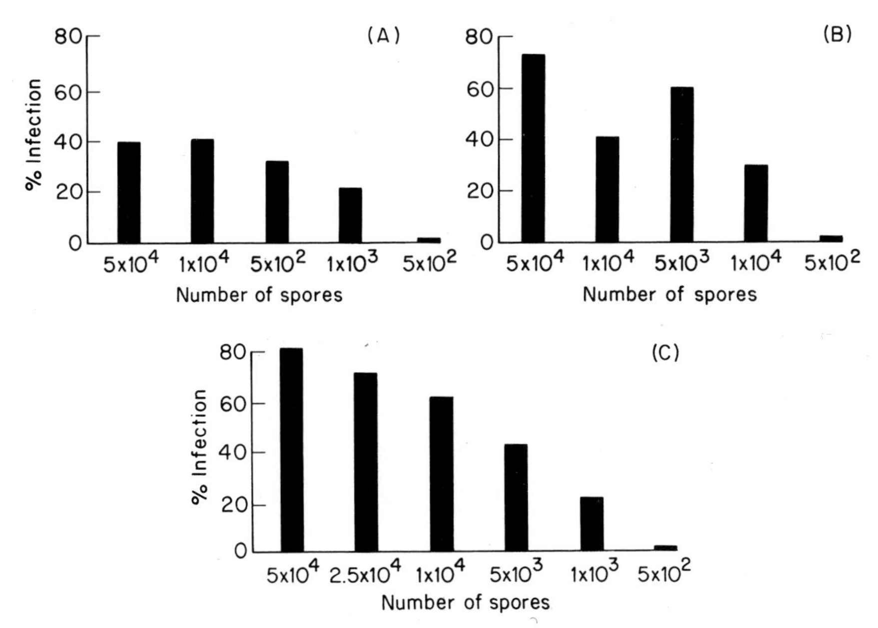

Fig. 6. Effects of spore number and relative humidity on the infection potential of the NAN aggressive subgroup of *Ophiostoma ulmi*. (A) and (B) Experiments carried out in consecutive years with feeding grooves subject to the prevailing climatic conditions. (C) Feeding grooves maintained at > 90% relative humidity for the duration of the experiment.

grows outward across the cambium and into the bark. This is often followed by the formation of a unique fungal structure, the "pressure pad", consisting of a pair of cushion-like structures formed back to back, one developing from the bark and the other from the wood (Fig. 7A). As they expand in thickness they raise and rupture the bark, thus creating a large oval cavity within which the sporulating mats develop, producing asexual spores (Fig. 7B and Leach *et al.*, 1952).

Still partially covered by the bark, the sporulating mats emit a fruity odour, attracting nitidulid beetles in much the same way that insects are attracted to rust pycnia (see Section I). The dorso-ventrally flattened bodies of these insects allows them to slip under the bark and crawl over the surface of the mats where they feed on mycelium and spores (Jewell, 1956). When they leave the mats they commonly carry on their body surface and in their gut viable spores of *C. fagacearum*; as many as 760,000 spores have been recorded on a single beetle (Jewell, 1954; Juzwick and French, 1983). Some of the nitidulids may also visit a number of mats on different trees and again, just as with the rusts, this can lead to fertilization if spores of one mating type are transferred onto a mat of the opposite mating type. Thus

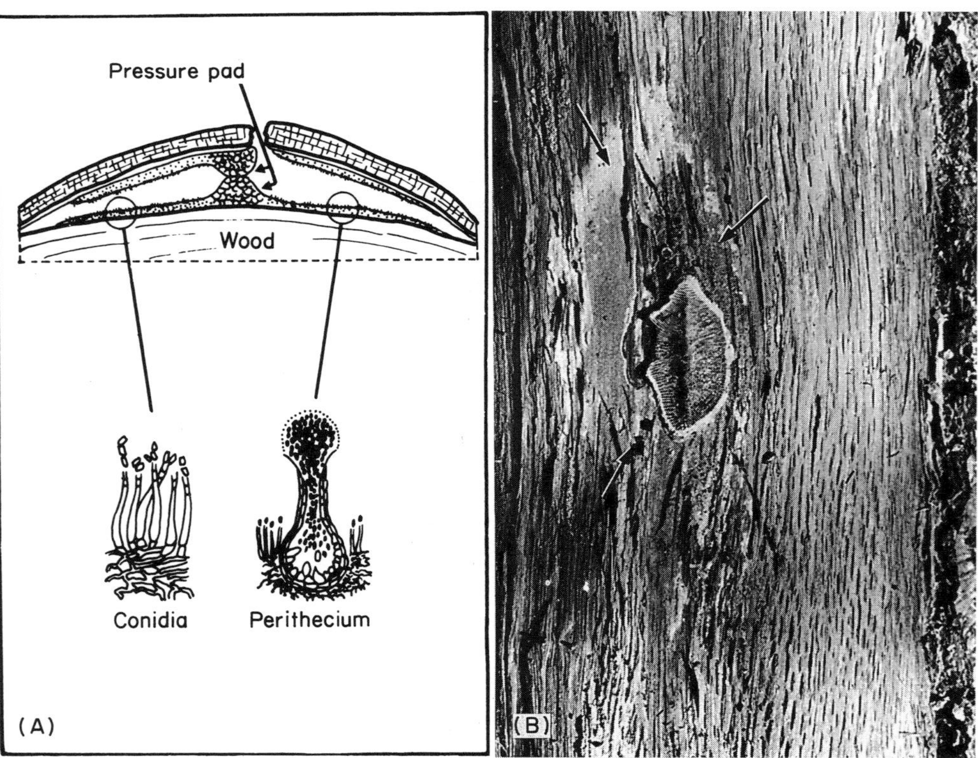

Fig. 7. Oak wilt pathogen, *Ceratocystis fagacearum*. (A) Diagrammatic representation of the pressure cushions and associated sporulating mats showing asexually produced spores (conidia) borne on the mycelium and sexually produced spores (ascospores) formed within a perithecium. (Adapted from True *et al.*, 1960.) (B) Pressure pad revealed after removal of the bark, surrounded by the sporulating fungus mat (arrowed).

nitidulids not only transfer spores from mats to wounds, thereby causing infection, but also from mat to mat, resulting in fertilization and a change from asexual to sexual spore production by *C. fagacearum*.

Within a matter of a week, however, the sporulating mats decline and become overgrown by various species of antagonistic fungi (Shigo, 1958). Even in the xylem below the mats *C. fagacearum* can be quickly replaced by other fungi. This poor competitive saprophytic ability of *C. fagacearum*, indicated by its rapid replacement, is the main reason why bark beetles (Scolytidae) have relatively little significance in the disease transmission of oak wilt (Gibbs and French, 1980). The small North American oak bark beetles, *Pseudopityophthorus pruinosus* and *P. minutissimus* (H25.63), typically breed in branches that are less than 10 cm in diameter, and these are the very branches from which *C. fagacearum* declines most rapidly under the combined influence of competing fungi and high summer temperatures (Gibbs, 1980b). Thus although the beetles make wounds suitable for infection during maturation while feeding on young twigs (Rexrode and Jones, 1970), so few of the beetles carry the fungus that infection rarely results.

This lack of emphasis on bark beetles as vectors of oak wilt could change, however, if the disease was ever introduced into Europe (Gibbs, 1978). In both Britain and Europe there is a native species of bark beetle, *Scolytus intricatus*, which is larger than the American oak bark beetles and consequently breeds in large diameter oak branches and in the main trunk (Yates, 1984). As *C. fagacearum* tends to persist for a much longer time in such material, this beetle species would probably have a much greater opportunity to come into contact with inoculum of the pathogen and so could become an effective vector.

C. Black-stain Root Disease

Over the past decade black-stain root disease has become recognized as a growing problem in western states of North America, attacking several *Pinus* species and also Douglas fir (*Pseudosuga menziesii*) (Hansen, 1985; Sinclair *et al.*, 1987). Like oak wilt, it is probably a native disease, but intensive forestry management has made large areas of young forests particularly vulnerable (Hansen *et al.*, 1986).

External symptoms include reduced shoot and leader growth and later, reddening of the foliage as the affected trees die (Witcosky and Hansen, 1985). The pathogen *Verticicladiella wageneri* (F6.2) has many similarities with vascular wilt fungi such as *Ophiostoma ulmi* and *Ceratocystis fagacearum* (F4.24) but unlike them, infects only via the roots and its spread up the stem in the xylem tracheids is restricted to a height of only a few metres. The resulting internal symptoms consist of bands of black-stained wood, usually widest at the root collar, but tapering into the trunk and roots. Local spread can be via root grafts, but *V. wageneri* is also unusual among vascular wilt pathogens in being able to grow through soil and thereby infect adjacent healthy roots (Hicks *et al.*, 1980; Hessburg, 1984). However, it cannot penetrate intact roots and requires exposed xylem for infection.

The involvement of insects in vectoring the disease was first suspected when the pathogen was observed sporulating in bark beetle galleries in ponderosa pine roots (Goheen and Cobb, 1978). Since then, two root weevil and one bark beetle species have been proven to act as vectors in experiments involving both artificially and naturally *V. wageneri* infested insects (Table IV) (Harrington *et al.*, 1985; Witcosky *et al.*, 1986).

As roots of infected trees gradually succumb to black-stain they are often sequentially colonized by bark beetle (H25.63) and weevil (H25.63) species over a period of 2–4 years. *Steremnius* and *Hylastes* species primarily colonize the roots, while *Pissodes* tends to favour the lower stem and root collar (Witcosky and Hansen, 1985). The behaviour of the larvae in etching into xylem during pupation also encourages the association between vector and pathogen by releasing *V. wageneri* from the tracheids, enabling it to colonize the pupal chambers and so contaminate the adult beetles. Transmission then occurs when the insects feed on the roots of healthy trees, thereby introducing the pathogen into fresh infection courts (Table IV). The action of the vectors in seeking out damaged or stressed roots for oviposition and breeding can also lead to the infection of previously black-stain-free trees.

However, transmission of the pathogen does appear to be a relatively rare event. In field experiments naturally infested *Hylastes* beetles succeeded in transmitting black-stain root disease to just 2% of Douglas fir seedlings (Table IV) and only a small proportion of all the vector species sampled within active disease foci appeared to carry *V. wageneri* (Table V). Such poor levels of vectoring may in part be due to loss of viable inoculum during dispersal, in much the same way that occurs with *O. ulmi*, especially as black-stain vectors may crawl or fly considerable distances in their search for suitable feeding and breeding sites.

Nonetheless, the role of the insect vectors should not be underestimated. New foci of black-stain root disease are being found increasingly in newly thinned plantations or beside recently installed forest roads (Hansen, 1978). These types of disturbance inevitably cause root damage and the insects rapidly respond to host volatiles such as alpha pinene produced by injured roots, and move into these new areas bringing with them the disease. Interestingly, roots colonized by *V. wageneri* are even more attractive to the

TABLE IV. Transmission of *Verticicladiella wageneri* to two-year-old Douglas fir seedlings.

Treatment	Insect species	No. of seedlings	% of seedlings infected with *V. wageneri*	Mean no. of feeding wounds per seedling
Insects artificially infested with *V. wageneri*	*Pissodes fasciatus*	97	68	13.6
	Steremnius carinatus	128	25	5.1
	Hylastes nigrinus	61	47	1.6
Insects collected from the field	*Hylastes nigrinus*	1000	2	1.5

From Witcosky *et al.* (1986).

TABLE V. *Verticicladiella wageneri* infestation of field-collected insects.

Insect species	Number examined	Percentage carrying *V. wageneri*
Hylastes nigrinus	173	2.3
Steremnius carinatus	433	0.5
Pissodes fasciatus	21	4.8

From Witcosky *et al.* (1986).

vector insects than merely damaged roots, and the insects actively seek them out (Witcosky, 1985).

D. The *Dendroctonus ponderosae*/pathogenic blue-stain complex

The mountain pine beetle *Dendroctonus ponderosae* (H25.63) is a very serious insect pest on lodgepole pine (*Pinus contorta*) and other pines in western North America. Major outbreaks occur at irregular intervals, usually in mature or over mature forests (Coulson and Witter, 1984). Two vascular stain fungi, *Ceratocystis clavigera* (syn. *Europhium clavigerum*) and *Ophiostoma montia* (F4.24), are closely associated with the beetle and are strongly implicated in overcoming host resistance.

Just a few weeks after the mass beetle attacks, distinct regions of blue-stain caused by the two pathogens can be seen adjacent to many of the egg galleries. These in turn, are surrounded by zones of sapwood with reduced moisture content (Fig. 8). Within only six weeks all the sapwood of affected trees has become blue stained (Reid *et al.*, 1967). Despite this dramatic evidence for host invasion, little has been discovered about the pathogenic ability of *C. clavigera* and *O. montia* in the absence of their bark beetle vector. The inoculation work that has been described has involved experiments on trees with successful defensive reactions. Under these circumstances fungal invasion was restricted to within a few centimetres of the innoculation point (Reid *et al.*, 1967) and there are no data equivalent to those from *Ophiostoma polonica* on spruce, where it has been shown that increasing the density of inoculation points can lead to host killing (Christiansen, 1985 and Section II, C). Recently, however, Owen *et al.* (1987) were able to kill two-year-old seedlings of *Pinus ponderosae* by inoculation with *C. clavigera*, indicating that this fungus has the intrinsic capacity to cause significant damage.

In contrast, the relationship between the two fungi and *D. ponderosae* during beetle breeding has been the subject of detailed study by Whitney

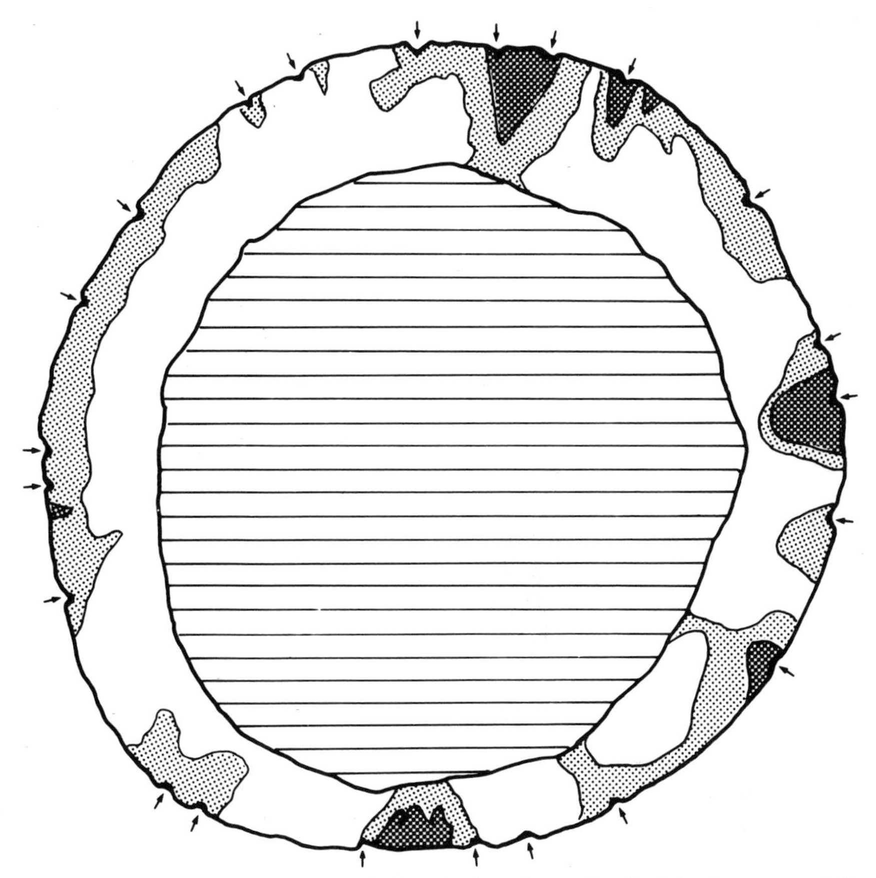

Fig. 8. Cross-section of the xylem of a lodgepole pine a few weeks after it has been successfully colonized by *Dendroctonus frontalis*. Arrows show the position of egg galleries. The zone of blue-stained wood and the surrounding zones of wood with reduced moisture are shown with different degrees of stippling. The heartwood is hatched. Some of the patterns are due to the presence of other egg galleries above or below the section. (Reproduced with kind permission from Reid *et al.*, 1967.)

(1971). It appears that *C. clavigera* and *O. montia* are always present in the mycangia of the beetle and are routinely introduced into the bark during gallery construction. First instar larvae readily yield both fungi on culturing, but these appear to be lost as the larvae then mine into fresh bark. Later larval stages are also typically free from fungal contamination and it is notable that the area of fungal colonization has a well-defined margin that usually lies several millimetres behind the feeding larvae. However, by the time the pupal chambers are excavated, the fungal associates have colonized the surrounding phloem tissue and contact with the insect is re-established.

Towards the end of the pupation period the chamber is lined with a 1–2 mm thick layer of closely packed conidia and conidiophores, but very little mycelium can be seen. Where pupae are in more or less continuous contact with it, the layer becomes much denser and somewhat thinner. The degree of asexual sporulation of blue stain fungi in the pupal cell is unequalled elsewhere in the gallery and strongly resembles the ambrosia associated with myceto-xylophagous beetles.

As they lie in the pupal chamber, pupae often rock their abdomens which causes them to brush against the fungi growing on the walls of the chamber. It has been suggested that this habit of abdominal flexing could prevent suffocation and entanglement in fungal growth during metamorphosis and also that it may stimulate the abundant sporulation. Young tenerals chew away the fungus material lining the pupal chambers and then enlarge the chambers by eating or chewing away the surrounding phloem and phellem. It is probably during this process that the maxillary mycangia become charged with spores and the outer body contaminated with the fungi.

The evidence clearly points to a well developed relationship between the insect and its fungi, their combined action enabling them to overwhelm a host tree. Although there is a need for more detailed studies, current evidence suggests that a similar kind of relationship exists between *Ips typographus* and *Ophiostoma pennicillata*/*O. polonica* on Norway spruce in Europe (Horntvedt *et al.*, 1983), between *Dendrotonus rufipennis* and *Ophiostoma piceae* on spruce in North America (Whitney, H.S., personal communication), and between various *Ips* species and *Ophiostoma ips* and *Ophiostoma pilifera* (Mathre, 1964a, b; Owen *et al.*, 1987).

E. *Dendroctonus frontalis* and *Ophiostoma minor*

Dendroctonus frontalis, the southern pine beetle (H25.63), is the most destructive insect pest to pine forests in thirteen south-eastern States of the USA and in parts of Mexico and Central America. It is an aggressive tree killer, attacking and overcoming healthy vigorous trees when its populations are large, but confined to weakened or dying trees when populations are low (Payne, 1980).

For many years it has been recognized that there is an association between *D. frontalis* and *Ophiostoma minor* (F4.24), and there is no doubt that *O. minor* has the ability to develop as a vascular stain pathogen in pine (Mathre, 1964b; Basham, 1970). However, it is now recognized that the association is fairly casual—unlike the *Dendroctonus ponderosae*/blue-stain fungi relationship. *O. minor* is not found in the mycangia but instead is principally located on phoretic mites which are carried under the beetle elytra and on the thoracic segments (Bridges and Moser, 1983; Moser and Bridges, 1986).

Furthermore, not all the beetle-colonized trees become blue-stained, and even where stain caused by *O. minor* is present, it is typically restricted in its distribution (Fig. 9). Under these circumstances it is important to know if other non-staining agents are being introduced to the tree by the beetle, and if so what they are. At present, evidence for another pathogen is lacking and it is still possible to argue, as did Mathre (1964b), that with this beetle some of the killing of trees results from the direct mechanical damage caused by the construction of the breeding galleries.

Another matter of interest is the relationship between *D. frontalis* and *O.*

Fig. 9. Black patches of wood colonized by *Ophiostoma minor* (arrowed) may occupy only a limited amount of the sapwood in loblolly pine killed by *Dendroctonus frontalis*. (Reproduced with kind permission from Bridges and Moser, 1983.)

minor in the breeding galleries. Barras (1970) showed that the beetle is much less successful at breeding in logs in which the bark is colonized by *O. minor* after artificial inoculation, than in uncolonized bark. Under these circumstances the female beetle can be seen turning away from fungus-colonized tissue during gallery construction, and any larvae feeding in the bark produce abnormal galleries.

Dendroctonus brevicomis, the western pine beetle, is also associated with *O. minor*, and here again the blue-stain in beetle-colonized trees is restricted to scattered patches and wedges. Whitney (1982) suggested that non-staining fungi present in the mycangia might be the primary cause of mortality in trees infested with *D. brevicomis*. However, Owen *et al.* (1987) have recently shown that while *O. minor* is capable of causing significant damage to ponderosa pine seedlings, the mycangial fungi from *D. brevicomis* are not pathogenic. Clearly much remains to be discovered about these associations.

IV. CONCLUSIONS

The relationship between fungal pathogens and their insect vectors can be mutually beneficial in many respects. For the fungus, the vector offers a means of transport which overcomes spatial discontinuities between host plants. It also frequently ensures that the pathogen is "targetted" not only to a host species susceptible to pathogenic attack, but also to a favourable infection court. The insect, in turn, may benefit from the steady provision of a suitable breeding habitat—often on a vast scale—and an abundant food supply of moribund host tissues which may sometimes be augmented by the mycelium of the fungal partner.

It is nevertheless important to recognize that there can be "weak links" within such an alliance. Firstly, there may be difficulties for the partners in establishing and maintaining the relationship. This is particularly evident in associations involving vascular wilt pathogens where, during disease development, the pathogen becomes completely separated from its vector and must be reunited with it for the disease cycle to be completed. Sometimes the developmental processes which have evolved to secure such reunions can be complex—as illustrated by the highly specialized pressure pads produced by *Ceratocystis fagacearum* (F4.24). Many of the vascular stain fungi, however, continue to maintain a close spatial relationship with their vectors throughout their life-cycles and are separated from them for only very limited periods, if at all. Thus in the case of *Dendroctonus ponderosae* (H25.63), separation from its blue stain associates occurs only transiently during larval development.

Secondly, there may be times when the pathogen and its vectors are in direct conflict. This is shown most strikingly with some of the phloeophagus bark beetles, where too early or too rapid invasion of the bark by the pathogen can lead to competition between fungus and beetle and disruption of normal larval development.

Thirdly, the behaviour of the insect can greatly influence its efficiency as a vector. Thus, for example, the polyphagus feeding habits of sap-feeding nitidulid beetles means that only a tiny proportion of these insects may actually visit a sporulating mat of *C. fagacearum*, and even fewer are then likely to move on to an oak infection court suitable for the pathogen. Even in host-specialized vectors such as bark beetles, apparently minor deviations in behaviour can significantly alter the proportion of insects carrying the pathogen and the number and condition of the propagules they deliver to a new host.

Defining where and how such weak links occur in pathogen—vector associations can often suggest what opportunities exist for breaking the disease cycle. In the future, this may also be a profitable basis for the development of disease control measures.

Acknowledgements

We are grateful to Drs Clive Brasier and Glen Kile for their helpful comments and to the many colleagues who kindly supplied photographs for this manuscript. We also wish to acknowledge the Pilkington Charitable Trust for the financial support of J. F. W. during the preparation of this paper.

REFERENCES

Anagnostakis, S. L. (1987). Chestnut blight: the classical problem of an introduced pathogen. *Mycologia* **79**, 23–37.

Atanasoff, D. (1920). "Ergot of Grains and Grasses." U.S. Dep. Agric., Washington.

Barras, S. J. (1970). Antagonism between *Dendroctonus frontalis* and the fungus *Ceratocystis minor*. *Ann. Entomol. Soc. Am.* **63**, 1186–1189.

Basham, H. G. (1970). Wilt of loblolly pine inoculated with blue-stain fungi of the genus *Ceratocystis*. *Phytopathology* **60**, 750–754.

Batra, L. R. (1963). Ecology of ambrosia fungi and their dissemination by beetles. *Trans. Kans. Acad. Sci.* **66**, 213–236.

Batra, L. R., and Batra, S. W. T. (1985). Floral mimicry induced by mummy-berry fungus exploits host's pollinators as vectors. *Science* **228**, 1011–1013.

Brasier, C. M. (1987). Recent genetic changes in the *Ophiostoma ulmi* population: the threat to the future of the elm. *In* "Populations of Plant Pathogens: their Dynamics and Genetics" (M. S. Wolfe and C. E. Caten, eds). pp. 213–226. Blackwell Scientific, Oxford.

Bridges, R. J., and Moser, J. C. (1983). Role of two phoretic mites in transmission of bluestain fungus, *Ceratocystis minor*. *Ecol. Entomol.* **8**, 9–12.

Christiansen, E. (1985). *Ceratocystis polonica* inoculated in Norway spruce: blue staining in relation to inoculum density, resinosis and tree growth. *Eur. J. For. Pathol.* **15**, 160–167.

Choudhury, J. H. (1979). Flight activity, flight orientation and elm-bolt infestation by *Scolytus multistriatus* Marsh. Ph.D. Thesis, Imperial College, London.

Cobb, F. W., Fergus. C. L., and Stambaugh, W. J. (1965). Factors affecting infection of red and chestnut oaks by *Ceratocystis fagacearum. Phytopathology* **55**, 1194–1199.

Coulson, R. N., and Witter, J. A. (1984). "Forest Entomology, Ecology and Management." J. Wiley and Sons, New York.

Coutts, M. P. (1969). The mechanism of pathogenicity of *Sirex noctilio* on *Pinus radiata* II. Effects of *S. noctilio* mucus. *Aust. J. Biol. Sci.* **22**, 1153–1161.

Craigie, J. H. (1931). An experimental investigation of sex in rust fungi. *Phytopathology* **21**, 1001–1040.

Devay, J. F., Lukezic, F. L., English, H., Trujillo, E. E., and Moller, W. I. (1968). *Ceratocystis* canker of deciduous fruit trees. *Phytopathology* **58**, 949–956.

Evans, H. C. (1971). Transmission of *Phytophthora* pod rot by invertebrates. *Nature (London)* **232**, 346–347.

Fisher, R. C. (1937). The genus *Scolytus* in Great Britain, with notes on the structure of *Scolytus destructor* 0L. *Ann. Appl. Biol.* **24**, 110–130.

Francke-Grosmann, H. (1967). Ectosymbiosis in wood-inhabiting insects. *In* "Symbiosis, Vol. 2—Associations of Invertebrates, Birds, Ruminants and Other Biota" (S. M. Henry, ed.) pp. 141–205. Academic Press, London.

Fransen, J. J. (1939). Iepenziekte, iepenspintkevers en beider bestrijding. Comite Inzake Bestudeering en Bestrijding en van de Iepenziekte—Mededeeling **32**. Wageningen Agric. Coll.

Garrett, S. D. (1956). "Biology of Root Infecting Fungi." Cambridge Univ. Press, Cambridge.

Gibbs, J. N. (1978). Oak wilt. *J. Arboric.* **3**, 351–356.

Gibbs, J. N. (1980a). The role of *Ceratocystis piceae* in preventing infection by *Ceratocystis fagacearum* in Minnesota. *Trans. Br. Mycol. Soc.* **74**, 171–174.

Gibbs, J. N. (1980b). Survival of *Ceratocystis fagacearum* in branches of trees killed by oak wilt in Minnesota. *Eur. J. For. Pathol.* **10**, 218–224.

Gibbs, J. N., and French, D. W. (1980). The transmission of oak wilt. *U.S. For. Serv. Res. Pap.* NC-185.

Goheen, D. J., and Cobb, F. W. (1978). Occurrence of *Verticicladiella wagenerii* and its perfect state, *Ceratocystis wagenerii sp. nov,* in insect galleries. *Phytopathology* **68**, 1192–1195.

Hansen, E. M. (1978). Incidence of *Verticicladiella wagenerii* and *Phellinus weirii* in Douglas-fir adjacent to and away from roads in western Oregon. *Plant Dis. Rep.* **62**, 179–181.

Hansen, E. M. (1985). Forest Pathogens of NW North America. *Forest Rec., Lon.* **129**.

Hansen, E. M., Goheen, D. J., Hessgurg, P. F., and Witcosky, J. J. (1986). Biology and management of black-stain root disease in Douglas-fir. *In* "Forest Pest Management in Southwest Oregon" (O. T. Helgerson, ed.), pp. 13–19. Oregon State Univ., Corvallis.

Harrington, T. C., Cobb, F. W., and Lownsbury, J. W. (1985). Activity of *Hylastes nigrinus*, a vector of *Verticicladiella wageneri*, in thinned stands of Douglas fir. *Can. J. For. Res.* **15**, 519–523.

Henry, B. W., Moses, C. S., Richards, C. A., and Riker, A. J. (1944). Oak wilt, its significance, symptoms and cause. *Phytopathology* **34**, 636–647.

Hessburg, P. H. (1984). Pathogenesis and intertree transmission of *Verticicladiella wageneri* in Douglas-fir (*Pseudosuga menziesii*). Ph.D. Thesis, Oregon State Univ., Corvallis.

Hicks, B. R., Cobb, F. W., and Gersper, P. J. (1980). Isolation of *Ceratocystis wageneri* from forest soil with a selective medium. *Phytopathology* **70**, 880–883.

Horntvedt, R., Christiansen, E., Solheim, H., and Wang, S. (1983). Artificial inoculation with *Ips typographus*-associated blue-stain fungi can kill healthy Norway spruce trees. *Medd. Nor. Inst. Skogforsk.* **38**, 1–20.

Hunt, R. S., and Cobb, F. W. (1982). Potential arthropod vectors and competing fungi of *Fomes annosus* in pine stumps. *Can. J. Plant Pathol.* **4**, 247–253.

Ingold, C. T. (1978). Dispersal of micro-organisms. *In* "Plant Disease Epidemiology" (P. R. Scott and A. Bainbridge, eds), pp. 11–22. Blackwell Scientific, Oxford.

Iton, E. F. (1961). Studies on a wilt disease of cacoa at River Estate. II. Some aspects of wind transmission. *In* "A Report on Cacoa Research, 1959–1960", pp. 47–58. Univ. College of West Indies, Imperial College of Tropical Agriculture, Trinidad.

Iton, E. F., and Conway, G. R. (1961). Studies on a wilt disease of cacoa at River Estate. III. Some aspects of the biology and habits of *Xyleborus* spp. and their relation to disease transmission. *In* "A Report on Cacoa Research, 1959–1960", pp. 59–65. Univ. College of West Indies, Imperial College of Tropical Agriculture, Trinidad.

Jennersten, O. (1983). Butterfly visitors as vectors of *Ustilago violacea* spores between Caryophyllaceous plants. *Oikos* **40**, 125–130.

Jewell, F. F. (1954). Viability of the conidia in *Endoconidiophora fagacearum* Bretz. in fecal material of certain nitidulidae. *Plant Dis. Rep.* **38**, 53–54.

Jewell, F. F. (1956). Insect transmission of oak wilt. *Phytopathology* **46**, 244–257.

Juzwik, J., and French, D. W. (1983). *Ceratocystis fagacearum* and *C. piceae* on the surfaces of free-flying and fungus-mat-inhabiting nitidulids. *Phytopathology* **73**, 1146–1168.

Juzwik, J., and French, D. W. (1986). Relationship between nitidulids and *Ceratocystis fagacearum* during late summer and autumn in Minnesota. *Plant Dis.* **70**, 424–426.

Kile, G. A., and Hall, M. F. (1988). An assessment of *Platypus subgranosus* (Coleoptera: Platypodinae) as a vector of *Chalara australis*, the causal agent of a vascular disease of *Nothofagus cunninghamii*. *N.Z. J. For. Sci.* (in press).

Kirby, S. G., and Fairhurst, C. P. (1983). The ecology of elm bark beetles in northern Britain. *Bull. For. Commn, Lond.* **60**, 29–39.

Kuntz, J. E., and Drake, C. R. (1957). Tree wounds and long distance spread of oak wilt. *Phytopathology* **47**, 22.

Lanier, G. N., and Peacock, J. W. (1981). Vectors of the pathogen. *In* "Compendium of Elm Diseases" (R. J. Stipes and R. J. Campana, eds), pp. 14–16. American Phytopathological Soc., Minnesota.

Leach, J. G. (1940). "Insect Transmission of Plant Diseases." McGraw-Hill, New York and London.

Leach, J. G., True, R. P., and Dorsey, C. K. (1952). A mechanism for the liberation of spores beneath the bark and for diploidization in *Chalara quercina*. *Phytopathology* **42**, 537–539.

Lekander, B., Bejer-Peterson, B., Kangas, E., and Bakke, A. (1977). The distribution of bark beetles in Nordic countries. *Acta Entomol. Fenn.* **32**, 1–37.

Madden, J. L., and Coutts, M. P. (1977). The role of fungi in the biology and ecology of woodwasps (Hymenoptera: Siricidae). *In* "Insect–fungus Symbiosis" (L. R. Batra, ed.), pp. 165–174. Allanheld, Osmun, Montclair, N.J.

Maslow, A. D. (1970). "Insects Harmful to Elm Species and their Control." Forest Industries Publications, Moscow.

Mathre, D. E. (1964a). Survey of *Ceratocystis* spp. associated with bark beetles in California. *Contrib. Boyce Thompson Inst. Plant Res.* **22**, 353–362.

Mathre, D. E. (1964b). Pathogenicity of *Ceratocystis ips* and *Ceratocystis minor* to *Pinus ponderosa*. *Contrib. Boyce Thompson Inst. Plant Res.* **22**, 363–388.

Moser, J. C., and Bridges, J. R. (1986). *Tarsonemus* (Acarina: Tarsonemidae) mites phoretic on the southern pine beetle (Coleoptera: Scolytidae): attachment sites and numbers of bluestain (Ascomycetes: Ophiostomatacae) ascospores carried. *Proc. Entomol. Soc. Wash.* **8**, 297–299.

Okaisabor, E. K. (1974). *Phytophthora* root infections from the soil. *In* "*Phytophthora* Disease of Cocoa" (P. H. Gregory, ed.), pp. 161–168. Longman, London.

Owen, D. R., Lindahl, K. Q., Wood, D. L., and Parmeter, J. R. (1987). Pathogenicity of fungi isolated from *Dendroctonus valens, D. brevicomis* and *D. ponderosae* to ponderosa pine seedlings. *Phytopathology* **77**, 631–636.

Parker, D. E. (1939). "Investigations on the relation of elm insects to the Dutch elm disease in Great Britain and other European countries during the years 1935–38." U.S. Dep. Agric., Bureau of Entomology and Plant Quarantine, Morristown, N.J.

Payne, T. L. (1980). Life history and habits. *In* "The Southern Pine Beetle" (R. G. Thatcher, J. L. Searcy, J. E. Coster and G. D. Hestel, eds), pp. 7–28. U.S. Dep. Agric., Washington.

Prior, C. (1986). Sudden death of cocoa in Papua New Guinea associated with *Phytophthora palmivora* cankers invaded by bark beetles. *Ann. Appl. Biol.* **109**, 535–543.

Readio, P. A. (1935). The entomological phases of Dutch elm disease. *J. Econ. Entomol.* **28**, 341–353.

Reid, R. W., Whitney, H. S., and Watson, J. A. (1967). Reactions of lodgepole pine to attack by *Dendroctonus ponderosae* Hopkins and blue stain fungi. *Can. J. Bot.* **45**, 1115–1126.

Rexrode, C. O., and Jones, T. W. (1970). Oak bark beetles, important vectors of oak wilt. *J. For.* **68**, 294–297.

Russin, J. S., Shain, L., and Nordin, G. L. (1984). Insects as carriers of virulent and cytoplasmic hypovirulent isolates of the chestnut blight fungus. *J. Econ. Entomol.* **77**, 838–846.

Savile, D. B. O. (1976). Evolution of the rust fungi (Uredinales) as reflected by their ecological problems. *Evol. Biol.* **9**, 137–207.

Saunders. J. L. (1965). The *Xyleborus-Ceratocystis* complex of cacao. *Cacao (Turrialba, Costa Rica)* **10**, 7–13.

Shigo, A. L. (1958). Fungi isolated from oak wilt trees and their effects on *Ceratocystis fagacearum*. *Mycologia* **50**, 757–769.

Sinclair, W. A., Lyon, H. H., and Johnson, W. T. (1987). "Diseases of Trees and Shrubs." Cornell Univ. Press, Ithaca and London.

Talbot, P. H. B. (1977). The *Sirex–Amylostereum–Pinus* association. *Annu. Rev. Phytopathol.* **15**, 41–54.

Taylor, B., and Griffin, M. J. (1981). The role and relative importance of different ant species in the dissemination of black pod disease of cocoa. *In* "The Epidemiology of *Phytophthora* on Cocoa in Nigeria" (P. H. Gregory and A. C. Maddison, eds), pp. 114–131. C.M.I., Kew, Surrey.

True, R. P., Barnett, H. L., Dorsey, C. K., and Leach, J. G. (1960). Oak Wilt in West Virginia. *W. Va. Agric. Exp. Stn Bull,* **448T**.

Webber, J. F. (1981). A natural biological control of Dutch elm disease. *Nature (London)* **292**, 449–501.

Webber, J. F. (1987). The influence of the d^2 factor on survival and infection by the Dutch elm pathogen *Ophiostoma ulmi*. *Pl. Pathol.* **36**, 531–538.

Webber, J. F. (1988). The effectiveness of three species of Dutch elm disease vectors. *Eur. J. For. Pathol.* (in press).

Webber, J. F., and Brasier, C. M. (1984). Transmission of Dutch elm disease: a study of the processes involved. *In* "Invertebrate–microbial Interactions" (J. Anderson, A. D. M. Rayner and D. Walton, eds), pp. 271–306. Univ. Press, Cambridge.

Webber, J. F., and Hedger, J. N. (1986). Comparison of interactions between *Ceratocystis ulmi* and elm bark saprobes *in vitro* and *in vivo*. *Trans. Br. Mycol. Soc.* **86**, 93–101.

Webber, J. F., Brasier, C. M., and Mitchell, A. G. (1988). The saprophytic phase of Dutch elm disease. In "Fungal Infection of Plants" (G. Pegg and P. G. Ayres, eds) (in press).

Whitney, H. S. (1971). Association of *Dendroctonus ponderosae* with blue stain fungi and yeasts during blood development in lodgepole pine. *Can. Entomol.* **103**, 1495–1503.

Whitney, H. S. (1982). Relationships between bark beetles and symbiotic organisms. *In* "Bark Beetles of North American Conifers" (J. B. Mitton and K. M. Sturgeon, eds), pp. 183–211. Univ. Texas Press, Austin.

Witcosky, J. J. (1985). The root-insect–black-stain root disease association in Douglas-fir: vector relationships and implications for forest management. Ph.D. Thesis, Oregon State Univ., Corvallis.

Witcosky, J. J., and Hansen, E. M. (1985). Root colonising insects associated with Douglas-fir in various stages of decline due to black-stain root disease. *Phytopathology* **75**, 399–402.

Witcosky, J. J., Schowalter, T. D., and Hansen, E. M. (1986). *Hylastes nigrinus* (Coleoptera: Scolytidae), *Pissodes fasciatus* and *Steremnius carinatus* (Coleoptera: Curculionidae) as vectors of black-stain root disease of Douglas-fir. *Environ. Entomol.* **15**, 1090–1095.

Wright, E. (1935). *Trichosporium symbioticum, N.Sp.,* a wood staining fungus associated with *Scolytus ventralis. J. Agric. Res. (Washington, D.C.)* **50**, 525–538.

Yates, M. G. (1984). The biology of the oak bark beetle, *Scolytus scolytus* (Ratzeburg) (Coleoptera: Scolytidae), in southern England. *Bull. Entomol. Res.* **74**, 569–579.

8

The Roles of the Bark Beetle *Ips cembrae*, the Woodwasp *Urocerus gigas* and Associated Fungi in Dieback and Death of Larches

D. B. REDFERN

I. INTRODUCTION

Associations between fungi and the phloeophagous bark beetles of coniferous trees are numerous, and mainly involve the beetle genera *Ips* and *Dendroctonus* (H25.63) and the fungus genera *Ceratocystis* and *Ophiostoma* (F4.24) (see Webber and Gibbs, this volume). Many of the fungus species merely cause blue-staining in the sapwood of dying trees or logs cut from freshly felled trees (Whitney, 1982), but some are pathogenic and in association with relatively aggressive bark beetles can cause severe damage to living trees. Examples have been reported from both North America (Bramble and Holst, 1940; Mathre, 1964a; Molnar, 1965) and continental Europe (Horntvedt *et al.*, 1983), but there have been no previous reports from Britain. However this is not surprising since all but one of the eight coniferous species planted commercially on a large scale are exotic and, for the most part, the bark beetles associated with them in their native habitats are absent.

Nonetheless there are some exceptions, and this paper describes a disease of larches (*Larix* spp.) which is primarily caused by one of them, the introduced beetle *Ips cembrae* (H25.63) associated with a previously undescribed species of *Ceratocystis* (F4.24). In addition, another insect–fungus association, the woodwasp *Urocerus gigas* (H33.1) and the wood-rotting basidiomycete *Amylostereum chailletii* (F5.8), can also be involved, forming a succession. This paper discusses the roles of these two associations in the development of damage.

II. DESCRIPTION OF DAMAGE AND INOCULATION EXPERIMENTS

Ips cembrae (H25.63) was introduced into Britain probably from continental Europe after the 1939–45 war in consignments of timber imported from Germany (Crooke and Bevan, 1957). It now occurs widely throughout east Scotland (Crooke and Kirkland, 1960; J. T. Stoakley, unpublished). In its native habitat the species is regarded as a secondary but important pest of European larch (*Larix decidua*), and causes damage by maturation feeding on young twigs, resulting in shoot pruning, and through breeding attacks on the main stems of standing trees. In Scotland it commonly breeds in larch logs, but has been recorded attacking live trees for breeding in a few instances (Crooke and Bevan, 1957; Crooke and Kirkland, 1960).

During an investigation of dieback in older stands (25–120 years) of European larch, *I. cembrae* galleries were found in dead and dying bark on main stems in the crowns of affected trees. However, the intensity of attack and the extent of gallery formation seemed insufficient to have caused the amount of damage observed by purely physical injury. It was also notable that the sapwood was stained blue and brown, suggesting that fungi might also be involved. In the event, isolations made from the stained sapwood did indeed yield a number of fungi, including a newly described species of *Ceratocystis* (F4.24), *C. laricicola* (Redfern *et al.*, 1987) and *Amylostereum chailletii* (F5.8).

In order to assess their significance, several of these fungi (see Table I) were inoculated into trees of similar size (15–20 m) to those affected by dieback, using methods described by Redfern *et al.* (1987). Trees were inoculated through four wounds, 3.5 cm diameter, made in the bark at 90° intervals round the stem at a point just above the lowest whorl of live branches, where examination of affected trees had shown that attack by *I. cembrae* was concentrated. The inoculum consisted of cultures grown on sterile rice grains; controls were inoculated with rice grains alone.

TABLE I. Effect of *Ceratocystis laricicola, Amylostereum chailletii* and other fungi on the bark and sapwood of 37-year-old *Larix decidua*.

	Mean extent of bark death (cm)	Mean % sapwood affected by drying and stain
C. laricicola	20.4	49.0
Ceratocystis sp.	1.4	0
A. chailletii	3.7	26.2
Unknown basidiomycete	0.4	0
Control	0	0

Only *C. laricicola* and *A. chailletii* killed significant areas of phloem and cambium, causing the formation of vertically orientated, lenticular lesions. The sapwood beneath the lesions was stained either blue by *C. laricicola*, or brown by *A. chailletii*. This stained wood extended a variable distance beyond the extremities of the lesions in a spindle shape. This in turn was enveloped in a narrow but similarly elongated zone of dry wood which extended longitudinally well beyond the area of stain. The extent of injury caused by the inoculated fungi after approximately 18 months is shown in Table I.

Sapwood affected by blue stain becomes non-conducting and this, rather than death of phloem and cambium, seems to be the primary cause of mortality in conifers attacked by bark beetles and pathogenic blue-stain fungi (Caird, 1935; Mathre, 1964b; Basham, 1970; Horntvedt *et al.*, 1983). Both *A. chailletii* and *C. laricicola* inoculations caused quite extensive vascular disruption (Table I), but as a proportion of the sapwood remained undamaged between each inoculation point, none of the trees showed crown symptoms.

III. ASSOCIATION OF *CERATOCYSTIS LARICICOLA* AND *AMYLOSTEREUM CHAILLETII* WITH *IPS CEMBRAE* AND *UROCERUS GIGAS*

In the naturally occurring cases of dieback examined, *Ceratocystis laricicola* (F4.24) was isolated so frequently from blue-stained wood associated with early stages of *Ips cembrae* (H25.63) gallery formation that it suggested the beetles might be acting as a vector of the fungus. To examine this possibility over 300 beetles were collected during six different sampling periods and plated on various agar media. *C. laricicola* was isolated from a maximum of 27% of the beetles in three of the six samples. Moreover, freshly caught beetles were used to inoculate 9-year-old trees through wounds made by removing 1 cm diameter circles of bark. Living beetles were crushed into the wounds which were resealed with the bark disc and foil. Two out of 12 such inoculations caused bark lesions and blue-staining of the sapwood from which *C. laricicola* was reisolated.

By contrast with this apparently "new" association, the relationship between woodwasps (H33.1) and species of *Amylostereum* (F5.8) has been known for many years. Oidia of the fungus are carried in specialized mycangia and introduced into wood during oviposition by the insects. *A. chailletii* is carried by a number of siricid species (Stillwell, 1966) including *Urocerus gigas* (Francke-Grosmann, 1967), and in the work described here it was isolated from the mycangia of *U. gigas* emerging from larch logs.

Woodwasp species vary in their preference for breeding material: host species and the condition of the material are both important (Spradbery and Kirk, 1981). Generally they are not regarded as forest pests (Chrystal, 1928; Hanson, 1939) but some species attack trees weakened by insect defoliation or other forms of stress (Stillwell, 1960; Neumann and Minko, 1981). Thus, in Canada, *A. chailletii* has been recorded as a cause of stain and decay in debilitated trees following attack by various siricids (Stillwell, 1960, 1966), but there have been no reports that it causes dieback.

In Britain there are no reported examples of significant attacks by *U. gigas* on weakened trees, but it is apparently able to oviposit in the exposed wood of wounds created at the bases of otherwise healthy trees during harvesting operations (Hanson, 1939). This allows *A. chailletii* to be introduced into healthy trees of several coniferous species where it then causes a sapwood decay (Redfern, unpublished).

In the work reported here the association between *U. gigas* and *A. chailletii* in the damaged trees appeared to be less consistent than that between *I. cembrae* and *C. laricicola*. Thus, whereas *A. chailletii* was isolated from affected trees in four dieback outbreaks, woodwasp larvae were only observed in the same trees in two cases. However, the most likely explanation for this discrepancy probably lies in the difficulty of detecting *U. gigas* during the early months of its three-year life-cycle (Hanson, 1939).

IV. DISCUSSION

The relative importance of the two associations as causes of dieback is difficult to assess. As the two insects appear to be the sole or principal means by which the fungi associated with them are introduced into trees, the development of damage will be influenced as much by their behaviour as by the ability of the fungi to kill living tissue. Three factors in particular may be important: (1) the timing of attacks by the two insects relative to

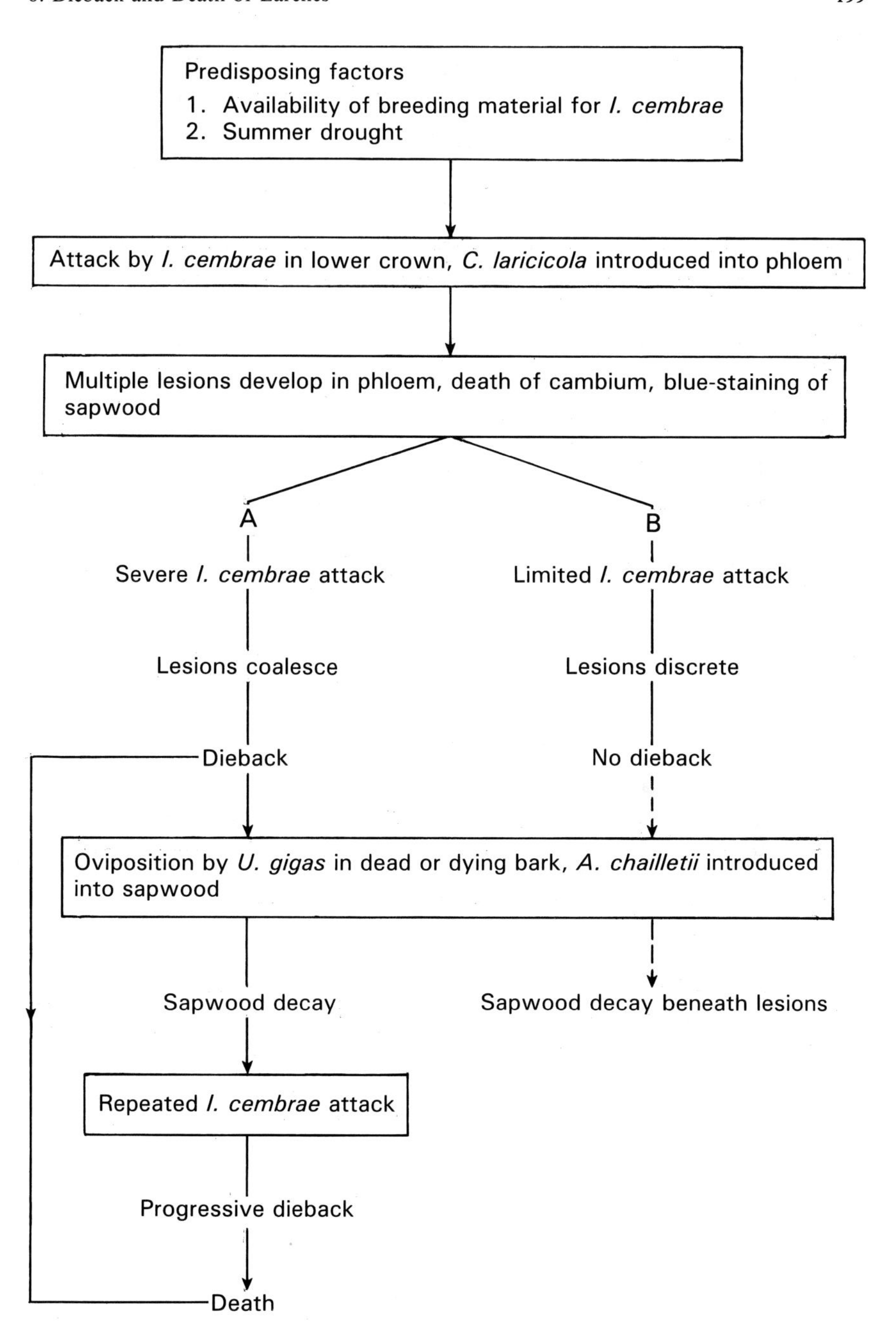

Fig. 1. The development of damage in larch caused by *Ips cembrae, Urocerus gigas* and their associated fungi *Ceratocystis laricicola* and *Amylostereum chailletii*.

each other, (2) the tissues into which the fungi are introduced, and (3) the method of introduction.

Figure 1 outlines the way in which damage may develop and indicates the possible roles of the two associations. This diagram summarizes observations and experiments described more fully elsewhere (Redfern *et al.*, 1987), but is also partly speculative (broken lines in figures).

The relative timing of attacks on live trees by the two insects has not been established by direct observation but examination of recently affected trees, in which lesions were always associated with newly excavated *Ips cembrae* mother galleries, suggests that attack by *I. cembrae* precedes that by *Urocerus gigas*. This is also supported by the dynamics of both insect populations. Almost all cases of dieback were associated with recent thinning or felling operations, which provided breeding material in the form of logs and large branches. Whilst this material would have been suitable for both insects, because of the extended siricid life-cycle the beetles would probably have responded more rapidly to this stimulus than the woodwasps. In the past, the most damaging siricid attacks on live trees (Stillwell, 1960) have occurred during a spruce budworm (*Choristoneura fumiferana*) outbreak which probably favoured a generally high woodwasp population because of progressive tree mortality.

In addition to the availability of breeding material at almost all sites where dieback occurred, there was also good circumstantial evidence that the standing trees had suffered drought stress during the growing season. As both high beetle populations and drought were associated with serious outbreaks of *I. cembrae* in continental Europe in the early part of this century (Nechleba, 1923; Schimitschek, 1930–31), it seems likely that both these factors are prerequisites for damaging attacks.

The evidence strongly suggests that *Ceratocystis laricicola* is introduced into larch during breeding attacks by *I. cembrae* and that it kills phloem and cambium around mother galleries causing the formation of narrow, vertical lesions (Figs 2 and 3). The sapwood beneath these lesions is occupied by the fungus and thus ceases to conduct water. Vascular disruption is further increased by drying of the sapwood beyond the wood occupied by *C. laricicola*. Tangential spread is restricted but numerous separate introductions, which might occur in the same season or over a period of time, could cause complete girdling of the bark and sapwood, leading to dieback. Horntvedt *et al.* (1983) induced dieback in Norway spruce with *Ceratocystis polonica* by a multiple inoculation technique which simulated a mass attack by bark beetles. A similar process could be expected for *C. laricicola* in larch.

The likelihood that attack will result in dieback is further increased by the way in which initial *I. cembrae* attacks on live trees are concentrated over a

Fig. 2. Lesion caused by *Ips cembrae* and *Ceratocystis laricicola* in the phloem and cambium of 51-year-old *Larix decidua*. The outer bark has been removed over a rectangular area to expose the white phloem. The upper end of a second lesion can be seen in the lower right-hand corner.

Fig. 3. Transverse section of 51-year-old *Larix decidua* through two 3-year-old lesions caused by *Ips cembrae* and *Ceratocystis laricicola*. The dead bark covering one of them has been removed to emphasize the formation of new wood at the margins. The sapwood beneath this lesion is blue-stained. The other three dark features on the same annual ring as the two lesions are other lesions above or below the plane of the section.

relatively short length of stem in the lower crown, thus maximizing the chance that stems will be girdled. Schimitschek (1930–31) observed similar behaviour in newly felled trees on which the crowns had been left intact. In those cases in which the girdled area extends below the lowest whorl of live branches the tree can be expected to die without passing through the stage of progressive dieback proposed in Fig. 1A.

In trees affected by dieback, a transition zone occurs between the live and dead portions of the main stem in which healthy bark interdigitates with strips of dead bark. This zone may extend for a distance of several metres. In subsequent seasons repeated attacks by *I. cembrae* may extend the area of dead bark, leading to progressive dieback and death.

The most likely means of entry for *Amylostereum chailletii* is during oviposition by *U. gigas* into bark killed by *C. laricicola*, and as noted above there is some evidence to support this for trees affected by dieback. In the absence of dieback, discrete lesions (Figs 1B, 2 and 3) or larger areas of dead bark formed by lesions coalescing may also be attractive as oviposition sites, but no examples have been found so far.

It appears that *U. gigas* is utilizing bark killed by a fungal pathogen as an oviposition site. Together with wood exposed at wounds (Hanson, 1939; Redfern, unpublished) and bark damaged by fire (Spradbery and Kirk, 1981), these sites represent a niche for the woodwasp which is independent of the general health of the tree.

Although *A. chailletii* can kill bark, it is unlikely to do so under the circumstances in which it is introduced into trees under natural conditions; that is through dead or dying bark directly into the sapwood. Thus its role in trees previously damaged by *I. cembrae* and *C. laricicola* is probably limited to that of a sapwood decay organism.

Since *I. cembrae* is an introduced beetle it is of particular interest to consider the nature of its relationship with *C. laricicola*. If the association developed in Britain the fungus is probably merely a casual component of the surface microflora which is transmitted only accidentally. On the other hand, if it evolved in continental Europe it might be more intimate—perhaps involving specialized mycangia on the insect. However, there are no previous records of an association between *I. cembrae* and *Ceratocystis* species and, as far as can be ascertained, no reference is made to blue staining of sapwood in attacked trees.

Whatever the status of the association, it may have ecological benefit for a beetle such as *I. cembrae* which has a limited ability to attack living trees. Repeated attacks by small numbers of beetles on trees which are becoming increasingly debilitated through the annual death of fresh areas of bark may contribute to the maintenance of a population in the absence of events which would permit mass attacks on stems.

Acknowledgements

I should like to thank my colleagues, Dr. S. C. Gregory and Mr J. E. Pratt, for helpful discussion of the manuscript and for taking the photographs in Figs 2 and 3.

REFERENCES

Basham, H. G. (1970). Wilt of Loblolly pine inoculated with blue-stain fungi of the genus *Ceratocystis*. *Phytopathology* **60**, 750–754.

Bramble, W. C., and Holst, E. C. (1940). Fungi associated with *Dendroctonus frontalis* in killing Shortleaf pines and their effect on conduction. *Phytopathology* **30**, 881–899.

Caird, R. W. (1935). Physiology of pines infested with bark beetles. *Bot. Gaz. (Chicago)* **96**, 709–733.

Crooke, M., and Bevan, D. (1957). Note on the first British occurrence of *Ips cembrae* Heer (Col. Scolytidae). *Forestry* **30**, 21–28.

Crooke, M., and Kirkland, R. C. (1960). Resurvey of distribution of the bark beetle *Ips cembrae. Rep. For. Res. Lond.* 167–169.

Chrystal, R. N. (1928). The Sirex wood-wasps and their importance in forestry. *Bull. Entomol. Res.* **19**, 219–247.

Francke-Grosmann, H. (1967). Ectosymbiosis in wood-inhabiting insects. *In* "Symbiosis, Vol. 2—Associations of Invertebrates, Birds, Ruminants and Other Biota" (S. M. Henry, ed.), pp. 141–205. Academic Press, London and New York.

Hanson, H. S. (1939). Ecological notes on the *Sirex* wood wasps and their parasites. *Bull. Entomol. Res.* **30**, 27–65.

Horntvedt, R., Christiansen, E., Solheim, H., and Wang, S. (1983). Artificial inoculation with *Ips typographus*-associated blue-stain fungi can kill healthy Norway spruce trees. *Medd. Nor. Inst. Skogforsk.* **38**, 1–20.

Mathre, D. E. (1964a). Survey of *Ceratocystis* spp. associated with bark beetles in California. *Contrib. Boyce Thompson Inst.* **22**, 353–362.

Mathre, D. E. (1964b). Pathogenicity of *Ceratocystis ips* and *Ceratocystis minor* to *Pinus ponderosa. Contrib. Boyce Thompson Inst.* **22**, 363–388.

Molnar, A. C. (1965). Pathogenic fungi associated with a bark beetle on Alpine fir. *Can. J. Bot.* **43**, 563–570.

Nechleba (1923). *Ips cembrae* als Bestandesverderber. *Z. Angew. Entomol.* **9**, 365–368.

Neumann, F. G., and Minko, G. (1981). The sirex wood wasp in Australian radiata pine plantations. *Aust. For.* **44**, 46–63.

Redfern, D. B., Stoakley, J. T., Steele, H., and Minter, D. W. (1987). Dieback and death of larches caused by *Ceratocystis laricicola sp. nov.* following attack by *Ips cembrae. Plant Pathol.* **36**, 467–480.

Schimitschek, E. (1930–31). Der achtzahnige Larchenborkenkafer *Ips cembrae* Heer. Zur Kenntris seiner Biologie und Okologie sowie seines Labensvereines. *Z. Angew. Entomol.* **17**, 255–344.

Spradbery, J. P., and Kirk, A. A. (1981). Experimental studies on the responses of European siricid woodwasps to host trees. *Ann. Appl. Biol.* **98**, 179–185.

Stillwell, M. A. (1960). Decay associated with Woodwasps in Balsam fir weakened by insect attack. *For. Sci.* **6**, 225–231.

Stillwell, M. A. (1966). Woodwasps (Siricidae) in conifers and the associated fungus, *Stereum chailletii*, in Eastern Canada. *For. Sci.* **12**, 121–128.

Whitney, H. S. (1982). Relationships between bark beetles and symbiotic organisms. *In* "Bark beetles of North American Conifers" (J. M. Mitton and K. B. Sturgeon, eds), pp. 183–211. Univ. Texas Press, Austin.

9

Mycopathogens of Insects of Epigeal and Aerial Habitats

H. C. EVANS

I. INTRODUCTION

The subject to be covered in this chapter is precisely defined in the title, but nature rarely categorizes with such precision. Many larval stages of aerial-inhabiting insects feed or pupate within the soil and are subject to fungal attack both inside and outside the soil environment. Thus, the borderline between a soil insect and an aerial or epigeal insect is often ill-defined. However, an attempt will be made to confine examples to non-anomalous situations to avoid any duplication or overlap with other authors.

A group of fungi which are not considered to be mycopathogens *sensu stricto* is included within this review. These are the Septobasidiales (F5.3), an order of essentially mutualistic Basidiomycotina—although pathogenic (*sensu lato*) representatives are known—which have developed highly complex interactions with their scale insect (Diaspididae (H20)) hosts. In evolutionary terms, they are much too important to be omitted from the proceedings of a symposium devoted exclusively to insect–fungus interactions. Since they are not covered by the chapters dealing with mutualism, their inclusion here may appear to be more by default than by design. Nevertheless, if the evolutionary end point of an entomopathogenic association is mutualism, there being a progressive advancement towards pure parasitism (Humber, 1984), then the Septobasidiales fit logically within the concept of the chapter. Apart from the latter fungi, the interactions between most mycopathogens and their living insect hosts are transient, death rapidly following fungal invasion of the haemocoel. During their brief post-infection, pre-necrotic phase, the parasitized insects often demonstrate behavioural changes which can either be construed as "altruistic", particularly in social insects, or, conversely, as entirely beneficial to the pathogen. The behavioural responses of the host to infection, and the corresponding adaptations in fungal morphology in order to overcome or exploit such behavioural patterns, will be the main thrust of this review. Interactions at the cellular level are outside the present brief and these have been both thoroughly and speculatively reviewed recently (Charnley, 1984).

Examples will be drawn from each natural subdivision of the fungal kingdom, wherever relevant, but the subdivisions will be treated autonomously since the evolution of the entomopathogenic habit seems to have arisen *de novo* within each group. A final section, specifically devoted to ant–fungus interactions, is included as this has been the field of interest of the author for a number of years.

II. GENERAL INTERACTIONS: THE FUNGI INVOLVED

A. Mastigomycotina (F2)

The mastigomycete fungi pathogenic to insects of epigeal and aerial habitats are poorly represented, almost certainly because such habitats are difficult to exploit for organisms heavily dependent upon free water for dispersal. An interesting exception is the chytrid, *Myiophagus ucrainicus* (Wize) Sparrow, a parasite of armoured scale insects (Diaspididae (H20)). This insect–fungus association, however, has been little studied (Karling, 1948) and basic data on pathogenesis are wanting. *M. ucrainicus* has been recorded mostly from diaspidid hosts colonizing citrus trees in subtropical regions. Infection probably takes place during or shortly after heavy rainfall, the motile zoospores being released from resting sporangia within dead scales and swimming in the water film which forms over "drip leaves" in such conditions. As in similar aquatic fungi, the zoospores must home-in on the potential host, guided by specific chemoreceptors, and may slip between the scale and leaf to infect via the more vulnerable ventral surface of the insect. There would seem to be no behavioural defence strategies that these essentially stationary hosts can offer to avoid or limit infection within the immediate population.

B. Zygomycotina (F3)

1. Mucorales

The representatives of the subdivision Zygomycotina which attack insects of epigeal and aerial habitats occur predominantly in the Entomophthorales (F3.1). In fact, there is only one substantiated example outside of this order, namely *Sporodiniella umbellata* (F3.1) Boedijn (Fig. 1A), a mucoraceous fungus which has been reported mainly on canopy-inhabiting membracids in tropical tree crops (Evans and Samson, 1977). Infected insects die freely exposed, typically with their stylets embedded in the host plant tissues, usually the midribs on the abaxial surface of the leaf. The insect is further fixed to the substrate by hyphal strands emerging from the body orifices and between the plates of the integument. As will be shown, this is an extremely common phenomenon amongst mycopathogens of above-ground insects, serving to anchor the host in an exposed aerial position, thereby greatly facilitating inoculum dispersal over both time and space. If the dying insect were to fall into the leaf litter or soil, not only would the efficiency of spore dissemination be significantly decreased, but the chances of secondary predation and scavenging would also be markedly increased. Thus, in this particular insect–fungus interaction, the behaviour of the insect would appear to be of direct benefit to the fungus with no obvious avoidance mechanism being exhibited by the membracid host. The strikingly formed umbellate network of sporangiophores, so characteristic of the fungus, are delicately interlaced and may be triggered to release spores mechanically as insects pass.

2. Entomophthorales (F3.1)

A similar, if not more exaggerated response, is shown by many insect hosts infected with species of Entomophthorales (Figs 1B and 2A–D), the prominent death positions adopted by the diseased insects earning the description of "summit disease". Perhaps the best known example is that of *Entomophaga grylli* (Fres.) Batko (Fig. 2B), a pathogen of numerous grasshopper and locust species (Brady, 1979), which has "... reduced large destructive outbreaks to negligible proportions in various countries throughout the

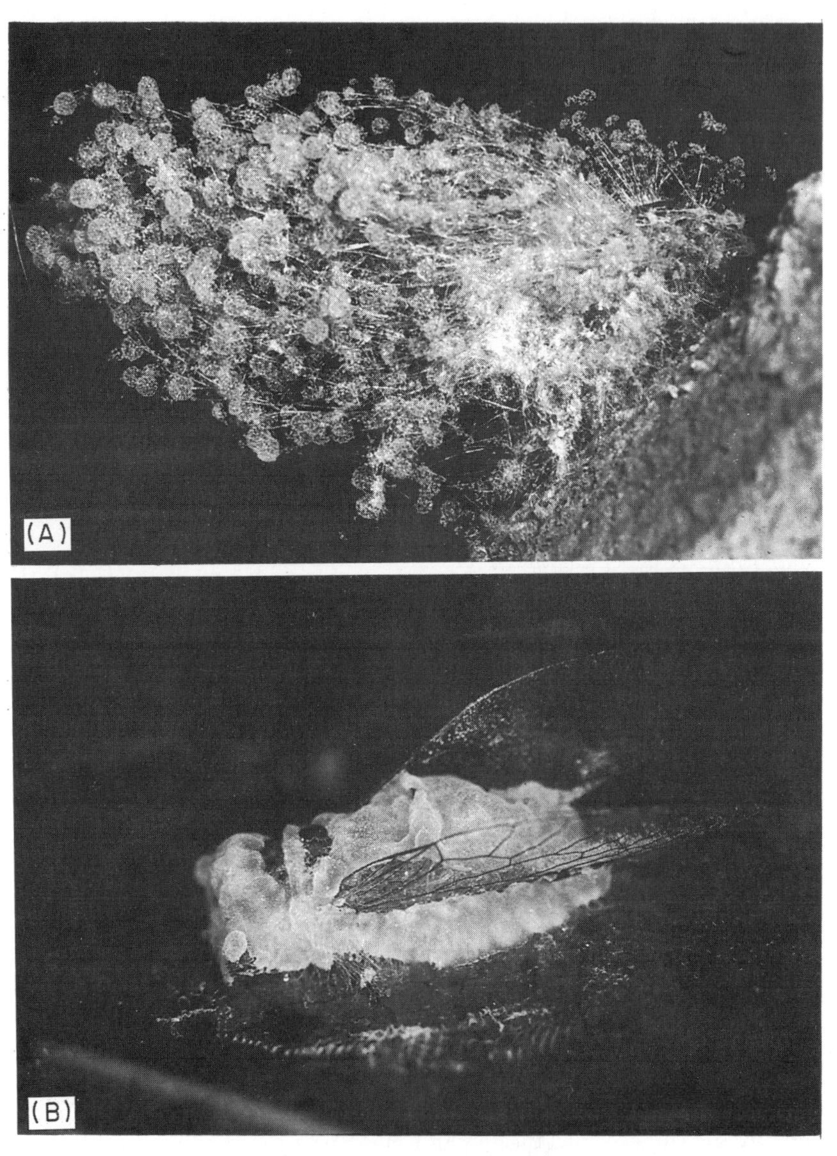

Fig. 1. Zygomycotina on cicada (H20) hosts. (A) *Sporodiniella umbellata* (Mucorales (F3.1)) on adult cicada attached to tree stem, Papua New Guinea. (B) *Entomophthora* sp. (Entomophthorales (F3.1)) on adult cicada attached to forest shrub leaf, Kenya.

world" (MacLeod, 1963). Indeed, because of its potential as a biological control agent of many important agricultural pests, *E. grylli* was one of the first mycopathogens to be evaluated in the field. The history of these pioneering biocontrol attempts, which ended somewhat ignominiously, has been well documented (Petch, 1925). Our knowledge of the life-cycle, epizootiology and pathogenesis of the fungus has increased considerably in recent years as interest in *E. grylli* as a biocontrol factor has been renewed (Nelson *et al.*, 1982; Krueger and Ramoska, 1985; Carruthers *et al.*, 1986). This is, therefore, a useful model of an insect–fungus interaction on which to build and integrate the other host–Entomophthorales examples.

Classification of the Entomophthorales is in a process of dynamic revision. Hence there is still considerable confusion in the literature, particularly concerning generic terminology. I have attempted to follow current nomenclature, where possible, but opinions are still strongly divided and the names used here may be subject to further emendations. For recent taxonomic viewpoints consult: Ben-Ze'ev and Kenneth (1982a, b), Humber (1981a, b, 1982), Humber and Ben-Ze'ev (1981), Remaudière and Keller (1980), Samson *et al.* (1988). In order to avoid adding to the confusion, only *Entomophaga* will be abbreviated in text.

a. Entomophaga (F3.1). MacLeod (1963) describes the "summit disease" syndrome in *Melanoplus bivittatus* in Canada, the infected grasshoppers congregating at the tops of plants, clasping the stems and each other in a death lock. This behaviour is similar to that reported much earlier for red locusts in South Africa (Skaife, 1925) and has been consistently noted not only for *E. grylli*, but also for most insect–entomophthoralean interactions. Rigor mortis ensures that the insects remain in this position until they are removed by weathering. A remarkable synchrony in the time of death has been observed in disease outbreaks amongst locust populations in South Africa, the majority of infected insects dying in the later afternoon (Skaife, 1925; Schaeffer, 1936). This behaviour optimizes both the formation of spores and their subsequent dispersal, which are normally favoured by increasing night-time humidity and dew formation. All these activities of diseased insects are abnormal; exposure, for example, will increase the chances of predation, particularly by birds. An hypothesis often voiced to explain this behaviour suggests that the host is being specifically programmed by the fungus, which either interferes with the nervous system directly or releases substances which induce behavioural changes indirectly. Other suggestions with a fungal-mediated element have been made, such as the insects becoming positively phototropic or seeking supplementary oxygen in the moving air layers above the ground vegetation as the spiracles become blocked by fungal hyphae. The true explanation, however, is almost

(A)
(B)
(C)
(D)

certainly much more complex than that propounded in these relatively simple, unifactorial theories. An attempt will be made to analyse host behaviour in more depth in the final section of this chapter.

In a recent study of *E. grylli* in populations of the grasshopper, *Camnula pellucida*, in North American rangeland, it was concluded that epizootics "... seem to be regulated through complex interactions between environmental conditions and the spatial patterns and dispersal of both host and pathogen populations" (Carruthers *et al.*, 1986). The movement of infected grasshoppers to, and their death in, elevated positions strongly favours dispersal of the fungus, especially horizontal transmission, by forcibly discharged primary spores. Some of the insects, probably older individuals, produce only resting spores within the body cavity, and such specimens are readily distinguishable since they rapidly disintegrate, scattering the spores in the process (MacLeod, 1963). Further indications as to the complexity of these Acrididae–Entomophthorales interactions have been revealed during an investigation of melanoploid grasshoppers attacked by a different pathotype of *E. grylli* which is incapable of producing primary spores on the host (Humber and Ramoska, 1986). Horizontal transmission is effected through an atypical spore state. Such spores ("cryptoconidia") are formed from hyphal bodies, but only when the latter are released from newly dead insects, the abdomens of which soften, elongate and then rupture in a streaming process, contaminating both plant and soil surfaces. Cryptoconidia also form in the more common pathotype, but only when the host suffers mechanical damage. Presumably, this is a form of insurance, ensuring that fungal dispersal would still be possible if the insect host died prematurely, either through predation or cannibalism.

Hutchinson (1962) investigated *Entomophaga*-induced epizootics amongst calyptrate flies (Calliphoridae (H28.42), Muscidae (H28.40), Sarcophagidae, Tachinidae (H28.41)) in the USA, and observed that: "... many diseased flies were flying or walking with abdominal parts missing". Gelatinous substances, containing fungal hyphal bodies, exuded from the insects following their "abdominal dismemberment" and glued the flies to the substratum, usually soil or mud in river beds. Further movements of the flies resulted in additional loss of bodyparts and the excretion

---

Fig. 2. Zygomycotina on various hosts. (A) *Erynia delphacis* (F3.1) on rice leafhopper (*Cofana spectra* (Cicadellidae (H20)), Sulawesi). The host has died towards the apex of the rice leaf with its stylet embedded in the leaf. (B) *Entomophaga grylli* (F3.1) on locustid (H15.2) host, showing classic "summit disease" position at top of grass inflorescence, Kenya. (C) *Entomophthora muscae* on onion fly (*Delia* sp., Anthomyiidae (H28.39)), dying away from host crop, freely exposed on asparagus plant, UK. (D) *Entomophthora muscae* on housefly (H28.40) attached to window pane, surrounded by white halo of discharged ballistospores, UK.

of viscid internal contents. Healthy flies were apparently attracted to diseased insects, possibly by the odour of decaying tissues. Laboratory observations showed that flies developed symptoms about four days after exposure to fungal inoculum, becoming sluggish and falling to the ground. Swelling of the abdomen followed rapidly and internal contents often leaked from between the sclerites, accompanied in many instances by jerky, uncoordinated leg movements. Sporulation occurred from the hyphal bodies contained within the host or in the gelatinous exudates. The process of host colonization is somewhat different from that described for other *Entomophaga* spp. The fungus preferentially invades the abdomen and progresses only slowly into the rest of the body. The glue-like substances seem to have a multifunctional role: anchoring the host, disseminating propagules and attracting healthy flies. Rhizoids would fulfill no function in this disease syndrome.

b. Erynia (F3.1). A system-modelling approach, similar to that being developed for *E. grylli* (F3.1) (Carruthers *et al.*, 1986), has also been adopted to study insect–fungus interactions in the case of entomophthoralean pathogens of eastern spruce budworm (*Choristoneura fumiferana* (H32)) in Canadian forest ecosystems (Perry and Whitfield, 1984). A synchronous development of the insect host and the fungi involved (*Erynia radicans* (Bref.) Humber, Ben-Ze'ev and Soper = *Entomophthora sphaerosperma* Fres., and *Entomophaga aulicae* (Reichardt in Bail) Humber = *Entomophthora egressa* MacLeod and Tyrrell) was demonstrated with temperature as a critical factor, determining not only quantitative interactions of host and pathogen, but also qualitative responses, particularly of the fungi. This investigation followed an earlier assessment that had been made of the same pathogen complex on the same host in north-eastern USA (Vandenberg and Soper, 1978). Significantly greater disease levels were observed in the lower crown of spruce trees compared with other parts of the plant. This was correlated with a combination of factors: higher spore concentrations, the spores tending to drop into this layer from the upper crown; a more favourable (buffered) microclimate; and insect behaviour. Late instar larvae usually abandon protected feeding sites on the upper shoots and spin down to the lower foliage, where they are more exposed to fungal inoculum. Such larvae later move down into the soil to pupate and infected pupae in the soil provide the means of long-term survival. Insects killed by *Erynia radicans*, as in many *Erynia*–insect interactions (see Fig. 2A), are securely attached to the aerial substrata (stems, branches, leaves) by holdfast-type structures, the rhizoids. Sawyer (1931) followed the development of *Erynia radicans* in lepidopteran larvae (*Rhopobota*, Olethreutidae (H32)) and found that rhizoids formed at approximately the same time as the host died, hyphae pushing through the ventral segments of the thoracic region and fanning out onto the substratum. He noted that this was the only morphological stage of the fungus which failed to develop *in vitro* and concluded that rhizoid formation marked the end of the vegetative growth phase. The author also observed that the larvae, although infected, "... show a remarkable indifference to the progress of the disease and death does not occur until most of the body tissues are in a marked state of disintegration". This is a consistent theme in Entomophthorales–insect interactions which, as will be shown, distinguishes them from other entomopathogens.

However, not all *Erynia* spp. possess rhizoids, and their presence or absence can be used as a specific or generic character. As suggested by Sawyer's work, the selection pressure to produce rhizoidal structures has probably been determined by the biology of the host involved. A case in point, once again in a North American forest, is that of *Erynia crustosa* (MacLeod and Tyrrell) Humber and Ben-Ze'ev and the forest tent caterpillar, *Malacosoma disstria* (Lasiocampidae (H32)). The infected larvae are found firmly anchored to tree branches by their prolegs, which stiffen just prior to or at death. No rhizoids are produced, presumably because the death position adopted by this host is sufficient to fix it in place, although it has been observed that undefined mycelia do grow out of the prolegs onto the substratum (MacLeod and Tyrrell, 1979). The position at death is characteristic; the anterior part of the larva droops backwards from the twig or branch. Dissolution of the internal tissues by the fungus precedes death and may cause visible flaccidity and subsequent collapse of the host. It may be expected that progress toward this condition in the dying host would result in loss of coordination so that it would simply drop from the tree: a behavioural response which would, of course, adversely affect the horizontal dispersal ability of the fungus through ballistospores but which would favour its survival in the form of resting spores. The latter develop later in the season in larvae which still remain attached but which shrivel up, become brittle and disintegrate to release the spores.

A similar life-cycle is seen in *Erynia bullata* Thaxter and MacLeod, a pathogen of blow-flies (Calliphoridae (H28.42)) in North America, with the resting spores falling into the soil from mummified and crumbling cadavers (MacLeod *et al.*, 1973). According to these authors, the dead flies were conspicuous in hardwood forests, attached to leaf tips and branches of understorey shrubs by well-developed rhizoids. Resting spores and primary spores never occurred together on the same insect. This phenomenon, as will be obvious from the examples still to be discussed, is common to many insect–Entomophthorales interactions. A plausible explanation was put forward by Wilding (1972) and corroborated by the findings from the blow-fly study. Physiological age of the host was thought to determine the type of spore produced by the fungus; resting spores were formed predominantly in the older flies, especially females which have a significantly longer life-span. Kramer (1979) has further endorsed this hypothesis as the results of laboratory experiments with *Erynia bullata* show. Dead blow-flies (*Phormia* sp.), examined 5–10 days after inoculation, were found to contain either primary spores (79%) or resting spores (18%) with only a 3% mixture. Of the young flies (less than 3 days old), 80% had primary spores, whilst resting spore formation was largely confined to older insects.

An intriguing disease syndrome has been observed in sugarbeet aphids (*Pemphigus betae* (H20)) attacked by *Erynia neoaphidis* (Harper, 1958). The aphids, which normally feed below ground on the lower stem and root system, emerge from their soil niche when infected and move to the upper vegetation, eventually dying on the foliage. Production of ballistospores from these aerially exposed cadavers would ensure horizontal transmission of the fungus to other aphid populations. This behavioural response will be discussed later in relation to insect–fungus interactions as a whole.

c. *Entomophthora* (F3.1). In the genus *Entomophthora sensu stricto*, there is a progressive advancement from sporulation shortly after the death of the host to sporulation on the still living insect. The former group of species all demonstrate rhizoid formation, whilst the latter lack these structures—obviously a non-essential feature in an insect–fungus interaction where the host becomes a mobile spore-disseminating unit. Amongst these fungi, *Entomophthora erupta* (Dustan) Hall has received most attention. Dustan (1924) described this species from the green apple bug (*Lygus communis* var *novascotiensis*, Miridae (H20.3)) in the USA. He noted that fungal colonization is confined to the abdomen and, although the internal contents and also the posterior leg and wing muscles are utilized by the fungus, the insect continues to survive and move about with relative ease. Indeed, Dustan reported that the insects exhibit rapid movement, even when the ballistospores are being discharged from the ruptured dorsum, such movement being possible because the muscles of the pre- and mesothorax remain uncolonized. Moreover, healthy nymphs and adults cluster around diseased mirids and apparently attempt to feed by inserting their stylets into the sporulating mass on the dorsal surface. However, those insects infected later in the year towards the end of the infection cycle behave differently: the swollen abdomen fails to rupture and the host becomes progressively more lethargic, eventually crawling beneath tree bark or into crevices before dying. The abdomens of these hosts contain resting spores which are released the following spring as the cadavers disintegrate (MacLeod *et al.*, 1976). Hall (1959) investigated the same fungus on the black grass mirid, *Irbisia solani* (H20.3), whilst Wheeler (1972) also recorded it on the alfalfa plant bug, *Adelphocoris lineolatus* (H20), in the USA, and both these workers confirmed the ability of this fungus to sporulate on the living host. Subsequently, Keller (1981) reported *Entomophthora erupta* from a mirid host (*Notostira elongata* (H20.3)) in Switzerland, but the pathogen was found to invade the entire host and to sporulate only after its death, the mirid dying on grass stems, head downwards, attached by its legs and stylet. It was concluded that these pathobiological differences resulted from different host–pathogen interactions. However, further taxonomic studies revealed that the species attacking *N. elongata* in Switzerland was a different, previously undescribed, species—*Entomophthora helvetica* Keller Ben-Ze'ev—and that pathobiological characteristics could be employed as criteria by which closely-related species could be separated (Ben-Ze'ev *et al.*, 1985). Several other *Entomophthora* taxa also exhibit restricted host colonization. Samson *et al.* (1979) described a species from Europe, *Entomophthora thripidum* Samson, Ramakers and Oswald (on *Thrips tabaci* (H21.3), Thysanoptera), which sporulates on the abdominal dorsum for as long as the host remains alive. Prior to sporulation, however, the infected thrips often move to an elevated position on their host plants, increasing spore dispersal efficiency.

Disease development in various hosts affected by the ubiquitous type species of the genus, *Entomophthora muscae* (Cohn) Fres. (Figs 2C, D), has been well documented. However, it is only recently that rhizoids have been positively identified in this species (Balazy, 1984). Berisford and Tsao (1974) closely monitored the symptoms in seedcorn maggot flies (*Delia platura*, Anthomyiidae (H28.39)), which were found to aggregate at the tops of plants and even on surrounding fences when infected. Apparently healthy active flies were found to contain fungal mycelium in the labellar lobes; within two hours such flies became lethargic and either spread their legs around, or grasped onto, the plant substrata. Sticky droplets were seen around the labellar lobes and several hours later the flies were dead and firmly attached by their probosces. Sporulation from the now swollen abdomen followed within four hours of host death. Attachment of the insect was explained entirely by the presence of glue-like substances around the labellar lobes and later by mycelial growth from the proboscis. Brobyn and Wilding (1983) subsequently carried out a detailed study of the infection process in *Musca domestica* (H28.40). It was concluded that, whilst the flies often showed proboscis attachment to the substrate, no specific structures were involved and there was no fungal-induced secretion of adhesive substances, although the tenacity of the attachment process was clearly evident as removal of flies resulted in proboscis rupture and detachment. It was postulated that at death the fly regurgitated food which helped to cement the proboscis in position. Contrasting results were obtained by Balazy (1984), however, when he examined a range of flies in the Anthomyiidae (including *Delia* spp. (H28.39)), Calliphoridae (H28.42), Empididae (H28.19), Muscidae (H28.40), Scatophagidae and Syrphidae (H28.23), infected by *Entomophthora muscae* and discovered specialized attachment structures in these insect–fungus interactions. These were described as lobate holdfasts, firmly securing the insect to the substrate despite the fact that, although dead, the insect continued to struggle and move its legs. Balazy recognized the evolutionary and biological significance of the rhizoids: "As the parasitic phase increases in length, the importance of rhizoids should diminish". This trait is clearly illustrated in the genus *Entomophthora*.

Epizootiological studies have also been undertaken to evaluate disease caused by *Entomophthora muscae* in agriculturally important insects. Recently, Carruthers *et al.* (1985) investigated disease outbreaks in the onion fly (*Delia antiqua* (H28.39)) and found that the majority of infected flies moved to elevated positions amongst the crop, characteristically in the late afternoon. After dying on and becoming attached to the aerial vegetation, the insects rapidly become covered by sporulating hyphae, ensuring that the ballistospores are released during the night and early morning when environmental conditions are optimal for infection. Rhizoids were said to be absent. A small proportion (less than 10%) of the insects died on or in the soil and disease symptomatology was unusual: the abdomens blackened, becoming brittle and fragmenting into the soil. Such flies contained resting spores which thus provide the means of intergeneration and interseasonal carryover. Primary host infection occurs as the fly emerges from the soil-buried pupal case and passes to the surface. Exaggerated behavioural responses of *Entomophthora muscae*-infected carrot flies (*Psila rosae*, Psilidae) have also been reported (Eilenberg, 1986). The infected adult females die at the tops of neighbouring hedgerow plants, but before doing so they lay their eggs there instead of in the soil around the crop plants.

There is a complex of species of Entomophthorales associated with Aphididae (H20) (Hall and Dunn, 1957; Hagen and Van den Bosch, 1968; Wilding and Perry, 1980). However, relatively few behavioural patterns of infected aphids have been monitored and the present review will be confined to those interactions where significant host responses to fungal infection have been noted. During a study of insects on plum trees in the UK, several aphid species (*Phorodon humuli* and *Brachycaudus helichrysi*) were found to be attacked by three species, *Erynia neoaphidis* Remaudière and Hennebert, *Entomophthora planchoniana* Cornu and *Conidiobolus obscurus* (Hall and Dunn) Remaudière and Keller (Byford and Ward, 1968). Collections of diseased aphids attached to leaves revealed that these were producing only ballistospores, whereas those found hidden in bark crevices contained resting spores only. These authors discussed the finding in relation to epizootiology and they considered that the resting spores were in an ideal position to overwinter, release ballistospores in spring and initiate the infection cycle on newly-hatched nymphs emerging from eggs laid the previous autumn around the leaf buds of the tree. The aphids which died on the leaves would, of course, have been shed into the leaf litter, thereby effectively removing the fungus from the infection court. This has been thought to happen in those species of the Entomophthorales which do not form rhizoids (Rockwood, 1950).

d. Massospora (F3.1). Species of the genus *Massospora* are obligate pathogens of Cicadidae (H20). Some of these insect–fungus interactions are amongst the most remarkable of all the relationships covered by this review. Perhaps the most bizzare is that of *M. cicadina* Peck and the periodical cicada, *Magicicada septendecim* (H20), in the USA. Speare (1921) found that the fungus invades the abdomen, confining its activities to the posterior segments only. The internal tissues become disorganized and the intersegmental membranes are destroyed, resulting in progressive sloughing-off of the abdominal sclerites. Despite this apparently damaging dismemberment, the insects continue to behave normally, flying and crawling about, and remain alive for considerable periods. The mass of fungal sporogenous structures, which by now occupy the abdomen, are gradually exposed and primary spores are disseminated as the infected insects move about. A diseased cicada still attempts to feed, and "... mingles promiscuously with its fellows". During the latter part of the season, most infected insects contain sulphur yellow to brown resting spores, both spore types never developing within the same host. These older, diseased cicadas usually return to and die in the soil, which thus becomes contaminated with resting spores. However, the mechanisms which trigger germination of these spores, 16 years and 9 months later, as the next generation emerges, defies speculation; the very nature of the life-cycle probably ensures that they will never be adequately determined. Soper (1963) has studied the interactions of *M. levispora* Soper with the cicada, *Okanagana rimosa*. As with *M. cicadina*, only the posterior abdominal region is colonized. An inverse relationship between age and primary spore production was determined: in cicadas infected early in adult life only primary spores formed; in those attacked later, towards the end of their flight period, only resting spores were produced. However, those containing resting spores can be distinguished externally, as there is no loss of abdominal sclerites. In both cases the cicadas remain active, the infected males even calling the females and attempting to copulate with them.

e. Strongwellsea (F3.1). Anthomyiid flies of the genus *Delia* (H28.39) are subject to another highly specialized entomophthoralean infection, the causal agent of which is sufficiently distinct not only morphologically, but also in its biology, to merit generic status (Humber, 1976). *Strongwellsea castrans* Batko and Weiser, which colonizes and sporulates within the abdominal cavity of the host, has been reported from both North America and Europe (Strong *et al.*, 1960; Lamb and Foster, 1986). However, unlike the Entomophthorales discussed previously, the fungus causes minimal host damage, neither digesting the contents of the abdomen nor causing disruption of the sclerites. Fungal hyphae grow principally within the abdominal haemocoel, although some may eventually penetrate the nervous system and reach the brain "... without affecting the host's behaviour ..." (Humber, 1976). A highly organized cavity lined by sporogenous structures develops in the abdomen and ballistospores are ejected through a neat, gaping hole which appears in the abdominal pleuron. Although affected flies continue normal activities (Nair and McEwen, 1973), they tend to die earlier than healthy flies once the abdominal wall is breached. This is probably due to increased exposure to abiotic factors, although the nutrient drain on the insect by the fungus must contribute to this weakening of the host. Death of the host is rapidly followed by that of the fungus. Production of resting spores is almost exclusively confined to adult females and, as noted previously, this is probably governed by the age of the host. This genus is probably the most highly advanced of the Entomophthorales, the fungi developing almost a parasitic, rather than a pathogenic, relationship with their hosts.

C. Ascomycotina (F4) (Figs 3 and 4)

This is not intended to be a comprehensive treatment and only the most relevant genera will be discussed. However, the entomopathogenic Deuteromycotina (F6) are included within this subdivision since many of them are either proven or purported anamorphs of Ascomycotina, and to consider them separately would disrupt the logical progression of the review. Unfortunately, field and laboratory data are wanting for most insect–Ascomycotina interactions, especially compared with those involving

Fig. 3. Ascomycotina on various hosts. (A) *Cordyceps* sp. (F4.4) on coleopteran larva (H25) buried in log, Kenya. The wood has been removed to reveal the hidden, mummified host and the two fungal fructifications emerging from within. (B) *Cordyceps tuberculata* on moth (H32), Brazil, dorsal view. Numerous narrow-cylindrical fructifications produce superficial perithecia (arrowed) laterally and the anamorph (*Akanthomyces pistillariaeformis*) towards the apex. (C) Aphids (H20) killed by *Verticillium lecanii* (F6.2), UK. This fungus, in common with many homopteran entomopathogens, produces mucilaginous or slime spores which move efficiently over leaf and stem surfaces in run-off water.

(A)

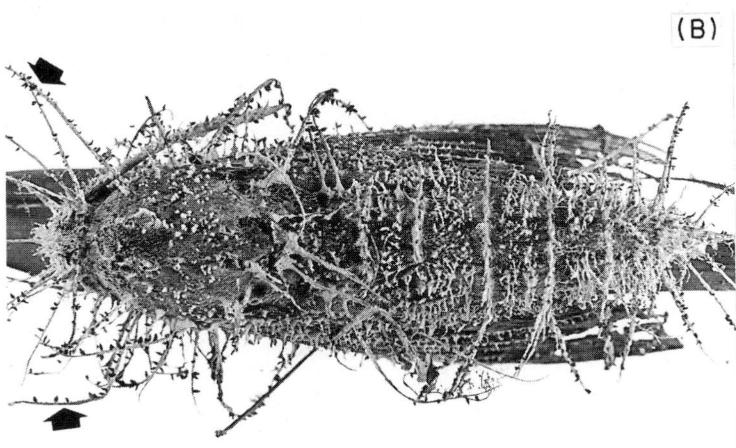
(B)

(C)

(A)

(B)

(C)

Entomophthorales, probably because they are largely confined to tropical and subtropical situations, whilst insect pathologists tend to have a predominantly temperate distribution.

1. Clavicipitales (F4.4)

a. Hypocrella. All representatives of this genus are pathogens of soft scale insects (Coccidae or Lecaniidae) and whiteflies (Aleyrodidae). Petch (1921a) described more than 40 species with their *Aschersonia* (Coelomycete (F6.1)) anamorphs. Most are strikingly coloured and common in tropical and subtropical ecosystems, particularly in humid forests (Evans, 1982). The conidia ooze from acervuli buried in a mycelial cushion which covers and effectively obliterates the host insect. These slime spores are best adapted for short-distance (vertical) dispersal by rain-splash or in run-off water on leaf surfaces. The ascospores are forcibly ejected and most probably function as long-distance dispersal units. It is possible that infected alates may further move the fungus around, as shown for some aphid–Entomophthorales interactions (Wilding and Perry, 1980). These fungi can devastate coccid or whitefly (H20) populations in the hot humid tropics, especially those colonizing rain forest trees or understorey shrubs (Petch, 1921b), which makes it difficult to envisage how these hosts overcame pathogen pressure. In an essentially stationary insect there is no scope for avoidance of infection, or at least of contact with inoculum, although humoral or cellular resistance may be important here. More selective feeding by homopterans and restriction of their plant host range, in essence a speciation pressure, may have reduced the chances of mycopathogens decimating contiguous populations of the same insect host, which in turn would have increased the necessity of the fungi to improve the efficiency of horizontal transmission. The often weird and wonderful shapes and bright colours of *Hypocrella* spp., and indeed of many of the entomopathogenic Clavicipitales (F4.4), may be a response to this pressure, perhaps in some way serving to attract "vector" insects.

b. Torrubiella. This genus shares much in common with *Hypocrella*, occurring predominantly in tropical forests and being well adapted to plant-feeding homopterans (H20). The anamorphs, however, are hyphomycetous and are usually formed superficially, together with perithecia, on

---

Fig. 4. Ascomycotina on ant hosts (H33.9). (A) *Cordyceps unilateralis* (F4.4) on *Camponotus* sp. (Formicinae), Brazil. Infected ants climb shrubs, then die grasping and biting foliage. (B) *Akanthomyces* sp. (F6.2) on *Cephalotes atratus* (Myrmicinae), Brazil. (C) *Cordyceps kniphofioides* on *Cephalotes atratus*, Brazil. These ants are usually found hidden beneath bark or epiphytes around the buttress roots or lower stems of forest trees.

the vividly coloured mycelial stroma. Typically they pertain to the genera *Hirsutella* and *Verticillium* (F6.2) with water-dispersed slime conidia.

c. Cordyceps. Because of its spectacular fruitbodies, this is perhaps the most frequently illustrated genus of entomopathogenic fungi and, therefore, the best known popularly (Figs 3A, B and 4A, C). Scientifically, "... however, the genus has been investigated little and the student must search extensively to find other than taxonomic descriptions of most species" (McEwen, 1963). *Cordyceps* is taxonomically close to *Torrubiella*, but the perithecia and conidiogenous cells are borne on or in highly organized aerial structures. Several hundred species have been described on a wide range of insect hosts. Coleoptera larvae (H25) are common hosts and most have an epigeal habit, as for example, those feeding in woody tissues or leaf litter. How these insects become infected, since they rarely come into contact with aerially-dispersed inoculum, is unclear, but *Cordyceps* species have successfully adapted to many such hidden targets. The well coordinated, positively phototropic fruitbodies ensure that the fungus is able to reach more suitable habitats for spore dissemination. Mathieson (1949) described a *Cordyceps* outbreak on *Aphodius* grubs (Scarabaeidae (H25.14)) in Australia. Scarabaeid larvae are, in fact, frequently subject to *Cordyceps* infections and, although theoretically these insects inhabit the soil, the above-ground fungal fruitbodies can be said to be truly epigeal. He studied the development of infection in field-collected larvae and found that the incubation period was long, somewhere between 2–8 weeks. Diseased larvae can be distinguished from healthy ones as the body becomes darker and the cuticle opaque and wrinkled prior to death. Shortly after death, the larvae begin to dehydrate or mummify as they become colonized internally by fungal mycelium and hyphal bodies. The insect is transformed into a tough, sclerotial-like structure in which the fungus can survive for extended periods and produce fruitbodies when conditions are favourable and whilst food reserves last. This slow mycelial colonization and subsequent conservation of the cadaver, which are characteristics of most *Cordyceps* infections, is in sharp contrast to the ephemeral nature of the post-mortem activity of the Entomophthorales. Observations of similar *Cordyceps* infections of scarabaeid larvae have shown that, as the fruitbodies develop, exudates or crystalline deposits form on the surface and attract an insect fauna. It is conceivable that these insects are either indirect or direct vectors of the sticky conidia of the anamorph (*Hirsutella*) which first develops on the immature fruitbodies. Several anamorphs may occur in various genera of the hyphomycetes (*Hymenostilbe, Paecilomyces* and *Verticillium*, for example, F6.2) (see Fig. 3C).

There are many gaps in our knowledge of insect–*Cordyceps* interactions, especially relating to their ecology. Most information on the behavioural responses of infected hosts has been obtained from *ad hoc* studies of the disease syndrome in ants. These are discussed separately.

d. Nectria. All the entomopathogenic species of *Nectria* are restricted to armoured scale insects (Diaspididae). The perithecia, as in *Torrubiella*, are brightly coloured and freely exposed on or around the insect host. The *Fusarium* (F6.2) anamorphs are distinctive and, because of macromorphological adaptations to an entomogenous habit, their generic affinity has often been overlooked. The stalked sporodochia of the anamorph are covered by pigmented hyphal sheaths which protect the conidia from desiccation. Rupture of this covering and the subsequent release of the mucilaginous conidial mass imparts a horn-like appearance to the structure (Petch, 1921c).

2. Pleosporales (=Dothideales—F4.9)

a. Podonectria. This is an obligately pathogenic genus attacking the Diaspididae (H20) (Petch, 1921c; Rossman, 1978), with many features in common with both *Nectria* and *Hypocrella*. The conspicuous perithecia are probably involved in long-distance dissemination of inoculum, as forcibly discharged ascospores, with the *Tetracrium* anamorph (F6.2) serving for local dispersal within and between diaspidid colonies. The anamorph forms distinctive stauroconidia, with elaborate arm-like processes, remarkably similar in form to the spore types of some Entomophthorales infecting aquatic hosts. Their function may be to catch onto potential hosts as the conidia are washed over foliage during heavy rainfall.

3. Suspected anamorphs

These comprise the more ubiquitous, and scientifically the most studied, genera of entomopathogenic fungi. *Beauveria* (F6.2) and *Metarhizium* (F6.2) species show a predeliction for insects of soil habitats; in order to avoid duplication with the following chapter they will only be discussed briefly here.

A similar initial symptomatology to that reported for the Entomophthorales has been reported for insects with *Beauveria* and *Metarhizium* infections, beginning with restlessness and loss of appetite, and followed by sluggishness and decreased irritability (Schaeffer, 1936; Madelin, 1963). There is some evidence of a "summit disease" syndrome (Madelin, 1963), although no mention was made of this behaviour in red locusts (H15.2) infected with *Beauveria bassiana* (Bals.) Vuill., whilst the same populations attacked by *Entomophaga grylli* were consistently reported to climb (Schaeffer, 1936). Similarly, leafhoppers (*Cofana spectra*, Cicadellidae

(H20)) infected with *Metarhizium album* Petch (Rombach *et al.*, 1987) in rice fields in Indonesia were usually found lower down the rice plants compared to hoppers attacked by an *Erynia* sp., which were always situated at the tops of the plants (see Fig. 2A). Moreover, whilst the *Metarhizium*-infected insects had their wing cases tightly appressed, those affected by *Erynia* died with their wings outstretched.

Traditionally, genera such as *Beauveria* and *Metarhizium* are not associated with phototropic fruitbodies (synnemata), possibly because there are no external pressures to produce them on freely exposed hosts. However, species of both genera can form synnemata when the host is hidden away. For example, in Brazilian forests subterranean locustids (H15.2) infected with *M. anisopliae* (Metsch.) Sorok. develop highly organized epigeal synnemata, whilst *B. amorpha* (Hohn.) Samson and Evans produces abundant synnemata which, although less well-structured, creep beneath tree bark from hidden beetle hosts (H25) and emerge through cracks and crevices (Samson and Evans, 1982). The latter authors also described epizootics of *B. velata* Samson and Evans on leaf-eating Lepidoptera larvae (H32). Infected insects fall from the tree canopy where they feed and die on the leaf litter; a similar situation is reported for *Pantorhytes* weevil pests of cocoa in Papua New Guinea, which drop from the canopy when infected with *B. bassiana* (Prior *et al.*, 1988). The possible significance of such behaviour will be discussed later.

D. Basidiomycotina (F5)

1. Septobasidiales (F5.3)

This order comprises two genera which are obligate parasites of armoured scale insects (Diaspididae (H20)). The genus *Septobasidium* Pat. contains more than 175 described species (Hawksworth *et al.*, 1983), although Steinhaus (1949) puts this nearer to 300 species, which are predominantly tropical or subtropical in distribution. Initial reports on *Septobasidium* species erroneously considered them to be pathogenic on the scale insects: "The fungus grows over and kills whole colonies of scale insects ..." (Petch, 1921a). This author subsequently described several species associated with diaspidid hosts on citrus and tea crops in Sri Lanka (Petch, 1927), although evidently still not appreciating the nature of the interaction with the insect. It was left to the classic investigations of J. N. Couch, published in a series of papers from 1929 to 1935 and later in monograph form (Couch, 1938), to clarify the relationship. These still remain the standard reference works on the genus *Septobasidium*, and there have been few studies conducted since.

a. Septobasidium (F5.3). Septobasidium species are always found in association with scale insects colonizing tree hosts, in which situation they form perennial colonies often possessing distinct annual growth rings. This interaction, therefore, is long-lasting and stable in contrast to those described previously. The fungal colony is highly organized, usually multi-layered with a tough outer layer of coarse interwoven hyphae. A labyrinth of chambers and tunnels, in which the insects live and move, permeates the lower stratum of mycelium. The shape and size of the chambers correspond to the morphology of the scale insect involved. The overall structure, therefore, is elaborately designed and adapted to a specific diaspidid host. The so-called insect houses (Couch, 1946) are usually colonized by only a single individual and that author relates observations on the behaviour of young scales (crawlers), which actively seek out unoccupied houses even though the chambers can accommodate a substantial number of immature insects. A mature female, however, completely fills the house. It would appear, therefore, that the crawlers are programmed to predict future space requirements. Only a proportion of the housed scales are parasitized by the fungus, and typically this varies with the season. Fine hyphae emerge from the natural orifices of infected insects and link into the more robust hyphal network of the fungal colony. Within the host, such hyphae form complex coiled structures in the haemocoel, seemingly without provoking phagocytosis, which perform a haustorial function, absorbing nutrients from the haemocoel and transferring them to the external fungal thallus. Infected insects fail to mature, remaining underdeveloped or dwarfed, and are incapable of reproduction. There is no evidence, however, that their life-span is foreshortened; indeed there is every indication that this is prolonged. The insects also seem to be relatively free to move about within the chamber and, of course, they continue to feed on the tree host.

The basidiospores of the fungus are dispersed within or on the fungal colony and young scales become infected when they pick up these spores as they crawl over the mycelium. Penetration and colonization of the insect host occurs at this stage. Apparently, those scales which escape infection when young become resistant with age. Infected crawlers either enter the labyrinth of tunnels within the colony until vacant or newly-formed houses are encountered, or they move away and colonize new areas of tree bark, thereby initiating new fungus–insect colonies. The fungus cannot survive in nature without the diaspidid, but the latter is capable of a separate existence. The benefits accruing to the fungus are obvious, being nutritional as well as those of dispersal, whilst those enjoyed by the insect are less obvious. Nevertheless, the interaction can be described as truly mutualistic since the scale insects are afforded a buffered environment in which they are protected from extremes of climate and the attacks of birds and predatory insects. The main enemies of diaspidids, however, are hymenopterous parasites which, although capable of locating potential hosts within the fungal thallus, can reach the insects within the chambers only rarely. The structure of the outer mycelial layer (150 to > 1000 μm in thickness) ensures that only those diaspidids at the very edge of the colony are at risk from the wasp's ovipositor (200–300 μm in length). Free-living scale insects are rapidly decimated by a combination of adverse weather conditions, parasites and predators. Thus, this complex interaction of fungus and insect benefits both partners to the detriment of the tree host. The very complexity of this mutualism, in terms of morphological adaptation on the part of the fungus and behavioural responses by the insect, suggests a long period of co-evolution. Indeed, the association may be even more sophisticated than outlined above. For example, the fungus can be grown *in vitro* but only simple mycelial colonies result, there being no formation of the structured thallus found in nature. Couch (1938) postulated that the insects excrete substances which affect or determine the growth of the fungal mycelium. These extraneous growth modifiers would be highly specific. The insects may also be attracted to the fungus by chemotactile stimuli.

b. Uredinella (F5.3). The closely related genus *Uredinella* was proposed by Couch (1937) to accommodate a similar diaspidid-parasitizing fungus, but with basidial characteristics close to primitive rust species. Unlike the mutualistic relationship established by species of *Septobasidium* with their insect hosts, in *Uredinella*–diaspidid interactions the benefits are strictly one-way. Only a single insect host is parasitized, essentially in the same manner as previously described, and consequently the fungal colony is restricted in size. Once again, it is in the interest of the parasite to keep the host alive as long as possible and there are no indications of premature death of the insect. Nevertheless, the parasitized diaspidid is effectively removed from the breeding population since it remains infertile. The fungus, therefore, has a negative effect on scale insect population dynamics and can be regarded as a mortality factor. Although only two species have been ascribed to the genus so far, it is considered that, because of their discrete habit, they have been under-collected and that many more remain to be discovered. More studies of this group are warranted especially in view of their suggested links with the rust fungi. Couch's initial views on this relationship have been largely ignored by most specialists concerned with the phylogeny of the fungi. Ultrastructural studies of the hyphae of five species of *Septobasidium* have now shown that septal morphology in all the taxa is typical of the Uredinales (Urediniomycetes), and not the Tremellales (Hymenomycetes) with which they are usually grouped (Dykstra, 1974). The phylogenetic significance of the Septobasidiales is clearly in need of further investigation and a better understanding of their unique interactions with scale insects may help to clarify evolutionary pathways.

III. ANT–FUNGUS INTERACTIONS

Bequaert (1922) concluded that: "From data collected it is evident that ants (H33.9) are remarkably immune from the attacks of parasitic fungi; only a few species of fungal ant parasites are known and these are rarely encountered". The same conclusion was reached by Allen and Buren (1974) and the grooming activities of ants have been thought to explain this phenomenon: "These habits of the ants must, in all probability, tend to suppress or render impossible the development of fungi except under unusual conditions" (Wheeler, 1910). The scattered but extensive mycological literature on ant–fungus interactions belies the dearth of entomological information. A few of these interactions have been analysed recently (Andrade, 1980; Evans and Samson, 1982, 1984), specifically involving species of *Cordyceps* (F4.4), the main groups of entomopathogenic fungi in tropical regions. There are also reports of ant–fungus interactions in temperate countries but the mycopathogens, in contrast, belong to the Entomophthorales (Balazy and Sokolowski, 1977; Humber, 1981b). The former authors described an epizootic of *Erynia formicae* Humber and Balazy (F3.1) on *Formica* spp. (H33.9) in pine forests in Poland. The ants were attached to understorey vegetation; rhizoids with discoidal holdfasts were noted by Humber (1981b), emerging from the ventral neck region. The same fungus was also reported from Switzerland, the ants being secured to pasture grasses both by fungal rhizoids and the grasping action of their legs and mandibles (Loos-Frank and Zimmerman, 1976); a classic death position for ants. The same workers also investigated the histopathology of diseased ants and concluded that the fungal hyphae found in the brain and nervous system disrupted the host's behavioural pattern, causing it to climb and die in an aerial situation, ideally suited for spore dispersal. Another species of *Erynia, E. myrmecophaga* Turian and Wuest, is also associated with an ant host (*Serviformica fusca*) in Swiss grasslands (Turian and Wuest, 1969). Once again, the ants grasp the vegetation with their mandibles and legs prior to death. The presence of rhizoids, however, has not been confirmed (Humber, 1981b).

Marikovsky (1962) studied several ant–fungus interactions in the USSR: one involving *Formica rufa* and a purported *Alternaria* sp. (F6.2); the other between *Tetramorium caespitum* and an unconfirmed species of *Tarichium*. It is possible that the resting spores of the latter within the ant cadavers represent those of an *Erynia* sp. Dead ants were again found attached by mandibles and legs to grass stems overhanging ant trails: Marikovsky speculated that they moved away from the main nest to avoid being eaten and that foraging workers carried back recently dead ants to the nest where they were consumed before sporulation occurred. He observed, furthermore, that diseased ants moved away from the nest and ascended the vegetation towards the latter part of the day and he considered that substances released by the fungus controlled this activity.

Tropical forest or tree crop ecosystems appear to have the richest and most diverse interactions between ants and fungi. Ponerine ants infected with *Cordyceps australis* (Speg.) Sacc. and formicine ants with *C. unilateralis* (Tul.) Sacc. (Fig. 4A) are frequently encountered in such habitats in both East and West Africa (Bequaert, 1922; Paulian, 1949; Evans, 1974; Samson *et al.*, 1982; Evans and Samson, 1984). The impulse to climb is strong, the infected ants clinging to vegetation and exhibiting twitching of the forelegs and antennae accompanied by abundant release of mucus from the everted mouthparts. Such ants move erratically when disturbed, constantly trying to groom themselves and spasmodically developing convulsions. Loss of orientation is also evident and ants falling from the vegetation will retrace their steps upwards, eventually selecting elevated sites, and grip the substratum with their mandibles and legs, often repeatedly biting and scarring the stem or leaf tissues. The ants die in this exposed position and fungal mycelium develops from the body openings and sutures, some 2–3 days later, further securing the cadavers to the vegetation. During this incubation period the ants may be scavenged by other ant species, and often the headparts fixed by the powerful mandibles are all that remain.

The ponerine ants, especially the principal host, *Paltothyreus tarsatus*, are essentially ground-dwelling in habit and rarely climb. Moreover, some of the infected formicine ants collected from understorey forest shrubs (Evans, 1974) were relatively uncommon, upperstorey forest dwellers, which were outside their normal territory. Evans also noted the presence of arboreal myrmicine ants, infected with various mycopathogens, fastened in abundance on the lower leaf surfaces of forest herbs. This provides further evidence of radical changes in behaviour by fungus-infected ants. However, the situation in the New World is not so clear-cut. Ponerine ants infected by *C. australis* are rarely exposed and usually seek refuge in forest leaf litter or beneath tree bark, the bright crimson heads of the fruitbodies thrusting out of these substrata. The *C. unilateralis*-infected formicines, although not obviously hidden, are not often found on understorey shrubs but predominantly occur in the herb layer.

During a detailed study of a *Cordyceps* complex associated with *Cephalotes* ants (Myrmicinae) in Amazonian forests, Evans and Samson

(1982) followed the behavioural patterns of diseased ants. These arboreal ants appear to move to predetermined death sites when infected. The ants of one species aggregated on or around tree boles, and numerous individuals bearing *Cordyceps* fruitbodies were collected regularly over a two-year period (Fig. 4C). As previously described, diseased ants were readily identified by their erratic, uncoordinated movements. Some concentrated around ant cadavers, hidden beneath epiphytes and tree bark, and attempted to remove them, clearing large areas of bark in the process. If successful, the ant then dropped the extracted cadaver to the ground, but itself often displayed abnormal activities and symptoms of infection. Observations of recently dead ants revealed that mycelial growth occurs 3–4 days after death, appearing from the orifices and joints and forming a vigorous attachment to the substrate. The dead ants were heavily scavenged during this period by other ants (*Solenopsis* spp.) and arachnids. Formation of the *Cordyceps* fruitbodies, however, was slow, and a 2–3 month overall maturation period was not uncommon. Synnemata of the anamorph formed more rapidly and these crept over the substrata, anchored by "rhizoids", remaining long after the host had been removed by scavengers or weathering. These synnemata produce a range of multifunctional anamorphs: some are adapted for rain-splash dispersal (slime spores), others to aerial dispersal (dry conidia), whilst others appear to be adapted more for long-term survival, probably as contact propagules, forming dark mucoid spore balls. Old cadavers were also found to regenerate new *Cordyceps* fruitbodies, further extending the survival period. The number of diseased ants collected per month was relatively constant over the sampling period and it was concluded, contrary to initial impressions, that disease is at an enzootic rather than an epizootic level in forest ant populations.

Other anamorphs have been regularly collected on ants in the tropics, especially members of the genus *Stilbella* (F6.2) (Samson *et al*., 1981). *S. burmensis* (Mains) Samson and Evans was commonly recorded on formicine ants both in cocoa farms and forests in Ghana. In an epizootic amongst a population of *Oecophylla longinoda*, infected worker ants were deeply lodged in cocoa tree bark (crevices, old pruning wounds) with the bicoloured synnemata emerging in abundance. Old queens, from the previous season, were found to be the source of infection and, although overgrown by moss and algae, were covered by fresh synnemata. It was concluded that: "The migration of infected alate ants presents an efficient means of long distance dispersal of the pathogen, especially as the queens die fully exposed and rarely hide away". A similar, *Camponotus–Stilbella* complex was reported by Katerere (1983) in an East African pine plantation. Another interaction, between ants of the genus *Technomyrmex* and the fungus *S. dolichoderinarum* Samson and Evans was also noted in

Ghanaian cocoa farms (Samson *et al.*, 1981). Numerous diseased ants were recovered from insect tunnels within tree bark and synnematal formation was adapted to this habitat. The synnemata always breaks out from the posterior abdominal or anal region and pass directly to the exterior of the tunnel, the pink fertile heads forming shortly after emergence. The positioning of the fruitbodies of these mycopathogens on their ant hosts is characteristic, each species invariably being associated with a particular part of the host cadaver. This specificity is also reflected in host range for all ant–fungus interactions, indicating an ancient co-evolution of host and pathogen.

IV. DISCUSSION AND CONCLUSIONS

The insect exoskeleton is a formidable barrier, the secret to the success of the insects in their colonization and domination of a broad spectrum of terrestrial habitats. Any organism which has managed not only to breach this defence, but also to overcome internal resistance factors, such as phagocytosis, must have co-evolved with their insect host over a considerable period. These cellular interactions have not been addressed here, nor have those spore–cuticle interactions which comprise both non-specific (electrostatic) and specific (chemical and structural) mechanisms (Boucias and Latgé, 1986; Samson *et al.*, 1988). Instead, emphasis has been upon the more neglected area of ecological interactions; how the insect hosts have attempted to limit the impact of mycopathogens and how the latter have overcome the considerable problems of dispersal and survival. Many of the data presented are observational in nature and, therefore, empirical, permitting only preliminary conclusions to be drawn. However, the most relevant and least controversial ones concerning the two major groups of fungi involved, the Entomophthorales (F3.1) and the Clavicipitales (F4.4), are summarized here.

The Entomophthorales reflect a continuum of sophistication and adaptation to their insect hosts. Humber (1984) discussed the evolution of the primary spore discharge mechanism which resulted ultimately in its abandonment in the genus *Massospora*. The mechanism became redundant as the activities of the living host served to disperse the spores. The ballistospores of the other genera were essential in the colonization of insect hosts, enabling the fungus to hit and stick onto difficult insect targets, initially from stationary hosts and subsequently from moving hosts. This increased parasitism at the expense of pathogenicity culminated in the genus *Strongwellsea*, the species of which induce both minimal damage and behavioural changes in their hosts. This was achieved by confining infection to the abdomen and effectively by-passing the host's alarm systems; a highly evolved parasitic state, in essence a biotrophic interaction. The dispersal strategy of the fungus demands that the host continues its normal behavioural pattern, thereby ensuring that healthy individuals come into contact with the inoculum. This is strikingly illustrated in *Massospora*-infected male cicadas, which still maintain their mating routine despite the loss of sexual organs.

It is significant that only the most primitive members of the Entomophthorales produce mycotoxins (Prasertphon and Tanada, 1969). These species all belong to the genus *Conidiobolus*, which is the least specialized of the order, having wide host ranges and killing their hosts rapidly. Field observations clearly show, however, that this rapid kill is positively disadvantageous since the tendency is for septicaemia to develop, the pathogen being out-competed by opportunistic microorganisms. It would appear, therefore, that without an antimicrobial back-up system, toxin production in entomopathogenic fungi offers no benefits. There is no evidence of antibiotic-producing systems in any of the Entomophthorales. The evolutionary option was to abandon toxins and keep the host alive until colonization or sporulation was completed. Thus, most Entomophthorales–insect interactions are characterized by internal mycelial colonization of parts, or all, of the living host. Death of the host in most species is followed, within a matter of hours, by the production of sporophores and subsequently of spores. The saprophytic phase is minimal or completely absent. Compare this situation with that in the Clavicipitales (and their purported anamorphs). These pathogens switch from a well-defined biotrophic (parasitic) phase, represented by yeast-like cells confined to the haemocoel, to an equally well-delimited saprophytic phase, involving mycelial colonization of the entire body only after death—a hemibiotrophic relationship comparable to that of some plant pathogens (Luttrell, 1974). The parallel development of powerful toxins, by the yeast phase, and antibiotics, probably by both phases, enabled the Clavicipitales to adopt this nutritional habit. There was no evolutionary pressure, therefore, for extended parasitism within this group, or the entomopathogenic Ascomycotina (Deuteromycotina) in general.

Long-term survival in the Clavicipitales is ensured by the mummification of the host and its conversion into a resting structure. The fungus is able to produce its fruitbodies over extended periods. This ability is vital in tropical forest ecosystems where potential hosts are widely dispersed and at a low density, but where, because of the buffered environment, there is almost constant insect activity throughout the year. Morphologically diverse anamorphs play an important role within the disease cycle, being adapted for both short- and long-distance dispersal, as well as for survival in certain cases. The Clavicipitales, particularly *Cordyceps* spp., advertise themselves by virtue of their striking form and colour and may serve the additional function of attracting potential victims or vectors, in addition to raising the sporogenous structures into aerial habitats.

The Entomophthorales are predominantly temperate in distribution, showing a "... morphogenetic dichotomy into conidial and resting-spore stages" (Madelin, 1966). The latter are essential for survival between seasonally determined host generations, particularly in northern temperate climates. Much available evidence suggests that resting-spore formation is directly related to host age: resting spores develop only in older insects, presumably dependent on the nutritional status of the host. It is also significant that these resting spore-producing insects behave differently to those bearing external spores, their activities generally being beneficial to the pathogen. For example, *Delia* flies (H28.39) which die in the soil and plum aphids (H20) which die beneath bark invariably contain resting spores, whilst those on the foliage produce only primary spores. The former are ideally placed to survive the winter and to infect newly emerging insects the following spring. However, rather than behaviour being controlled by the fungus, as is often postulated, this may simply be a normal response of older insects. The behavioural patterns of aphids may, in fact, be considerably more complex than previously thought. Data to support a host suicide hypothesis has recently been obtained from studies of pea aphids attacked by a parasitoid wasp (McAllister and Roitberg, 1987). Parasitized individuals throw themselves from their host plants into the soil, where there is little chance of survival. In effect, the insect kills itself and the parasitoid and, therefore, prevents subsequent parasitization of its kin. This response would, of course, only be applicable to insects which aggregate and are close genetically. The sugarbeet aphid moves upwards, away from its normal soil habitat, when infected with *Erynia*, and into the potentially more dangerous foliage. Could the aphid be removing itself and the pathogen from the colony by deliberately exposing itself to more unfavourable climatic conditions and to predators? Such activity would reduce the chances of vertical transmission of the fungus to its kin but, by dying in exposed aerial positions, actually enhancing horizontal transmission of the fungus. Closer analysis of aphid–fungus interactions would seem to be justified.

The "summit disease" syndrome, typified by *Entomophaga grylli* (F3.1), could also be explained by this theory, at least in colonial or aggregating insects. Swarming locusts or grasshoppers (H15.2), which are genetically very similar, have consistently been shown to climb when infected by *E. grylli*. Most theories to explain this behaviour favour a fungus-directed component. However, it is now considered more likely that invasion of the nerve tissues by fungal hyphae alerts the alarm systems and the diseased host takes evasive action, removing itself from its kin. In the summit position, the insect is exposed to predators, particularly birds. The behavioural responses are extremely strong; the insect is positively programmed to climb. Schaeffer (1936) detailed this behaviour in locusts, observing them to ascend and decend vegetation when infected, continuously searching for more elevated sites. Those removed from these sites and placed on the ground, simply crawled back up again. Shortly before death, the locusts underwent a complex routine, stretching and unstretching their legs and finally wrapping them around the plant stems. Some of the hypotheses to explain this activity have already been outlined: the suicide theory can be added to the list. There is some evidence against the theory, since the same locust species attacked by *Beauveria bassiana* (F6.2) did not show this behaviour, but tended to hide away. Nevertheless, these were found to be weakened or predisposed insects (Schaeffer, 1936) and the suicide theory, therefore, is still considered to be tenable.

The activities of diseased ants in tropical forests have also been interpreted as being of benefit to the colony or altruistic in nature (Evans, 1982). Modern ecological concepts, however, do not recognize altruism in animals and the term kin selection is probably more correct. The behaviour of *Cordyceps*-infected ants is directly opposite to their normal activities: arboreal ants descend and hide in lower storey vegetation, while terrestrial ants climb or seek refuge within various substrata. Once again, it can be supposed that the ant is removing itself and a potentially infectious disease from the colony and from its normal foraging areas. The well-being of the colony is protected. This activity, of course, effectively isolates the fungus and the ant-associated *Cordyceps* spp. (F4.4) have developed a battery of spore forms to maximize its dispersal efficiency in time and space. There is, therefore, a constant battle between host and pathogen in such forest ecosystems which results in a compromise or a balance; the disease constantly present but at a low (enzootic) level never gaining ascendency because of the behavioural responses of the host to infection. The balance in disturbed or agricultural ecosystems is lost and disease oscillates dramatically in insect populations, depending on environmental conditions.

REFERENCES

Allen, G. E., and Buren, W. F. (1974). Microsporidian and fungal diseases of *Solenopsis invicta* Buren in Brazil. *J. N.Y. Entomol. Soc.* **82**, 125–130.

Andrade, C. F. S. De (1980). Epizootia natural causada por *Cordyceps unilateralis* (Hypocreales, Euascomycetes) em adultos de *Camponotus* sp. (Hymenoptera, Formicidae) na regiao de Manaus, Amazonas, Brasil. *Acta Amazonica* **10**, 671–677.

Balazy, S. (1984). On rhizoids of *Entomophthora muscae* (Cohn) Fresenius (Entomophthorales: Entomophthoraceae). *Mycotaxon* **19**, 397–407.

Balazy, S., and Sokolowski, A. (1977). Morphology and biology of *Entomophthora myrmecophaga. Trans. Br. Mycol. Soc.* **68**, 134–137.

Ben-Ze'ev, I., and Kenneth, R. G. (1982a). Features-criteria of taxonomic importance in the Entomophthorales. I. A revision of the Batkoan classification. *Mycotaxon* **14**, 393–455.

Ben-Ze'ev, I., and Kenneth, R. G. (1982b). Features-criteria of taxonomic importance in the Entomophthorales. II. A revision of the genus *Erynia* Nowakowski 1881 (= *Zoophthora* Batko, 1964). *Mycotaxon* **14**, 456–475.

Ben-Ze'ev, I., Keller, S. and Ewen, A. B. (1985). *Entomophthora erupta* and *Entomophthora helvetica* sp. nov. (Zygomycetes: Entomophthorales), two pathogens of Miridae (Heteroptera) distinguished by pathological and nuclear features. *Can. J. Bot.* **63**, 1469–1475.

Bequaert, J. (1922). Ants in their diverse relations to the plant world. *Bull. Am. Mus. Nat. Hist.* **45**, 333–583.

Berisford, Y. C., and Tsao, C. H. (1974). Field and laboratory observations of an entomogenous infection of the adult seedcorn maggot, *Hylemya platura* (Diptera: Anthomyiidae). *J. Georgia. Entomol. Soc.* **9**, 104–110.

Boucias, D. G., and Latgé, J. P. (1986). Adhesion of entomopathogenic fungi on their host cuticle. *In* "Fundamental and Applied Aspects of Invertebrate Pathology" (R. A. Samson, J. M. Vlak and D. Peters, eds), pp. 432–434. Foundation of the Fourth Int. Colloq. Invertebr. Pathol., Wageningen.

Brady, B. L. K. (1979). *Entomophthora grylli. CMI Descriptions of Pathogenic Fungi and Bacteria* No. 606. Farnham Royal: Commonwealth Agricultural Bureaux.

Brobyn, P. J., and Wilding, N. (1983). Invasive and developmental processes of *Entomophthora muscae* infecting houseflies (*Musca domestica*). *Trans. Br. Mycol. Soc.* **80**, 1–8.

Byford, W. J., and Ward, L. K. (1986). Effect of the situation of the aphid host at death on the type of spore produced by *Entomophthora* spp. *Trans. Br. Mycol. Soc.* **51**, 598–600.

Carruthers, R. I., Haynes, D. L., and MacLeod, D. M. (1985). *Entomophthora muscae* (Entomophthorales: Entomophthoraceae) mycosis in the onion fly, *Delia antiqua* (Diptera: Anthomyiidae). *J. Invertebr. Pathol.* **45**, 81–93.

Carruthers, R. I., Soper, R. S., and Feng, Z. (1986). Epizootiology of *Entomophaga grylli* in populations of the clear-winged grasshopper, *Camnula pellucida*. *In* "Fundamental and Applied Aspects of Invertebrate Pathology" (R. A. Samson, J. M. Vlak and D. Peters, eds), p. 237. Foundation of the Fourth Int. Colloq. Invertebr. Pathol., Wageningen.

Charnley, A. K. (1984). Physiological aspects of destructive pathogenesis in insects by fungi: a speculative review. *In* "Invertebrate–Microbial Interactions: (J. M. Anderson, A. D. M. Rayner and D. H. Walton, eds). Br. Mycol. Soc. Symp. **6**, pp. 229–270. Cambridge Univ. Press.

Couch, J. N. (1937). A new fungus intermediate between the rusts and *Septobasidium*. *Mycologia* **29**, 665–673.

Couch, J. N. (1938). "The Genus *Septobasidium*". Univ. of North Carolina Press, Chapel Hill.

Couch, J. N. (1946). Two species of *Septobasidium* from Mexico with unusual insect houses. *J. Elisha Mitchell Sci. Soc.* **62**, 87–94.

Dustan, A. G. (1924). Studies on a new species of *Empusa* parasitic on the green apple bug (*Lygus communis* var. *novascotiensis* Knight) in the Annapolis Valley. *Proc. Acadian Entomol. Soc.* **9**, 14–36.

Dykstra, M. J. (1974). Some ultrastructural features in the genus *Septobasidium. Can. J. Bot.* **52**, 971–972.

Eilenberg, J. (1986). Effect of *Entomophthora muscae* (C) Fres. on egg-laying behaviour of female carrot flies (*Psila rosae* F.). *In* "Fundamental and Applied Aspects of Invertebrate Pathology" (R. A. Samson, J. M. Vlak and D. Peters, eds), p. 235. Foundation of the Fourth Int. Colloq. Invertebr. Pathol., Wageningen.

Evans, H. C. (1974). Natural control of arthropods, with special reference to ants (Formicidae), by fungi in the tropical high forest of Ghana. *J. Appl. Ecol.* **11**, 37–49.

Evans, H. C. (1982). Entomogenous fungi in tropical forest ecosystems: an appraisal. *Ecol. Entomol.* **7**, 47–60.

Evans, H. C., and Samson, R. A. (1977). *Sporodiniella umbellata*, an entomogenous fungus of the Mucorales from cocoa farms in Ecuador. *Can. J. Bot.* **55**, 2981–2984.

Evans, H. C., and Samson, R. A. (1982). *Cordyceps* species and their anamorphs pathogenic on ants (Formicidae) in tropical forest ecosystems. I. The *Cephalotes* (Myrmicinae) complex. *Trans. Br. Mycol. Soc.* **79**, 431–453.

Evans, H. C. and Samson, R. A. (1984). *Cordyceps* species and their anamorphs pathogenic on ants (Formicidae) in tropical forest ecosystems. II. The *Camponotus* (Formicinae) complex. *Trans. Br. Mycol. Soc.* **82**, 127–150.

Hagen, K. S., and Van den Bosch, R. (1968). Impact of pathogens, parasites and predators on aphids. *Annu. Rev. Entomol.* **13**, 325–384.

Hall, I. M. (1959). The fungus *Entomophthora erupta* (Dustan) attacking the black grass bug, *Irbisia solani* (Heidemann) (Hemiptera: Miridae), in California. *J. Insect Pathol.* **1**, 48–51.

Hall, I. M., and Dunn, P. H. (1957). Entomophthorous fungi parasitic on the spotted alfalfa aphid. *Hilgardia* **27**, 159–181.

Harper, A. M. (1958). Notes on behaviour of *Pemphigus betae* Doane (Homoptera: Aphididae) infected with *Entomophthora aphidis* Hoffm. *Can. Entomol.* **90**, 439–440.

Hawksworth, D. L., Sutton, B. C., and Ainsworth, G. C. (1983). "Ainsworth & Bisby's Dictionary of the Fungi". CMI, Kew.

Humber, R. A. (1976). The systematics of the genus *Strongwellsea* (Zygomycetes: Entomophthorales). *Mycologia* **68**, 1042–1060.

Humber, R. A. (1981a). An alternative view of certain taxonomic criteria used in the Entomophthorales (Zygomycetes). *Mycotaxon* **13**, 191–240.

Humber, R. A. (1981b). *Erynia* (Zygomycetes: Entomophthorales): validations and new species. *Mycotaxon* **13**, 471–480.

Humber, R. A. (1982). *Strongwellsea* vs. *Erynia*: the case for a phylogenetic classification of the Entomophthorales (Zygomycetes). *Mycotaxon* **15**, 167–184.

Humber, R. A. (1984). Foundations for an evolutionary classification of the Entomophthorales (Zygomycetes). *In* "Fungus–Insect Relationships" (Q. Wheeler and M. Blackwell, eds), pp. 167–183. Columbia Univ. Press, New York.

Humber, R. A. and Ben-ze'ev, I. (1981). *Erynia* (Zygomycetes: Entomophthorales): emendation, synonymy and transfers. *Mycotaxon* **13**, 506–516.

Humber, R. A., and Ramoska, W. A. (1986). Variations in entomophthoralean life cycles: practical implications. *In* "Fundamental and Applied Aspects of Invertebrate Pathology" (R. A. Samson, J. M. Vlak and D. Peters, eds), pp. 190–193. Foundation of the Fourth Int. Colloq. Invertebr. Pathol., Wageningen.

Hutchinson, J. A. (1962). Studies on a new *Entomophthora* attacking calyptrate flies. *Mycologia* **54**, 258–271.

Karling, J. S. (1948). Chytridiosis of scale insects. *Am. J. Bot.* **35**, 246–254.

Katerere, Y. (1983). Notes on *Stilbella burmensis* (Mains) Samson & Evans, an entomogenous fungus from Zimbabwe. *Trop. Pest Manage.* **29**, 195–196.

Keller, S. (1981). *Entomophthora erupta* (Zygomycetes: Entomophthoraceae) als Pathogen von *Notostira elongata* (Heteroptera: Miridae). *Mitt. Schweiz. Entomol. Ges.* **54**, 57–64.

Kramer, J. P. (1979). Interactions between blow-flies (Calliphoridae) and *Entomophthora bullata* (Phycomycetes: Entomophthorales). *J. N.Y. Entomol. Soc.* **87**, 135–140.

Krueger, S. R., and Ramoska, W. R. (1985). Purification and infectivity of *Entomophaga grylli* (Fresenius) Batko pathotype 2 against *Melanoplus differentialis* (Thomas) [Ort.: Locustidae]. *Entomophaga* **30**, 293–302.

Lamb, D. J., and Foster, G. N. (1986). Some observations on *Strongwellsea castrans* [Zygomycetes: Entomophthorales], a parasite of root flies (*Delia* spp.) in the South of Scotland. *Entomophaga* **31**, 91–97.

Loos-Frank, B., and Zimmermann, G. (1976). Uber eine dem *Dicrocoelium*–Befall analoge Verhaltensanderung bei Ameisen der Gattung *Formica* durch einen Pilz der Gattung *Entomophthora*. *Z. Parasitenkd.* **49**, 281–289.

Luttrell, E. S. (1974). Parasitism of fungi on vascular plants. *Mycologia* **66**, 1–15.

MacLeod, D. M. (1963). Entomophthorales Infections. *In* "Insect Pathology: An Advanced Treatise" (E. A. Steinhaus, ed.), Vol. 2, pp. 189–231. Academic Press, London and New York.

MacLeod, D. M., and Tyrrell, D. (1979). *Entomophthora crustosa* n. sp. as a pathogen of the forest tent caterpillar, *Malacosoma disstria* (Lepidoptera: Lasiocampidae). *Can. Entomol.* **111**, 1137–1144.

MacLeod, D. M., Tyrrell, D., Soper, R. S., and de Lyzer, A. J. (1973). *Entomophthora bullata* as a pathogen of *Sarcophaga aldrichi*. *J. Invertebr. Pathol.* **22**, 75–79.

MacLeod, D. M., Muller-Kogler, E., and Wilding, N. (1976). *Entomophthora* species with *E. muscae*-like conidia. *Mycologia* **68**, 1–29.

Madelin, M. F. (1963). Diseases caused by hyphomycetous fungi. *In* "Insect Pathology: An Advanced Treatise" (E. A. Steinhaus, ed.), Vol. 2, pp. 233–271. Academic Press, London and New York.

Madelin, M. F. (1966). Fungal parasites of insects. *Annu. Rev. Entomol.* **11**, 423–448.

Marikovsky, P. I. (1962). On some features of behaviour of the ant *Formica rufa L.* infected with fungus disease. *Insectes Soc.* **9**, 173–179.

Mathieson, J. (1949). *Cordyceps aphodii*, a new species, on pasture cockchafer grubs. *Trans. Br. Mycol. Soc.* **32**, 113–136.

McAllister, M. K., and Roitberg, B. D. (1987). Adaptive suicidal behaviour in pea aphids. *Nature (London)* **328**, 797–799.

McEwen, F. L. (1963). *Cordyceps* infection. *In* "Insect Pathology: An Advanced Treatise" (E. A. Steinhaus, ed.), Vol. 2, pp. 273–290. Academic Press, London and New York.

Nair, K. S. S., and McEwen, F. L. (1973). *Strongwellsea castrans* [Phycomycetes: Entomophthoraceae], a fungal parasite of the adult cabbage maggot *Hylemya brassicae* [Diptera: Anthomyiidae]. *J. Invertebr. Pathol.* **22**, 442–449.

Nelson, D. R., Valovage, W. D., and Frye, R. D. (1982). Infection of grasshoppers with *Entomophaga (= Entomophthora) gryll)* by injection of germinating resting spores. *J. Invertebr. Pathol.* **39**, 416–418.

Paulian, R. (1949). "Un Naturaliste en Cote d'Ivoire". Editions Stock, Paris.

Perry, D. F., and Whitfield, G. H. (1984). The interrelationships between microbial entomopathogens and insect hosts: a system study approach with particular reference to the Entomophthorales and the eastern spruce budworm. *In* "Invertebrate–Microbial Interactions" (J. M. Anderson, A. D. M. Rayner and D. H. Walton, eds), pp. 307–331. Cambridge Univ. Press.

Petch, T. (1921a). Fungi parasitic on scale insects. *Trans. Br. Mycol. Soc.* **7**, 18–40.

Petch, T. (1921b). Studies in entomogenous fungi. II. The genera *Hypocrella* and *Aschersonia*. *Ann. R. Bot. Gdns. Peradeniya* **7**, 167–278.

Petch, T. (1921c). Studies in entomogenous fungi. 1. The Nectriae parasitic on scale insects. *Trans. Br. Mycol. Soc.* **7**, 89–167.

Petch, T. (1925). Entomogenous fungi and their use in controlling insect pests. *Bull. Dep. Agric. Ceylon* **71**, 1–40.

Petch, T. (1927). *Septobasidium rameale. Trans. Br. Mycol. Soc.* **12**, 276–282.

Prasertphon, S., and Tanada, Y. (1969). Mycotoxins of entomophthoraceous fungi. *Hilgardia* **39**, 581–600.

Prior, C., Jollands, P., and le Patourel, G. (1988). Infectivity of oil and water formulations of *Beauveria bassiana* (Deuteromycotina: Hyphomycetes) to the cocoa weevil pest *Pantorhytes plutus* (Coleoptera: Curculionidae). *J. Invertebr. Pathol.* **55**, 66–72.

Remaudière, G., and Keller, S. (1980). Révision systématique des genres d'Entomophthoraceae à potentialité entomopathogène. *Mycotaxon* **11**, 323–338.

Rockwood, L. P. (1950). Entomogenous fungi of the family Entomophthoraceae in the Pacific North West. *J. Econ. Entomol.* **43**, 704–707.

Rombach, M. C., Humber, R. A., and Evans, H. C. (1987). *Metarhizium album*, a fungal pathogen of leaf and planthoppers of rice. *Trans. Br. Mycol. Soc.* **88**, 451–459.

Rossman, A. Y. (1978). *Podonectria*, a genus in the Pleosporales on scale insects. *Mycotaxon* **7**, 163–182.

Samson, R. A., and Evans, H. C. (1982). Two new *Beauveria* spp. from South America. *J. Invertebr. Pathol.* **39**, 93–97.

Samson, R. A ., Ramakers, P. M. J., and Oswald, T. (1979). *Entomophthora thripidum*, a new fungal pathogen of *Thrips tabaci. Can. J. Bot.* **57**, 1317–1323.

Samson, R. A., Evans, H. C., and Van de Klashorts, G. (1981). Notes on entomogenous fungi from Ghana V. The genera *Stillbella* and *Polycephalomyces. Proc. K. Ned. Akad. Wet.* **84**, 289–301.

Samson, R. A., Evans, H. C.,and Hoekstra, E. S. (1982). Notes on entomogenous fungi from Ghana. VI. The genus *Cordyceps. Proc. K. Ned. Akad. Wet.* **85**, 589–605.

Samson, R. A., Evans, H. C., and Latgé, J. P. (1988). "An Atlas of Entomopathogenic Fungi". Springer Verlag, Berlin.

Sawyer, W. H. (1931). Studies on the morphology and development of an insect-destroying fungus, *Entomophthora sphaerospora. Mycologia* **23**, 411–432.

Schaeffer, E. E. (1936). The white fungus disease (*Beauveria bassiana*) among red locusts in South Africa and some observations on the grey fungus disease (*Empusa grylli*). *Sci. Bull. Dep. Agric. For. Un. S. Afr.* **160**, 1–28.

Skaife, S. H. (1925). The locust fungus *Empusa grylli* and its effect on its host. *S. Afr. J. Sci.* **22**, 298–308.

Soper, R. S. (1963). *Massospora levispora*, a new species of fungus pathogenic to the cicada, *Okanagana rimosa. Can. J. Bot.* **41**, 875–878.

Speare, A. T. (1921). *Massospora cicadina* Peck. A fungus parasite of the periodical cicada. *Mycologia* **13**, 72–82.

Steinhaus, E. A. (1949). "Principles of Insect Pathology". McGraw-Hill, New York.

Strong, F. E., Wells, K., and Apple, J. W. (1960). An unidentified parasite on the seed-corn maggot. *J. Econ. Entomol.* **53**, 478–479.

Turian, G., and Wuest, J. (1969). Mycoses à entomophthoracées frappant des populations de fourmis et des drosophiles. *Mitt. Schweiz. Entomol. Ges.* **42**, 197–201.

Vandenberg, J. D., and Soper, R. S. (1978). Prevalence of Entomophthorales mycoses in populations of spruce budworm, *Choristoneura fumiferana. Environ. Entomol.* **7**, 847–853.

Wheeler, A. G. (1972). Studies on the arthropod fauna of alfalfa. III. Infection of the alfalfa

plant bug, *Adelphocoris lineolatus* (Hemiptera: Miridae) by the fungus *Entomophthora erupta*. *Can. Entomol.* **104**, 1763–1766.
Wheeler, W. M. (1910). "Ants, their Structure, Development and Behaviour". Columbia Univ. Press, New York.
Wilding, N. (1972). Entomophthoraceae. Field incidence in wheat bulb fly. *Rep. Rothamsted Exp. Stn 1971, Pt I*, pp. 227–228.
Wilding, N., and Perry, J. N. (1980). Studies on *Entomophthora* in populations of *Aphis fabae* on field beans. *Ann. Appl. Biol.* **94**, 367–378.

10

Mycopathogens of Soil Insects

S. KELLER AND G. ZIMMERMANN

I. INTRODUCTION

Considering the enormous importance of the soil as a "life support system", one might imagine that we possess a thorough knowledge of this habitat. This is indeed partly true. For example we know much about the formation of soils, their chemistry and physics and we also recognize a huge number of soil organisms. On the other hand our knowledge of the soil as a biotope is less extensive. Thus although certain autecological relationships have been elucidated in detail, we know rather less about demecological and synecological processes as far as short-term interactions are concerned, and even less when we consider long-term interactions. Likewise, although particular interactions between entomogenous fungi and soil arthropods are well understood, we are comparatively ignorant of the importance of that interaction in the ecosystem.

Such investigations are rendered more difficult by the large number and variability of soil types and their plant communities and consequently the soil organisms and their different interactions. The nature of soils used for agricultural purposes underlie additional man-made influences such as husbandry, crop rotation, fallow periods and the application of fertilizers and pesticides.

Compared with the other main habitats, air and water, the soil is characterized by the absence of light and by effective buffering against extreme biotic and abiotic influences. The soil is an excellent habitat for microorganisms and arthropods. Consequently the number of species and individuals are high relative to that in other habitats. The majority of these organisms lives exclusively in the soil, but others such as many pterygote insects spend only a part of their life in the soil, usually the larval or pupal stage or the overwintering period. Corresponding with this rich variety of organisms there exists a richness of interactions between species and groups of organisms: antagonism, commensalism, synergism, predation, parasitism, pathogenicity and competition dominate the soil life. From this complex of interactions, insects and their fungal antagonists are considered here.

II. MAIN PROPERTIES OF SOIL COMPONENTS AFFECTING SUBTERRANEAN LIFE

The interactions between the microflora and fauna of the soil are numerous and greatly influenced by the physicochemical factors of the habitat. For a better understanding of these phenomena, the major physicochemical characteristics affecting microorganisms as well as soil-inhabiting insects are outlined briefly (Lal, 1984). The absence of visible and ultra-violet light, major factors of the aerial environment, excludes any profound influence on growth, reproduction, behaviour and viability by either.

A. Texture

The texture of soils depends on the size and distribution of particles (of sand, silt and clay) and is an important component in soil classification. It also plays a significant role in determining microbial activity. For example, clay particles are colloidal in nature and may adsorb microorganisms, so restricting their distribution and movement. Clay may also function as a protectant against biodeterioration and thus may increase the stability and the longevity of conidia or blastospores (Fargues *et al.*, 1983).

B. Water

The survival, germination and growth of soil fungi are greatly influenced by water potential. The effect of soil moisture on the ecology of soil fungi was discussed in detail by Griffin (1963).

High moisture reduces the availability of oxygen and therefore leads to an increase of carbon dioxide. Conversely, with progressive evaporation of water the soil dries out and the resulting desiccation kills small arthropods and many microorganisms. Some fungi, however, modify themselves and survive as resting spores, sclerotia, resting cells or hyphae. Soil water not only affects the growth and survival of microorganisms and insects, but also profoundly affects their movement.

C. Atmosphere

In normal, well-drained soils, the voids are filled with air. The total volume of air, however, is largely determined by particle size and water content. The concentration of oxygen and carbon dioxide in soils plays a significant role in determining the activity, density and diversity of organisms in different soil types. Most of the fungi are strictly aerobic. Little is known about the effect of carbon dioxide on mycopathogens.

D. pH and Chemistry

The soil microflora is highly influenced by the soil pH. In general, high acidity decreases the growth of bacteria and increases that of soil fungi. In contrast, alkaline soils are more suitable for bacteria. Fungi are also more tolerant of very acid conditions and several species can grow even at pH 3.

In addition, the pH of soils is intimately associated with the availability of many nutrients. For example, at low pH several ions such as P and Ca become more soluble and different metal ions (Al, Mn, Ni, Fe) may become toxic as they pass into solution at low pH.

E. Temperature

Soil temperature is one of the key factors influencing the composition and activity of soil microflora and fauna. It depends greatly on the geographical location, aspect and gradient of surface slopes, exposure, soil colour and the nature and density of plant cover. In general, the temperature at the surface is directly related to the ambient air temperature. However, a steep temperature gradient occurs within the first few centimetres of the soil, while deeper layers are comparatively buffered.

F. Agricultural Practices

Agricultural practices such as ploughing, rotavation and application of pesticides may radically alter the population of soil organisms. This may be due to desiccation, a change in aeration of soil, a direct effect of pesticides on target or non-target organisms, or by the addition of undecomposed or partially decomposed organic material.

III. ENTOMOPATHOGENIC FUNGI AND THEIR SUBTERRANEAN HOSTS

Examples of entomopathogenic fungi occur throughout the fungus kingdom, but especially in the Zygomycotina (F3), the Ascomycotina (F4) and the Deuteromycotina (F6). In general, these microorganisms occur in all habitats in which insects of their developmental stages are present. Many insects of nearly all orders spend their life permanently or, more usually, temporarily in the soil. Thus, soil is used as a habitat for each stage in the life-cycle from the egg to the adult, for overwintering and diapause, for pupation or merely for concealment. In all these cases, insects may have contact with mycopathogens. During the last 25 years, many fungus species have been found on diseased soil-inhabiting insects throughout the world. The major species and their host insects are listed in Table I.

The usual ways of collecting entomopathogenic fungi on soil insects include searching for diseased insects or collecting samples of living insects which are then reared in the laboratory and observed for incidence of mycoses. Recently, Zimmermann (1986) showed that mycopathogens can

TABLE I. Important species of entomopathogenic fungi and their hosts in the soil.

Fungus	Host insects	Reference
Zygomycotina (F3)		
Conidiobolus coronatus (F3.1)	Many species, mainly Lepidoptera (H32)	Domsch *et al.* (1980)
Entomophthora sp. (F3.1)	*Tipula paludosa* (H28) (Diptera, Tipulidae)	Müller-Kögler (1957)
Massospora cicadina (F3.1)	Cicadas (Homoptera, Cicadidae—H20)	Soper *et al.* (1976)
M. levispora	*Okanagana rimosa* (Homoptera, Cicadidae)	Soper *et al.* (1976)
Ascomycotina (F4)		
Cordyceps spp. (F4.4)	Lepidoptera (H32)	Mains (1958)
C. militaris	*Tipula paludosa* (H28) (Diptera, Tipulidae)	Müller-Kögler (1965a)
C. aphodii	*Aphodius tasmaniae* (H25.14) (Coleoptera, Scarabaeidae)	Coles (1979)
Deuteromycotina (F6)		
Beauveria bassiana (F6.2)	Many species of several orders	Different authors
B. brongniartii (= *B. tenella*)	*Melolontha* spp. (H25.14) (Coleoptera, Scarabaeidae)	Ferron (1978)
Hirsutella spp. (F6.2)	*Haplodiplosis marginata* (H28.10) (Diptera, Cecidomyiidae)	Zimmermann (unpubl.)
	Dasyneura brassicae (do)	Müller-Kögler (unpubl.)
	Nematodes	Sturhan and Schneider (1980)
Isaria japonica (F6.2)	*Selatosomus aeneus* (H25) (Coleoptera, Elateridae)	Mietkiewski and Balazy (1982)
Metarhizium anisopliae (F6.2)	Many species of several orders, but mainly Coleoptera (H25)	Different authors
M. flavoviride	*Otiorhynchus sulcatus* (H25.63) (Coleoptera, Curculionidae)	Marchal (1977)
Paecilomyces farinosus (F6.2)	Lepidoptera (H32)	Different authors
P. fumosoroseus	Lepidoptera	Different authors
P. tenuipes	Lepidoptera, mainly Noctuidae	Zimmermann (1980)
	Bibio sp. (H28) (Diptera, Bibionidae)	
Tolypocladium spp. (F6.2)	*Galleria*-bait method	Zimmermann and Hokkanen (unpubl.)
Verticillium lecanii (F6.2)	Species of several orders	Different authors

also be detected in the soil by using bait insects, such as the larvae of the greater wax moth, *Galleria mellonella* (H32). This method was originally developed for trapping entomoparasitic nematodes in soil samples (Bedding and Akhurst, 1975) but also proved very useful for detecting entomopathogenic fungi, especially in studies on their occurrence and distribution. Records of the occurrence of some important species in the genera *Conidiobolus* (F3.1) *Beauveria* (F6.2), *Metarhizium* (F6.2) and *Paecilomyces* (F6.2) are reviewed by Domsch *et al.* (1980).

IV. ENVIRONMENTAL INFLUENCES AFFECTING MYCOPATHOGENS

A. Abiotic Factors

Mycopathogens in the soil, just as subterranean insects, are affected by abiotic factors, mainly temperature, relative humidity or soil water content, agrochemicals, and composition of soil types. In the following subsections, the effects of these factors are discussed in detail.

1. Temperature

Temperature is a key factor in the development and activity of all organisms. For mycopathogens, the mean optimum temperature normally is between 20 and 25°C, the maximum is about 35°C and the minimum 5–10°C (Roberts and Campbell, 1977). These temperature limits are similar to those for most of the soil-inhabiting insects but the optima may well differ, i.e. a suboptimal temperature for a pathogen may be optimal for the insect and *vice versa*. These differences may influence the numbers of hosts that succumb and the rate of development of disease within an host individual. For example, the fungus *Metarhizium anisopliae* (F6.2) is one of the most promising candidates for control of subterranean pest insects. The optimum temperature for germination and growth is about 25°C. In our regions of Europe, however, soil temperature from May to September is generally between 15 and 20°C. Therefore, low temperatures may be a limiting factor for the widespread use of the fungus in biological control. Recently, however, *Metarhizium* strains with lower temperature thresholds have been selected and isolated (Soares *et al.*, 1983; Latch and Kain, 1983).

The effect of temperature on the stability and survival of fungal conidia often depends on soil humidity. The half-life of conidia of *Beauveria bassiana* ranged from 14 days at 25°C and 75% water saturation to 276 days at 10°C and 25% water saturation (Lingg and Donaldson, 1981). Conidia held at −15°C exhibited little or no loss in viability regardless of water content, relative humidity or pH. However, conidia were not recoverable after 10 days from soils held at 55°C. In tropical and subtropical regions, temperatures in the upper soil layer may reach 30 and 40°C. Temperatures as high as 70°C occur commonly in compost heaps, an ideal breeding place for *Oryctes rhinoceros* whose larvae are often attacked by *M. anisopliae* (Johnpulle, 1938). In this connection it is interesting that the resistance of *Metarhizium* conidia to temperatures from 40°C to about 80°C increases with decreasing humidity, i.e. with desiccation (Zimmermann, 1982).

2. Moisture and Water

Soil moisture and water are of great importance: (1) for viability or activity; and (2) for the migration or movement of mycopathogens and insects. At the surface, moisture is frequently in equilibrium with atmospheres at less than 95% RH. In this case, growth of most species will be restricted or completely inhibited. In deeper layers and in our temperate zones, the relative humidity is nearly always at about 99% even if the soil is at the permanent wilting point of pF 4.2 (Griffin, 1963). Nevertheless, a higher water content in the soil may favour epizootics. For example the incidence of *Cordyceps aphodii* (F4.4) in the pasture cockchafer, *Aphodius tasmaniae* (H25.14), was 1–3% during May and June and 21–38% in July and August, coinciding with the rainfall recorded during these months (Coles and Pinnock, 1982).

Besides the effect of soil moisture and water content on viability mentioned above, water in the form of rain has a great influence on the vertical movement of entomopathogenic fungi. This is important with respect to the use of mycopathogens in biological control. If conidia applied to the soil accumulate at or near the surface, only those insects living in or coming to this region may be exposed to the fungus; pest insects living deeper in the soil will only be reached by fungal propagules washed down in water. Recent investigations revealed that the movement of conidia of *Nomuraea rileyi*, *Beauveria bassiana* and *Metarhizium anisopliae* (F6.2) differs according to soil type (Ignoffo *et al.*, 1977a; Storey and Gardner, 1987; Zimmermann, unpubl.). Trials made by Storey and Gardner (1987) with formulated *B. bassiana* conidia showed that the percentage of propagules recovered from the upper 5 cm of the soil columns was positively correlated with sand composition and negatively with clay or silt. However, unformulated *N. rileyi* conidia were retained by a silt-loam soil, probably because they were adsorbed on clay or organic particles (Ignoffo *et al.*, 1977a). It seems that the structure of spores, their formulation and probably the addition of a wetting agent may interfere with the ease with which they percolate through soils.

3. Agrochemicals

Agrochemicals include pesticides and fertilizers. During recent years much emphasis has been put on determining the effect of different pesticides on target and non-target organisms. Pesticides are applied on or in the soil for control of plant pathogenic fungi (fungicides), pest insects (insecticides) or weeds (herbicides). Their use may affect mycopathogens of insects. In general, fungicides are the most harmful substances, but some insecticides and even herbicides may also have an inhibitory effect on entomopathogenic fungi. For example, Ignoffo *et al.* (1975) tested the sensitivity *in vitro* of conidia of *Nomuraea rileyi* (F6.2) to 44 pesticides. Development of the fungus was inhibited by 7 of 8 fungicides, 13 of 25 insecticides and 4 of 11 herbicides. In integrated pest management systems, these side-effects should be considered. Pesticides may also favour fungal development and even enhance the development of epizootics. Ferron (1970, 1971) found that insecticides increased the susceptibility of *Melolontha melolontha* (H25.14) to *Beauveria brongniartii* (F6.2) and in field trials, more white grubs succumbed to this fungus in plots treated with one formulation of fonofos than in untreated plots, although another formulation and three other insecticides had no effect (Ramser and Keller, unpubl.).

Herbicides, too, may indirectly influence the interaction between entomopathogenic fungi and soil-inhabiting arthropods. As leaves and roots of many undesirable plants in the field are food for many insects, their elimination by the action of herbicides interferes profoundly with the soil fauna and, consequently, with its associated mycopathogens.

B. Biotic Factors

Besides abiotic factors, the mycopathogen–soil insect interaction is also influenced by biotic factors such as other soil microorganisms, soil arthropods and even plants.

1. Soil Microorganisms

Investigations by several authors showed that conidia of entomopathogenic fungi, e.g. *Beauveria bassiana* (F6.2), were unable to germinate in normal, non-sterile soil. However, in sterile soil, germination and mycelial growth were observed (Walstad *et al.*, 1970; Lingg and Donaldson, 1981). Even unsterilized aqueous extracts of soil inhibited germination of conidia and germ-tube growth in the species *B. bassiana* and *Paecilomyces farinosus* (F6.2) (Clerk, 1969). Trials revealed that the fungistatic effect in non-sterile soil is due to actinomycetes (Huber, 1958) or other saprophytic microorganisms, for example *Penicillium urticae* (F6.2), which was regularly isolated from *B. bassiana*-suppressive soils and which produces a water-soluble inhibitor of *B. bassiana* (Lingg and Donaldson, 1981; Shields *et al.*, 1981). Formerly, these observations led to the opinion that the soil microflora has a stabilizing and preserving effect on *B. bassiana* (Wartenberg and Freund, 1961). However, Lingg and Donaldson (1981) reported that conidial survival in non-sterile soil to which carbon and/or nitrogen sources had been added was greatly decreased and loss was often complete in less than 22 days. In contrast, the number of *B. bassiana* propagules increased dramatically in sterile soil treated in the same manner. Blastospores of the same species were also inactivated by soil bacteria and protozoa (Fargues *et al.*, 1983) and fewer conidia were produced by germinating resting spores of *Conidiobolus obscurus* (F3.1) on the surface of soil than on moist filter paper (Perry *et al.*, 1982). This was explained by a rapid degradation of resting spore germ tubes by soil microorganisms. While non-germinated resting spores of *C. obscurus* are remarkably resistant to biodegradation, germ tubes may be lysed during microbial attack. Conversely *B. bassiana* and *Metarhizium anisopliae* are known to exhibit antibiosis towards some soil fungi, including *Pythium* sp. (F2.3), *Phytophthora cactorum* (F2.3) or *Pullularia* sp. (F6.2) (Domsch *et al.*, 1980; Walstad *et al.*, 1970) and it seems likely that other less studied entomogenous species may have similar properties.

A variety of fungi and bacteria occurring naturally on the cuticle of insects also inhibit the activity of entomogenous fungi. In infection experiments with *M. anisopliae* (F6.2) and the curculionid *Hylobius pales* (H25.63), the abundance of microorganisms on the insect cuticle was reflected in either extremely low or completely suppressed germination of spores of *M. anisopliae* (Schabel, 1976). Surface-sterilizing the insect resulted in high germination percentages probably through the elimination of the antibiosis.

Insect pathogenic fungi may themselves be attacked by parasitic fungi (Müller-Kögler, 1961) but their practical importance appears to be low.

2. *Soil Arthropods*

Small soil arthropods, especially mites and Collembola (H1), may be very important in distributing entomopathogenic fungi. Using a modified Berlese–Tullgren funnel, Zimmermann and Bode (1983) demonstrated that mainly Acari and Collembola can transport conidia of *Metarhizium anisopliae* passively through a vertical soil layer of about 15 mm.

Several investigations have shown that phoretic mites may spread the conidia of *Beauveria bassiana* or *M. anisopliae* over the soil surface or through the soil (Samšiňák, 1964; Samšiňáková and Samšiňák, 1970; Schabel, 1982). Two species, *Tyrophagus putrescentiae* and *Sancassania phyllognathi*, carried spores from *Beauveria*-infected larvae of *Galleria mellonella* (H32) to healthy ones, while the mites themselves were resistant to fungal infection (Samšiňák, 1964; Samšiňáková and Samšiňák, 1970). In contrast, Schabel (1982) observed that an undescribed species of *Macrocheles* and the phoretic mite *Histiogaster anops* which live in depressions of the exoskeleton of *Hylobius pales* can transfer spores of *M. anisopliae* to their host, but at the same time become infected by the fungus. Thus, these mite species not only act as a vector, but also contribute to the build-up of an inoculum. A true fungal pathogen of soil mites and Collembola was described by Mankau (1968) as an *Acremonium* sp. (F6.2).

3. *Plants*

Little is known about the effect of plant roots and root exudates on the activity of insect mycopathogens. However, in experiments on control of the black vine weevil, *Otiorhynchus sulcatus* (H25.63), with *Metarhizium anisopliae*, the pathogen was less effective in cyclamen culture than in that of other plant species (Zimmermann, 1984). Further unpublished trials revealed that the fungus is directly inhibited by secondary plant products apparently diffusing into the soil from the cyclamen roots.

C. Morphological Adaptations and Saprophytic Abilities of Mycopathogens

In contrast to the movement of fungal spores in aerial and aquatic habitats, that in the soil is strongly limited. This appears to be the main reason why mycopathogens of soil-inhabiting arthropods are relatively scarce. We may conclude that those fungi which are successful in the soil environment have developed special adaptations.

The life-cycle of entomopathogenic fungi comprises: (1) a parasitic and (2) a saprophytic phase. The latter commences when the infected insect dies and generally ends when vegetative or sexual reproductive organs form outside the host. The saprophytic colonization of the dead host may temporarily cease on the formation of a pseudosclerotium; a dormant stage which enables the fungus to persist and withstand unfavourable conditions. If external factors, particularly temperature and humidity, are suitable, hyphae then emerge through the cuticle. Several entomopathogenic species, e.g. *Paecilomyces* spp. (F6.2), *Beauveria bassiana*, *Beauveria brongniartii* (F6.2) or *Cordyceps* spp. (F4.4), produce hyphal strands, synnemata or stromata outside host insect (Fig. 1). We may consider these long hyphal elements as a morphological adaptation to soil conditions, i.e. they may increase the probability of infecting a new host in and above the soil.

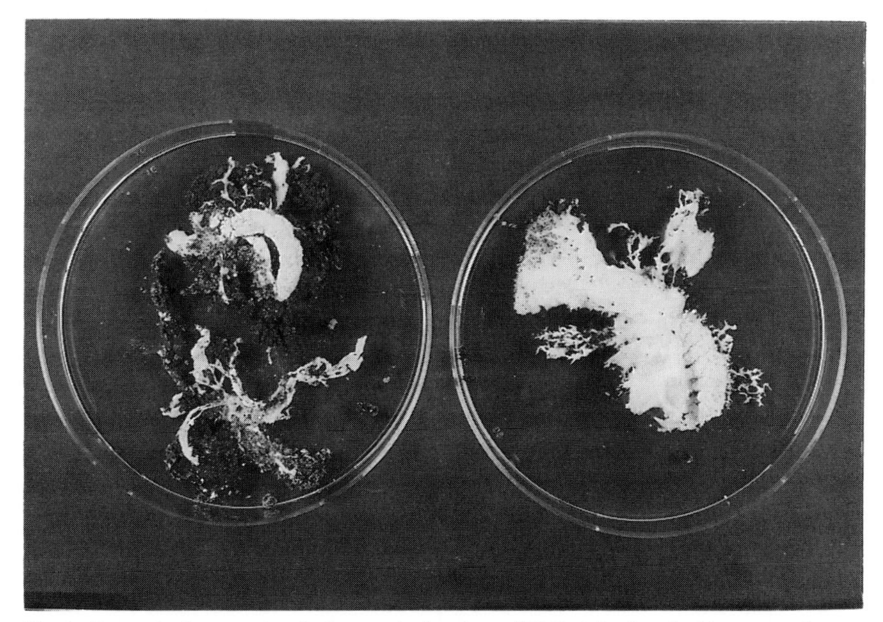

Fig. 1. Saprophytic growth of *Beauveria bassiana* (F6.2) infecting lepidopterous larvae (*Hepialus* sp. (H32)) in the soil (0.63 ×; BBA/Darmstadt).

Hyphae of *B. bassiana* from colonized larvae of the pecan weevil (H25.63) spread through the surrounding soil and formed colonies up to 8.5 cm in diameter (Gottwald and Tedders, 1984). In this way, adjacent, healthy weevil larvae were sometimes infected. Similar observations were also made on *B. brongniartii* infecting larvae of the cockchafer *Melolontha melolontha* (H25.4) (Ferron, 1978). In *Cordyceps aphodii* the branching stromata developing from pseudosclerotia grow up to 11 cm long in the soil. Although these stromata are produced in underground feeding chambers in the dark, they are positively phototropic (Coles, 1979). Growth of *Metarhizium anisopliae* is normally restricted to larval cadavers. However, an interesting observation was made by Coles and Pinnock (1982). Late third instar larvae of *Aphodius tasmaniae* (H25.14) infected with *M. anisopliae* developed hyphal strands 0.2–2 mm in diameter and 10–40 mm long, growing radially from the dead insect through the soil. These hyphae terminated in conidiophores which developed a green conidial layer. On early third instar larvae, however, the fungus produced only the typical, green spore layer on the surface of the dead host. Species of the genera *Paecilomyces* and *Cordyceps* produce synnemata and stromata, respectively, which extend above the soil from the dead, subterranean insect.

Conidia and ascospores are then distributed in the aerial environment. A special adaptation to soil life is the production of microconidia in the entomophthoralean species *Conidiobolus coronatus* (F3.1), which is a common opportunistic subterranean mycopathogen. These propagules produced as "satellites" on typical primary conidia, improve the chances for distribution and host-finding.

In other cases, the soil functions as a reservoir for fungi which, in general, infect insects on plant parts above the ground. For example, *Nomuraea rileyi* (F6.2) causes epizootics in populations of lepidopteran larvae feeding on foliage. The majority of the spores thus produced is washed onto and into the soil. Laboratory studies show that conidia then adhere to the leaves of plant seedlings as they germinate and emerge through the soil (Ignoffo *et al.*, 1977b). Similarly, a biological test showed that an active inoculum of the aphid attacking fungi *Conidiobolus obscurus* and *Erynia neoaphidis* persisted in the surface layers of the soil where an aphid (H20) population on the previous crop had been severely attacked by those species (Latteur, 1977). However, the mechanisms whereby other fungi, such as *Tarichium megaspermum* (F3.1), spend part of their life in the soil and attack insects on leaves, are still unknown.

The saprophytic phase is essential for the completion of the developmental cycle of entomogenous fungi. However, the saprophytic capacity of mycopathogens in the absence of a suitable host insect may also be very important. With the exception of a few Entomophthorales, all entomopathogenic species can be cultured on artificial media. Strictly, therefore, they are facultative pathogens. However, little is known about their free-living capabilities in the field. In a former section it was noted that the spores of some better known species are unable to germinate in normal, non-sterile soil. Furthermore, preliminary experiments reveal that *M. anisopliae* is a poor competitive saprophyte (Zimmermann, unpubl.). Living larvae of *Galleria mellonella* (H32) treated with conidia died within one week in sterile and non-sterile soil alike, whereas heat-killed and treated larvae developed the disease only in sterile soil. Similar results were published for the nematode-attacking fungus *Hirsutella rhossiliensis* (F6.2) (Jaffee and Zehr, 1985), which behaves similarly to entomopathogenic species of the same genus. The fungus colonized dead nematodes and autoclaved wheat seeds only in the absence of other soilborne fungi. Joussier (1977) noted saprophytic growth of *B. brongniartii* in sterile soil. A certain adaptation to saprophytism could explain the observation of Chudare (1982) that strains of entomophthoralean fungi isolated from the soil were less pathogenic than those isolated from insects.

Discussing fungal symbionts, Cooke (1978) considered that fungi are obligate symbionts, in the ecological sense, if they have no capacity for a free-living existence, other than as propagules, in the absence of a suitable host. In contrast, facultative symbionts have a well-developed free-living capability. According to this definition, it seems that most or all entomopathogenic fungi are ecologically more obligate than facultative pathogens.

D. Persistence and Biodegradation of Mycopathogens

Knowledge of the persistence of mycopathogens in the soil is important, not only with respect to their epizootiology, but also in the development of research programmes and strategies in biological control.

In general, the persistence of fungi in the soil depends on the species, strain and morphological stage of the fungus and on the different biotic or abiotic factors discussed above. Entomopathogenic fungi may persist as mycelium within mummified, fungus-killed cadavers, as conidia, as resting spores (Entomophthorales (F3.1)) or—following artificial application—as blastospores or hyphae. There are few published records on the viability of entomogenous fungi within mummified cadavers in soil. Pseudosclerotia of *Cordyceps aphodii* (F4.4) infecting larvae of *Aphodius tasmaniae* (H25.14) survive for 18 months or more in the soil and develop branching stromata with viable conidia when soil moisture levels increase (Coles, 1979; Coles and Pinnock, 1982). The winter survival of laboratory-produced pseudosclerotia of *Nomuraea rileyi* (F6.2) infecting *Heliothis virescens* (H32) was investigated as part of a study of the initiation of epizootics (Sprenkel and Brooks, 1977). Cadavers held on the surface of soil produced infectious conidia for the entire sampling period of 281 days. Infectious conidia were also present on cadavers buried 10 cm in the soil from 14 days to the last recorded date at 194 days. In this way *N. rileyi* may persist from one season to the next. After growing out from infected larvae of the pecan weevil, *Curculio caryae* (H25.63), *Beauveria bassiana* colonies threatened adjacent weevil larvae 90 to 120 days after the primary larvae were infected (Gottwald and Tedders, 1984).

Several workers have investigated the viability of isolated conidia, resting spores or blastospores on the soil surface or in the soil. The environmental persistence of conidia and resting spores of the Entomophthorales was reviewed in detail by Perry *et al.* (1982). The survival of Entomophthoralean conidia which have long been considered short-lived is strongly dependent on abiotic factors such as temperature and humidity. For example, the retention of the infectivity of conidia of *Conidiobolus obscurus* (F3.1) on the surface of soil may last several days at room temperature, but several months at 5°C. At higher temperatures, biodegradative microorganisms caused rapid conidial decomposition.

Blastospores of entomopathogenic Deuteromycetes (F6) generally biodegrade more rapidly than conidia. Fargues *et al.* (1983) found that naked blastospores from *B. bassiana* were inactivated after 3 weeks incubation in soil at 20°C, while several authors observed a viability of conidia of several months or even years. However, there are differences in the survival of conidia between species and even strains. In greenhouse experiments with the curculionid *Sitona lineatus* (H25.63) the number of viable conidia of *B. bassiana* in the soil—using the soil dilution plate method—decreased by a factor of more than 10 during less than one year (Müller-Kögler and Stein, 1970). A similar decrease of viable propagules by a factor of about 10^2 during one year was reported by Müller-Kögler and Zimmermann (1986). In another experiment, however, the viability of conidia of *Metarhizium anisopliae* dropped only to 30–50% during two years (Müller-Kögler and Stein, 1976). For *N. rileyi*, the half-life of conidia on or in the soil varied from 40–90 days in different experiments (Ignoffo *et al.*, 1978). After about 350 and 450 days, about 99% of the original infectivity was lost in buried and soil-surface samples, respectively.

The persistence of conidia of *B. bassiana*, *M. anisopliae*, *N. rileyi*, and *Paecilomyces fumosoroseus* at 19°C was compared using a trap technique (Fargues and Robert, 1985). Inocula of all species were substantially degraded after 6 months incubation to 70–80% dry weight loss. In contrast, the viability of another strain of *M. anisopliae* remained at the initial level even after 21 months. According to the authors, this different behaviour in the persistence of strains may be related to their ability to microcycle conidiation.

In the presence of host insects, mycopathogens persist for a long period in the soil. The activity of *B. bassiana* and *Paecilomyces farinosus* applied for control of Colorado beetle, *Leptinotarsa decemlineata* (H25), can last for two years (Wojciechowska *et al.*, 1977). Similarly, *B. brongniartii* was still active against the cockchafer, *Melolontha melolontha* (H25.14), several years after application; mortality was high in the second generation (Ferron, 1978; Keller, 1983).

V. EPIZOOTIOLOGY

The fundamentals of epizootiology of entomopathogenic fungi were considered in detail by Müller-Kögler (1965b), though without special reference to subterranean insect-fungus relationships. There have been very few long-term investigations in this field; most of them are limited to lists of fauna and their antagonists (see Section III).

A. Factors Affecting Epizootiology

1. The Host

a. Numbers and density of susceptible host species. A mycopathogen may infect one or more insect species, the susceptibilities of which may be similar but more commonly differ. This suggests the concept of primary and secondary hosts. The latter may be important for the survival of the pathogen, particularly when primary hosts are absent.

Metarhizium anisopliae (F6.2) is one of the most frequent mycopathogens of soil insects. It attacks larvae of Tipulidae (H28), Elateridae (H25), Carabidae (H25.1) and Scarabaeidae (H25.14) of the genera *Hoplia*, *Amphimallon* and *Melolontha*. Epizootics however, have been found affecting only wireworms (*Agriotes* spp.) and larvae of *Amphimallon solstitialis*; on Tipulidae and particularly on *Melolontha* the infection rate is always low (Keller, unpubl.). Because of this broad host spectrum a continuous presence of this fungus in the soil may be anticipated, but there may be different pathotypes with different host preferences (Ferron *et al.*, 1972; Fargues, 1976). Within wireworm populations a long development period and overlapping generations also create favourable conditions for the pathogen. Nevertheless the possibilities for prolonged survival are probably better when the pathogen is adapted to a combined pathogenic/saprophytic existence of the kind known for *Conidiobolus coronatus* (F3.1) (Papierok, 1985), though an adaptation to saprophytism may concomitantly impair pathogenicity.

On the other hand, *Beauveria brongniartii* (F6.2) seems to be highly host specific; during long-term field investigations only *Melolontha* sp. was infected with this fungus. This close host–pathogen relationship eventually leads to mutual control, expressed in periodic fluctuations in the host density as observed during the last two centuries (Keller, 1986c). Also, in recent years, epizootics coincided with collapses in the host population.

It is widely accepted that high host densities favour the occurrence of epizootics whereas lower ones inhibit them. More *Sitona lineatus* (H25.63) overwintering at high densities were attacked by *Beauveria bassiana* than at low densities (Hans, 1959), and similar observations were made by Payah and Boethel (1986) for *Cerotoma trifurcata*. Keller (1986a) found increased *B. brongniartii* infection rates in relatively dense *Melolontha* populations and immediately following their collapse. The continuous use of insecticides to control the cockchafer in eastern Switzerland between 1950 and 1960 indirectly suppressed the natural enemies of this insect, apparently by reducing the host density (Keller, 1981).

For many host–pathogen interactions, a certain density threshold seems to be necessary for the initiation of an epizootic. However, Aeschlimann

(1985) observed epizootics in populations of *Sitona discoideus* at two localities where the densities were usually about 100 larvae m^{-2}, whereas at another site the infection rate was low in spite of densities above 2000 larvae m^{-2}. These observations emphasize that host density is only one factor affecting the development of epizootics.

b. Susceptibility of host stadia. Eggs, larvae, pupae and adults may all be susceptible to infection by a certain fungus, though frequently one stage or more is not attacked. Under natural conditions only larval *Sitona humeralis* are attacked by *Beauveria bassiana* (Aeschlimann, 1979), whereas larvae, pupae and adults of *M. melolontha* are susceptible to *Beauveria brongniartii.* Eggs of this insect were often infected in breeding containers but only exceptionally under field conditions (Keller, unpubl). The susceptibility of adult insects may be an advantage for survival and distribution of the pathogen, especially if they leave the soil and disperse.

Not only may some stadia be resistant to infection, but there may also be quantitative differences in susceptibility between stadia and particularly between instars. For example, second and third instar larvae of *Melolontha melolontha* became more sensitive to *B. brongniartii* with increasing age (Ferron, 1967). The susceptibility of adults may differ from that of larvae (Ferron, 1978) or may be comparable as was demonstrated with *Metarhizium anisopliae* infecting *Oryctes rhinoceros* (H25.14) (Ferron *et al.*, 1975). Likewise, the susceptibility of larvae and adults of *Agriotes obscurus* to *M. anisopliae* differed little (Fox, 1961).

c. Duration of host development. The longer the development of an insect, the longer they are exposed to the environment and thus to attack by pathogenic fungi. In insects with a long development, the duration of each instar is correspondingly prolonged and ecdyses, resulting in the shedding of adhering spores or penetrating germ hyphae (Ferron, 1967), take place at longer intervals. Prolonged development or contact with the soil are therefore considered to favour the occurrence of fungus diseases (Ferron, 1978).

d. Host defence reactions. Ecdysis can be considered a defence reaction as mentioned above. In larvae of *Melolontha*, a black spot develops at the point of a penetrating germ hypha of *Beauveria brongniartii.* This melanization of the cuticle together with an aggregation of haematocytes at the infection site are further defence reactions. The so-called black spot disease described by Niklas (1958) and Hurpin and Vago (1958) is therefore not a disease *per se*. Since some of these white grubs survive it must be assumed that they can sometimes resist fungal infections, the degree of success depending on the pathogenic characteristics of the fungal species or strains involved.

e. Host vitality. Numerous physiological factors respecting the host can influence its susceptibility to a mycopathogen, e.g. nutrition, sublethal results of pesticides, injury, parasitization, previous infections or genetically based predisposition. White grubs infected with *Bacillus popilliae, Entomopoxvirus melolonthae* or *Rickettsiella melolonthae* were more susceptible to *Beauveria brongniartii* than uninfected ones (Ferron *et al.*, 1969; Ferron and Hurpin, 1974; Wille, 1959). Treatment with low doses of BHC (gamma HCH) and organophosphates also resulted in an increased susceptibility (Ferron, 1970, 1971). Injuries may facilitate infection not only by pathogenic organisms but also by non-specific pathogens and saprophytes.

f. Behaviour of infected insects. Epigeal insects infected with fungi, especially Entomophthorales (F3.1), often behave in a particular way that increases the dissemination of the spores. Comparable behaviour of fungus-infected subterranean insects is unknown. *Melolontha* larvae killed by *B. brongniartii* occur throughout the colonized profile. A possible accumulation of fungus propagules in upper soil levels may be the result of more deaths occurring between spring and autumn than during the winter when the larvae inhabit lower soil zones.

2. *The Pathogen*

a. Density of the infective material. The amount of the existing infective material has a decisive epizootiological significance. The shorter the developmental time (infection, colonization, sporulation) of a fungus in relation to the development of its host, the quicker the enrichment of infective material and the more rapidly the epizootic develops. In this connection toxins may be important in that they may cause the host to die quickly, or at least prevent reproduction and reduce or eliminate its defence reaction (Roberts, 1981).

The density of the infective material, usually spores or mycelial fragments, can be determined by: (1) bioassay using primary (Latteur, 1980) or secondary hosts (the "bait method") (Zimmermann, 1986); or (2) isolation of the fungi on selective substrates (Veen and Ferron, 1966; Joussier and Catroux, 1976).

The density of infective material necessary to initiate infection depends largely on the host and pathogen under investigation. Nevertheless it is interesting to note that for numerous insect–fungus combinations the LD_{50} lies in the range of 10^5–10^6 spores g^{-1} soil, e.g. *Melolontha melolontha*

(H25.14)/*Beauveria brongniartii* (F6.2) (Ferron, 1978); *Artipus floridanus* (H25.63)/Beauveria bassiana (F6.2) (McCoy *et al.*, 1986); *Otiorhynchus sulcatus* (H25.63)/*Metarhizium anisopliae* and *B. bassiana* (F6.2) (Prado, 1980; Zimmermann, 1981). Of course there are differences according to fungus strains, slope of dose-mortality-response, etc.

b. Number of pathogenic fungus species. Some soil insects are known to be attacked by only one pathogenic fungus species while others are attacked by more, resulting in interspecific competition. However, mixed infections by different pathogenic fungus species among soil insects are unusual or even, as for *Melolontha* spp., unknown (Niklas, 1958; Hurpin and Vago, 1958). In some cases the primary fungal pathogen may be usurped by a secondary faster-growing invader. For example, *Conidiobolus coronatus* (F3.1) can quickly overgrow insects already infected with fungi or other pathogens (Papierok, 1985).

c. Pathogenicity and virulence. Ferron (1978) found obvious differences in virulence between numerous strains of *Beauveria brongniartii* for *Melolontha melolontha* and *Acanthoscelides obtectus* (H25). The LD_{50} ranged from 6×10^6 to more than 3×10^{10} spores g^{-1} peat for *M. melolontha* and from 2×10^5 to more than 2.5×10^8 spores g^{-1} peat for *A. obtectus*. Field data suggest that strains of the same fungus exist which are pathogenic only for adults of *M. melolontha* and others which attack only the larvae (Keller, 1986a).

d. Persistence. The mechanisms developed by mycopathogens to persist in the soil are discussed in Sections IV, C and IV, D.

e. Vitality-reducing effects. During investigations on the progeny of females of *M. melolontha* infected with *B. brongniartii* a high percentage of the young larvae succumbed to mycosis and a significantly larger number than from uninfected females died with non-specific symptoms (Keller, 1978). This phenomenon was considered to be a result of a reduction in the vitality of the progeny due to the infection of the mother individuals.

Mycopathogens may also affect the vitality of host populations by decreasing fecundity and disturbing diapause (Müller-Kögler and Stein, 1970, 1976).

3. The Environment

Abiotic factors (see Section IV, A), biotic factors (see Section IV, B) and human influences (see Section IV, A, 3) have a significant role in epizootiology.

B. The Development of Epizootics

1. Principles

The study of epizootics needs long-term investigations over numerous host generations. Such investigations on soil insects are rare. More or less periodic long-term density fluctuations occur in cockchafer populations, the records based on the numbers of swarming adults (Keller, 1986c). These observations indicate the existence of density-dependent regulating mechanisms, which are probably mainly pathogenic microorganisms.

The hypothesis represented diagramatically in Fig. 2 suggests that a mycopathogen causes little infection during periods of low host and pathogen densities, and as a consequence the density of the host increases (phase 1). When that density passes a certain threshold, the mycopathogen begins to cause infections (phase 2). As the host population grows the fungus finds progressively improved conditions for establishment and spread, finally resulting in an epizootic (phase 3). The host population then

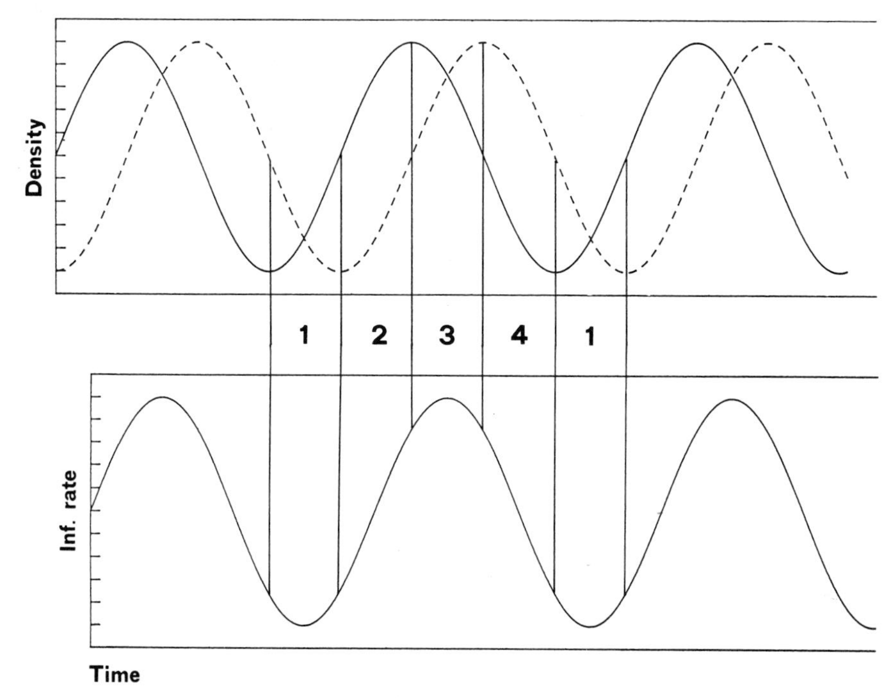

Fig. 2. Schematic drawing of the regulation of a *Melolontha* (H25.14) population by *Beauveria brongniartii* (F3.1). The upper frame shows the development of the host density (solid line) and of the pathogen density (broken line). The lower frame shows the resulting infection rate. The numbers 1–4 correspond to the four phases of an epizootic (see Table II).

collapses and the fungal density, enriched during the epizootic phase, exerts an infection pressure until the host population reaches such a low level that the infection chain is interrupted (phase 4). The population of the mycopathogen then decreases and finally allows the host population to recover.

Hypotheses describing the spatial distribution and spread of a mycopathogen are represented in Fig. 3. Within an area containing a host population (Situation A) an epizootic develops in a favoured site (Stage 1). On the

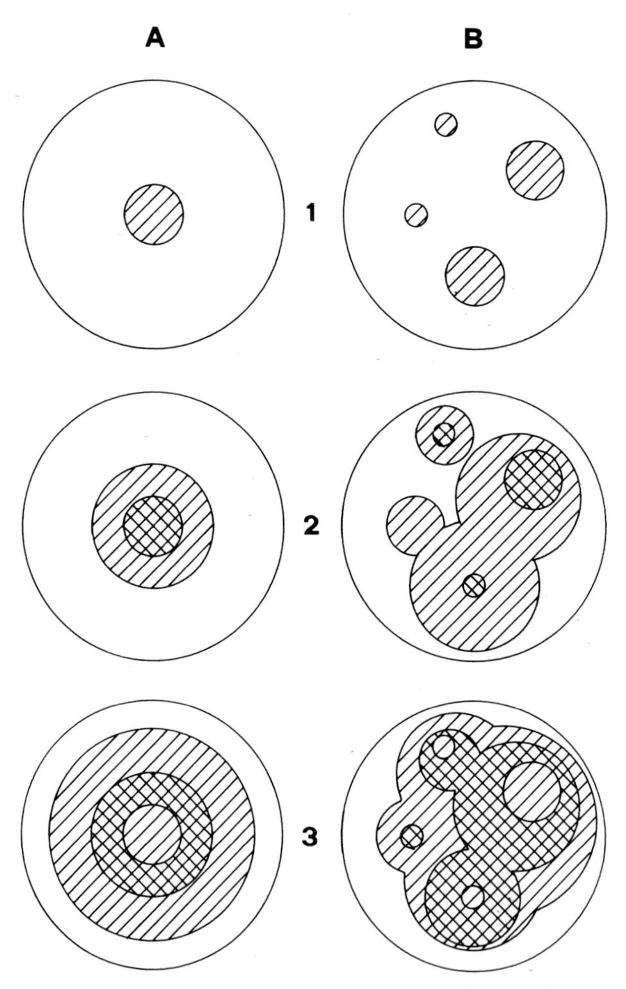

Fig. 3. Schematic drawing of the development of an epizootic in the field: (A) in an isolated population; (B) in a compound population (flight area). Cross-hatched, epizootic; hatched, enzootic; open, latency (healthy).

borders of that site, infection is at the enzootic level (Stage 2) developing to an epizootic. At the same time the host population in the centre decreases followed by a decrease in the infection rate (Stage 3). Situation A occurs in isolated populations or discrete parts of larger host breeding areas. Situation B occurs in areas consisting probably of more or less independent subpopulations of the host, each with an individual development of infection. Parts of these hypotheses were verified experimentally and by field observations of undisturbed populations of *Melolontha melolontha* (H25.14) (Keller, 1986b, c, and unpubl.).

Hitherto most investigations into epizootiology have emphasized host and disease development. The fungus, the other main component of the host–pathogen interaction, has often been neglected.

2. *The Development of Host Population and Disease*

Investigations of the population dynamics of cockchafers in eastern Switzerland strongly indicate that *Beauveria brongniartii* (F6.2) acts as the main regulating factor. One population was maximum in density and distribution in 1972–75 when the trees on which the beetles feed were completely or partially defoliated for a length of some 140 km along the borders of forests. The population density then began to decrease and in 1978 damaged forests were noticed on a borderlength of only 51 km, continuously diminishing to 9.6 km by 1987. The mean rate of infection with *B. brongniartii* at two sites with decreasing population density was 53% in 1981, 21% in 1984 and 12% in 1987, suggesting the post-epizootic phase. The corresponding values for two sites with a stable population density were 18%, 15% and 31% respectively, indicating the enzootic (pre-epizootic?) phase. Although these data demonstrate that an epizootic occurred during the decline of the population, they are not in themselves proof that the fungus was the cause.

A comparable relationship was noted in a smaller, rather isolated population. The first obvious decrease (from 17 to 6 adults m^{-2}) in the population occurred in the generation lasting from 1982–85. During the same period the infection rate increased from 14 to 52%. An even more marked population decrease (from 23.6 2nd instar larvae m^{-2} to 1.8 3rd instar larvae m^{-2}) occurred in the subsequent generation while the corresponding infection rates increased from 25 to 50%. Again, the beginning of the epizootic was not observed, and it is not clear whether the fungus itself induced the population decline or another cause favoured the activity of the fungus. Nevertheless, it was obvious that the collapse of both populations was accompanied by epizootics.

These data and those of experimentally induced population collapses (Ferron, 1978, 1983; Keller, 1986b) fit, quite well, the hypothesis outlined above (Fig. 2); the peak in disease incidence appears after the peak host density and the epizootic phase corresponds with the host population decline. According to the few observations available, the health of the host population recovers during the period of low density.

3. The Development of Fungus Population and Disease

To explain the progress of an epizootic it is important to know not only about the population dynamics of the host, but also about that of the fungus. Its density can be monitored either by employing a bioassay using the primary or a secondary host (Zimmermann, 1986) or by isolating the fungus from soil samples on selective media (Veen and Ferron, 1966; Joussier and Catroux, 1976).

The use of a bioassay provides an estimate of the number of infective propagules per unit of soil relative to that of a standard. The isolation technique gives the fungal density as the number of colony-forming units (CFU) per unit of soil. Without corresponding bioassay, however, this method gives no information about the pathogenicity of the isolates. Neither method indicates the nature of the fungal propagule existing in the soil.

Using the isolation technique, Joussier (1977) mapped the natural occurrence and density of *Beauveria brongniartii* in a breeding site of *Melolontha melolontha* and demonstrated the heterogeneous, insular distribution of the fungus. The author concluded that sample sites with high fungus density (*c.* 10^5 CFU g^{-1} soil) corresponded to sites where infected white grubs had died and the fungus had sporulated, but no close relationship was found between the number of CFUs and the number of infected grubs.

In a control trial of Ferron (1978), there was an increase in the fungus population in the treated plots during the two-year observation period and a corresponding increase in the infection rate of the white grubs. In the untreated plots the density of the fungus population and the proportion of diseased grubs decreased, suggesting that the naturally existing inoculum insufficient to cause an epizootic at the given host density.

Investigations by Fornallaz (in prep.) using the isolation technique in an area where ovipositing females of *M. melolontha* had been treated at their feeding sites with blastospores of *B. brongniartii* (Keller *et al.*, 1986) also demonstrated the clumped distribution of the fungus, reflecting that of the dead, infected females. From these foci the fungus spreads horizontally and vertically within a few months. Corresponding investigations in an area where the white grubs suffered from a naturally induced epizootic demonstrated a relatively homogeneous distribution of the fungus at a high density of about 10^5 CFU g^{-1} soil.

TABLE II. Characterization of the different epizootiological phases.

Epizootiological phase	Host population	Fungus population	Disease incidence
1 Latent	Healthy or: Disease not detectable: density too low sufficient defence reaction	Absent or: Unable to cause infections: density too low low pathogenicity	Zero
	Suffering from sporadic infections restricted to single individuals. Population development not affected.	Low density. Limited distribution. Low pathogenicity.	Very low
2 Enzootic (Pre-epizootic)	Small numbers of diseased individuals always present, distributed locally or throughout the population. Population development affected. Density tends to stabilize.	Low to moderate density. Wide or overall distribution. Inducing low to moderate infection pressure.	Low to moderate
3 Epizootic	High proportion of diseased individuals distributed throughout population. Population development severely affected. Density decreases.	High density, increasing. Overall distribution. Inducing high infection pressure.	High
4 Enzootic (Post-epizootic)	Moderate proportion of diseased individuals, decreasing, distributed throughout population. Population tends to stabilize at low density.	High density, decreasing. Overall distribution. Inducing high to moderate infection pressure, decreasing.	Moderate to low

Based on these results, the changes in fungus population during the development of an epizootic can be summarized as follows (Table II, Fig. 2): the enzootic phase is characterized by a clumped distribution of the fungus related to where infected hosts died. From these foci the fungus is spread by abiotic and biotic factors, including the host. The soil thus becomes progressively enriched with fungus material resulting in an epizootic. During this process the mycopathogen becomes distributed more or less homogeneously and reaches relatively high densities.

VI. SOIL FUNGI AS BIOCONTROL AGENTS

A. Use of *Beauveria brongniartii* (F6.2) to Control *Melolontha melolontha* (H25.14)

One of the first target pests for control by the application of a mycopathogen was the cockchafer, *Melolontha melolontha*. The earliest trials to control white grubs with *Beauveria brongniartii* were done probably in 1884 in Odessa (Russia) by Metchnikoff. Le Mould in northern France started control trials in 1890 and Dufour in Switzerland followed a year later. The soil was treated by introducing fungus-killed larvae or cultures (Dufour, 1894). In Poland in 1934 Karpinski (1950) failed to contaminate the breeding sites by dusting the adults with spores, and between 1950 and 1960 several further attempts to control white grub populations with *Beauveria* were also unsuccessful.

Nevertheless, systematic basic research using several promising pathogens resulted in the conclusion that *B. brongniartii* should be considered as the most promising agent for microbial control of *M. melolontha* populations (Hurpin and Robert, 1972). Subsequent trials demonstrated that the fungus introduced into populations caused epizootics and population collapse (Ferron, 1978; Keller, 1986b). These studies showed that the fungus acts slowly and long-term observations are necessary to follow the regulating process. It was noted that the population decrease did not occur until the generation following the treated one. During this period the breeding sites became enriched with fungus material able to control at least two further generations. In 1985 swarming cockchafers were treated with blastospores at their feeding sites over an area of some 80 ha, corresponding to a breeding area of about 4000 ha (Keller *et al.*, 1986). The disease became established but conclusions regarding the control success cannot yet be drawn.

B. Use of *Metarhizium anisopliae* (F6.2) to Control *Oryctes rhinoceros* (H25.14)

First attempts to control the rhinoceros beetle, *Oryctes rhinoceros*, were done by Friederichs 1913, who contaminated artificial breeding sites consisting mainly of rotten shells of cacao with spores of *Metarhizium anisopliae*. Later authors used similar methods with varying success. Only long-spore isolates, *M. anisopliae* var. *major*, are highly virulent, while the short-spore form, *M. anisopliae* var. *anisopliae*, has little or no virulence for *O. rhinoceros*. It was found that spores sprayed on breeding sites resulted in a certain mortality within 3 months, but 5–6 months later no infected larvae and no trace of the fungus was found. When the fungus was grown on oat grains and applied to natural breeding sites most of the *O. rhinoceros* larvae were killed during the following 3 months, and the spores remained viable for at least 24 months (Latch and Falloon, 1976).

C. Use of *Beauveria* and *Metarhizium* (F6.2) to Control Curculionidae (H25.63)

The Curculionidae (Coleoptera) include several pests of importance throughout the world. In many species, the adults feed on leaves of different ornamental plants and crops while their larvae live in the soil, damaging root systems. Because of problems with insecticidal control, research has focused on biological control methods. Under natural conditions curculionid adults and larvae are attacked mainly by two species of mycopathogens, *Beauveria bassiana* and *Metarhizium anisopliae*. Comprehensive greenhouse experiments with these two fungi against *Sitona lineatus* demonstrated that the application of mycopathogens to the soil may be successful even for a very small insect which lives for a comparatively short time (about 7 weeks) in the soil (Müller-Kögler and Stein, 1970, 1976). However, in a preliminary trial under field conditions the results with *B. bassiana* against larvae of *Sitona lineatus* were disappointing (Bailey and Milner, 1985).

A further curculionid candidate for biological control with mycopathogens in the soil is the black vine weevil. *Otiorhynchus sulcatus*. On different potted plants, a prophylactic treatment with a spore suspension of *M. anisopliae* resulted in an efficacy of 80–100% (Prado, 1980; Zimmermann, 1981, 1984). In the field, the application of the fungus gave results which varied according to soil temperature, formulation and, possibly, also soil conditions (Zimmermann, unpubl.). Recently, *B. bassiana* and *M. anisopliae* have also gained considerable attention as potential biological control agents for the pecan weevil, *Curculio caryae* (Gottwald and Tedders, 1983). The authors consider that *B. bassiana* is superior to *M. anisopliae* because of its higher pathogenicity, the production and release of 10 to 200 times more conidia per cadaver and its ability to spread saprophytically through the soil.

D. Use of Mycopathogens to Control other Pest Insects

In southern Australia, *Metarhizium anisopliae* (F6.2) was used as a control agent for larvae of the pasture cockchafer, *Aphodius tasmaniae* (H25.14) (Coles and Pinnock, 1982). The fungus was applied: (1) as a spore suspension, and (2) in a wheat bran bait. To ensure high larval mortalities, fungal baits were spread on pastures before the winter rains, when larvae came to the soil surface and foraged for food. After 4–5 months the average larval mortality caused by *M. anisopliae* was 55%. A similar application method was used in New Zealand for controlling porina caterpillars, *Wiseana* spp. (H32), which also cause severe damage to pastures. At 38.5 kg ha^{-1} of chipped wheat bran inoculated with *M. anisopliae*, 90% of larvae died from fungus infection in the soil (Latch and Kain, 1983).

Considerable efforts have been undertaken by several Polish authors to control adults of the Colorado beetle (H25) during hibernation in the soil using *Beauveria bassiana* and *Paecilomyces farinosus* (Fedorko *et al.*, 1977). The introduction of one or both pathogens into the soil resulted in an increased mortality of the adults, in a retarded emergence in the spring and, additionally, in a reduced viability of Colorado beetle populations in later generations (Bajan *et al.*, 1977).

VII. CONCLUSIONS

Although the soil appears a favourable habitat for insect-pathogenic fungi, only a few species have been recorded there. This may be at least partly because infected insects remain hidden in the soil habitat. Investigations on the life of soil insects and their interactions with mycopathogens are particularly time consuming, which may further explain our limited knowledge, restricted mainly to examples of practical interest. These include scarabaeid and curculionid larvae, which are the principal targets for microbial control of soil insects using mycopathogens. Consequently, to understand more about the importance of these fungi, studies on their interactions with beneficial and other non-target soil arthropods are necessary. Furthermore, ecological studies on the natural occurrence and distribution of mycopathogens in different soil types and in different geographical regions independent of specific target insects are needed.

The genus *Cordyceps* (F4.4) offers some fascinating opportunities for more theoretically oriented research. Some species of this genus are known to exist also in the imperfect stage, being classified in other taxa, e.g. *Hirsutella* (F6.2). Does this "double life" offer any advantages for survival? Are there differences in occurrence and host specificity of the two forms?

With respect to epizootiology, more research on the host–pathogen interaction is needed, particularly on the population dynamics of the mycopathogen. How far is a saprophytic life possible? Does the saprophytic ability vary only between species or also between strains? How does the fungus spread? Are there possibilities of improving the rate of its multiplication and its survival? Are there interactions between the mycopathogens themselves? Answers to these and similar questions are of great interest both fundamentally and practically with respect to soil insect–mycopathogen relationships.

Acknowledgement

The authors are greatly indebted to Dr N. Wilding for critically reviewing the manuscript and correcting the English phraseology.

REFERENCES

Aeschlimann, J. P. (1979). Sampling methods and construction of life tables for *Sitona humeralis* populations (Col., Curculionidae) in mediterranean climatic areas. *J. Appl. Ecol.* **16**, 405–415.

Aeschlimann, J. P. (1985). Occurrence and pathogenicity of *Beauveria bassiana* infesting larval *Sitona discoideus* (Col., Curculionidae) in the mediterranean region. *Entomophaga* **30**, 73–82.

Bailey, P., and Milner, R. (1985). *Sitona discoideus*: A suitable case for control with pathogens? *In* "Proc. 4th Aust. Conf. Grassland Invertebr. Ecol." (R. B. Chapman, ed.), pp. 210–214. Caxton Press, Christchurch, New Zealand.

Bajan, C., Kmitowa, K., Wojciechowska, M., and Fedorko, A. (1977). The effect of entomopathogenic microorganisms introduced into the soil on the development of successive generations of the Colorado beetle. *Pol. Ecol. Stud.* **3**, 157–165.

Bedding, R. A., and Akhurst, R. J. (1975). A simple technique for the detection of insect parasitic rhabditid nematodes in soil. *Nematologica* **21**, 109–116.

Chudare, Z. P. (1982). Pathogenicity of entomophthoraceous fungi. *Latv. PSR. Zinat. Akad. Vestis.* **10**, 88–105 (in Russian).

Clerk, G. C. (1969). Influence of soil extracts on the germination of conidia of the fungi *Beauveria bassiana* and *Paecilomyces farinosus*. *J. Invertebr. Pathol.* **13**, 120–124.

Coles, R. B. (1979). The biology of *Cordyceps aphodii* (Sphaeriales; Clavicipitaceae). *In* "Proc. 2nd Aust. Conf. Grassland Invertebr. Ecol." (T. K. Crosby and R. P. Pottinger, eds), pp. 207–212. Government Printer, Wellington.

Coles, R. B., and Pinnock, D. E. (1982). Control of the pasture cockchafer with the fungal pathogen *Metarhizium anisopliae. In* "Proc. 3rd Aust. Conf. Grassland Invertebr. Ecol." (K. E. Lee, ed.), pp. 191–198. S. A. Government Printer, Adelaide.

Cooke, R. (1978). "The Biology of Symbiotic Fungi." John Wiley, Chichester.

Domsch, K. H., Gams, W., and Anderson, T.-H. (1980). "Compendium of Soil Fungi," Vols 1 and 2. Academic Press, New York and London.

Dufour, J. (1894). Ueber die mit *Botrytis tenella* zur Bekämpfung der Maikäferlarven erzielten Resultate. *Forstl.-naturwiss. Zeitschr.* **6**, 249–255.

Fargues, J. (1976). Spécificité des champignons pathogènes imparfaits (Hyphomycetes) pour des larves de coléoptères (Scarabaeidae et Chrysomelidae). *Entomophaga* **21**, 313–323.

Fargues, J., and Robert, P. H. (1985). Persistance des conidiospores des hyphomycètes entomopathogènes *Beauveria bassiana* (Bals.) Vuill., *Metarhizium anisopliae* (Metsch.) Sor., *Nomuraea rileyi* (F.) Samson et *Paecilomyces fumoso-roseus* Wize dans le sol, en conditions contrôlées. *Agronomie (Paris)* **5**, 73–80.

Fargues, J., Reisinger, O., Robert, P. H., and Aubart, C. (1983). Biodegradation of entomopathogenic hyphomycetes: Influence of clay coating on *Beauveria bassiana* blastospores survival in soil. *J. Invertebr. Pathol.* **41**, 131–142.

Fedorko, A., Bajan, C., Kmitowa, K., and Wojciechowska, M. (1977). Effectiveness of the use of selected microorganisms to control the Colorado beetle during hibernation. *Pol. Ecol. Stud.* **3**, 127–134.

Ferron, P. (1967). Étude en laboratoire des conditions écologiques favorisant le développement de la mycose à *Beauveria tenella* du ver blanc. *Entomophaga* **12**, 257–293.

Ferron, P. (1970). Augmentation de la sensibilité des larves de *Melolontha melolontha* L. (Coléoptère Scarabaeidae) à *Beauveria tenella* (Delacr.) Siemaszko au moyen de quantités réduites de HCH. *Proc. Int. Colloq. Insect Pathol., 5th 1970 College Park, USA*. pp. 66–79.

Ferron, P. (1971). Modification of the development of *Beauveria tenella* mycosis in *Melolontha melolontha* larvae by means of reduced doses of organophosphorus insecticides. *Entomol. Exp. Appl.* **14**, 457–466.

Ferron, P. (1978). Etiologie et épidémiologie des muscardine. Thèse, Univ. P. and M. Curie, Paris.

Ferron, P. (1983). Induction artificielle d'une épizootie à *Beauveria brongniartii* dans une population de *Melolontha melolontha* L. *Symbioses.* **15**, 75–83.

Ferron, P., and Hurpin, B. (1974). Effets de la contamination simultanée ou successive par *Beauveria tenella* et par *Entomopoxvirus melolonthae* des larves de *Melolontha melolontha* (Col. Scarabaeidae). *Ann. Soc. Entomol. Fr.* **10**, 771–779.

Ferron, P., Hurpin, B., and Robert, P. H. (1969). Sensibilisation des larves de *Melolontha melolontha* L. à la mycose à *Beauveria tenella* par une infection préalable à *Bacillus popilliae. Entomophaga* **14**, 429–437.

Ferron, P., Hurpin, B., and Robert, P. H. (1972). Sur la spécificité de *Metarhizium anisopliae* (Metsch.) Sorokin. *Entomophaga* **17**, 165–178.

Ferron, P., Robert, P. H., and Deotte, A. (1975). Susceptibility of *Oryctes rhinoceros* adult to *Metarhizium anisopliae. J. Invertebr. Pathol.* **25**, 313–319.

Fox, C. J. S. (1961). The incidence of green muscardine in the European wireworm, *Agriotes obscurus* (Linnaeus), in Nova Scotia. *J. Insect Pathol.* **3**, 94–95.

Friederichs, K. (1913). Über den gegenwärtigen Stand der Bekämpfung des Nashornkäfers (*Oryctes rhinoceros* L.) im Samoa. *Der Tropenpflanzer* **17**, 660–675.

Gottwald, T. R., and Tedders, W. L. (1983). Suppression of pecan weevil (Coleoptera: Curculionidae) populations with entomopathogenic fungi. *Environ. Entomol.* **12**, 471–474.

Gottwald, T. R., and Tedders, W. L. (1984). Colonization, transmission, and longevity of *Beauveria bassiana* and *Metarhizium anisopliae* (Deuteromycotina: Hyphomycetes) on pecan weevil larvae (Coleoptera: Curculionidae) in the soil. *Environ. Entomol.* **13**, 557–560.

Griffin, D. M. (1963). Soil moisture and the ecology of soil fungi. *Biol. Rev. Cambridge Philos. Soc.* **38**, 141–166.

Hans, H. (1959). Beiträge zur Biologie von *Sitona lineatus* L. *Z. Angew. Entomol.* **44**, 343–386.

Huber, J. (1958). Untersuchungen zur Physiologie insektentötender Pilze. *Arch. Mikrobiol.* **29**, 257–276.

Hurpin, B., and Robert, P. H. (1972). Comparison of the activity of certain pathogens of the cockchafer *Melolontha melolontha* in plots of natural meadowland. *J. Invertebr. Pathol.* **19**, 291–298.

Hurpin B., and Vago, C. (1958). Les maladies du hanneton commun (*Melolontha melolontha* L.). *Entomophaga* **3**, 285–330.

Ignoffo, C. M., Hostetter, D. L., Garcia, C., and Pinnell, R. E. (1975). Sensitivity of the entomopathogenic fungus *Nomuraea rileyi* to chemical pesticides used on soybeans. *Environ. Entomol.* **4**, 765–768.

Ignoffo, C. M., Garcia, C., Hostetter, D. L., and Pinnell, R. E. (1977a). Vertical movement of conidia of *Nomuraea rileyi* through sand and loam soils. *J. Econ. Entomol.* **70**, 163–164.

Ignoffo, C. M., Garcia, C., Hostetter, D. L., and Pinnell, R. E. (1977b). Laboratory studies of the entomopathogenic fungus *Nomuraea rileyi*: Soil-borne contamination of soybean seedlings and dispersal of diseased larvae of *Trichoplusia ni*. *J. Invertebr. Pathol.* **29**, 147–152.

Ignoffo, C. M., Garcia, C., Hostetter, D. L., and Pinnell, R. E. (1978). Stability of conidia of an entomopathogenic fungus *Nomuraea rileyi* in and on soil. *Environ. Entomol.* **7**, 724–727.

Jaffee, B. A., and Zehr, E. I. (1985). Parasitic and saprophytic abilities of the nematode-attacking fungus *Hirsutella rhossiliensis*. *J. Nematol.* **17**, 341–345.

Johnpulle, A. L. (1938). Temperatures lethal to the green muscardine fungus, *Metarrhizium anisopliae* (Metch.) Sorok. *Trop. Agric.* **90**, 80–83.

Joussier, D. (1977). Recherches sur *Beauveria tenella* (Delacr.) Siem., agent de la muscardine des larves du hanneton et son emploi dans la lutte contre cet insecte. Thèse, Univ. Dijon. 125 pp.

Joussier, D., and Catroux, G. (1976). Mise au point d'un milieu de culture pour le dénombrement de *Beauveria tenella* dans les sols. *Entomophaga* **21**, 223–225.

Karpinski, J. J. (1950). The problem of controlling the beetle *Melolontha* by the fungus *Beauveria densa* Pic. *Ann. Univ. M. Curie-Skladowska, Sect. E., Lublin* **5**, 29–75 (in Polish, English summary).

Keller, S. (1978). Infektionsversuche mit dem Pilz *Beauveria tenella* an adulten Maikäfern (*Melolontha melolontha* L.). *Mitt. Schweiz. Entomol. Ges.* **51**, 13–19.

Keller, S. (1981). Früheres und gegenwärtiges Auftreten der Maikäfer (*Melolontha melolontha* L.) im Kanton Thurgau. *Mitt. Thurg. Naturf. Ges.* **44**, 75–89.

Keller, S. (1983). Die mikrobiologische Bekämpfung des Maikäfers (*Melolontha melolontha* L.) mit dem Pilz *Beauveria brongniartii*. *Mitt. Schweiz. Landwirtsch.* **31**, 61–64.

Keller, S. (1986a). Quantitative ecological evaluation of the May beetle pathogen, *Beauveria brongniartii*, and its practical application. *In* "Fundamental and Applied Aspects of

Invertebrate Pathology" (R. A. Samson, J. M. Vlak and D. Peters, eds), pp. 178–181. Foundation of the Fourth Int. Colloq. Invertebr. Pathol., Wageningen.

Keller, S. (1986b). Control of May beetle grubs (*Melolontha melolontha* L.) with the fungus *Beauveria brongniartii* (Sacc.) Petch. *In* "Fundamental and Applied Aspects of Invertebrate Pathology" (R. A. Samson, J. M. Vlak and D. Peters, eds), pp. 525–528. Foundation of the Fourth Int. Colloq. Invertebr. Pathol., Wageningen.

Keller, S. (1986c). Populationsdynamik. *In* "Neuere Erkenntnisse über den Maikäfer". *Beiheft Mitt. Thurg. Naturf. Ges.*, pp. 25–40.

Keller, S., Keller, E., and Auden, J. A. L. (1986). Ein Grossversuch zur Bekämpfung des Maikäfers (*Melolontha melolontha* L.) mit dem Pilz *Beauveria brongniartii* (Sacc.) Petch. *Mitt. Schweiz. Entomol. Ges.* **59**, 47–56.

Lal, R., ed. (1984). "Insecticide Microbiology." Springer Verlag, Berlin.

Latch, G. C. M., and Falloon, R. E. (1976). Studies on the use of *Metarhizium anisopliae* to control *Oryctes rhinoceros. Entomophaga* **21**, 39–48.

Latch, G. C. M., and Kain, W. M. (1983). Control of porina caterpillar (*Wiseana* spp.) in pasture by the fungus *Metarhizium anisopliae, N.Z. J. Exp. Agric.* **11**, 351–354.

Latteur, G. (1977). Sur la possibilité d'infection directe d'Aphides par *Entomophthora* à partir de sols hébergeant un inoculum naturel. *C. R. Hebd. Seances Acad. Sci., Sér D* **284**, 2253–2256.

Latteur, G. (1980). The persistence of infectivity of conidia of *Entomophthora obscura* at different temperatures on the surface of an unsterilised soil. *Acta Oecol., Oecol. Appl.* **1**, 29–34.

Lingg, A. J., and Donaldson, M. D. (1981). Biotic and abiotic factors affecting stability of *Beauveria bassiana* conidia in soil. *J. Invertebr. Pathol.* **38**, 191–200.

Mains, E. B. (1958). North American entomogenous species of *Cordyceps. Mycologia* **50**, 169–222.

Mankau, R. (1968). *Acremonium* sp. attacking soil mites and collembola. *J. Invertebr. Pathol.* **12**, 463–464.

Marchal, M. (1977). Fungi imperfecti isolé d'une population naturelle d'*Otiorrhynchus sulcatus* Fabr. (Col., Curculionidae). *Rev. Zool. Agric. Pathol. Vég.* **76**, 101–108.

McCoy, C. W., Beavers, G.M., and Tarrant, C.A. (1986). Susceptibility of *Artipus floridanus* to different isolates of *Beauveria bassiana. Fla. Entomol.* **68**, 402–409.

Mietkiewski, R., and Balazy, S. (1982). On the fungus *Isaria nipponica* Kobayasi 1939 discovered in Poland. *Bull. Acad. Pol. Sci., Sér. Sci. Biol.* **30**, 89–93.

Müller-Kögler, E. (1957). Über eine Mykose der Larven von *Tipula paludosa* Meig. durch *Empusa* sp. *Z. Pflanzenkr. (Pflanzenpathol.) Pflanzenschutz* **64**, 529–534.

Müller-Kögler, E. (1961). *Melanospora parasitica* Tul. als Parasit der insektenpathogenen *Beauveria tenella* (Delacr.) Siem. *Z. Pflanzenkr. Pflanzenschutz* **68**, 600–605.

Müller-Kögler, E. (1965a). *Cordyceps militaris* (Fr.) Link: Beobachtungen und Versuche anläßlich eines Fundes auf *Tipula paludosa* Meig. (Dipt., Tipul.) *Z. Angew. Entomol.* **55**, 409–418.

Müller-Kögler, E. (1965b). "Pilzkrankheiten bei Insekten". Parey, Berlin.

Müller-Kögler, E., and Stein, W. (1970). Gewächshausversuche mit *Beauveria bassiana* (Bals.) Vuill. zur Infektion von *Sitona lineatus* (L.) (Coleopt., Curcul.) im Boden. *Z. Angew. Entomol.* **65**, 59–76.

Müller-Kögler, E., and Stein, W. (1976). Gewächshausversuche mit *Metarhizium anisopliae* (Metsch.) Sorok. zur Infektion von *Sitona lineatus* (L.) (Col., Curculionidae) im Boden. *Z. Pflanzenkr. Pflanzenschutz* **83**, 96–108.

Müller-Kögler, E., and Zimmermann, G. (1986). Zur Lebensdauer von *Beauveria bassiana* in kontaminiertem Boden unter Freiland- und Laboratoriumsbedingungen. *Entomophaga* **31**, 285–292.

Niklas, O. F. (1958). Auftreten und Periodik verschiedener Krankheiten und Parasiten bei Larven des Maikäfers (*Melolontha* spec.). *Entomophaga* **3**, 71–88.

Papierok, B. (1985). Données écologiques et expérimentales sur les potentialités entomopathogènes de l'entomophthorales *Conidiobolus coronatus*. *Entomophaga* **30**, 303–312.

Payah, W. S., and Boethel, D. J. (1986). Impact of *Beauveria bassiana* (Balsamo) Vuillemin on survival of overwintering bean leaf beetles, *Cerotoma trifurcata* (Forster), (Coleoptera, Chrysomelidae). *Z. Angew. Entomol.* **102**, 295–303.

Perry, D. F., Latteur, G., and Wilding, N. (1982). The environmental persistence of propagules of the Entomophthorales. *Proc. Int. Colloq. Invertebr. Pathol., 3rd, 1982, Brighton*, pp. 325–330.

Prado, E. (1980). Bekämpning av öronvivellarver (*Otiorrhynchus sulcatus*) med hjälp av de insektspatogena svamparna *Beauveria bassiana, Metarrhizium anisopliae* och *Metarrhizium flavoviride*. *Växtskyddsnotiser* **44**, 160–167.

Roberts, D. W. (1981). Toxins of entomopathogenic fungi. *In* "Microbial Control of Pests and Plant Diseases 1970–1980" (H. D. Burges, ed.), pp. 441–464. Academic Press, London.

Roberts, D. W., and Campbell, A. S. (1977). Stability of entomopathogenic fungi. *In* "Environmental Stability of Microbial Insecticides" (D. L. Hostetter and C. M. Ignoffo, eds), *Misc. Publ. Ent. Soc. Am.* **10**, 19–76.

Samšiňák, K. (1964). Die Milben als Überträger von Insektenkrankheiten. *Vestn. Cesk. Spol. Zool.* **28**, 234–236.

Samšiňáková, A., and Samšiňák, K. (1970). Milben (Acari) als Verbreiter des Pilzes *Beauveria bassiana* (Bals.) Vuill. *Z. Parasitenkd.* **34**, 351–355.

Schabel, H. G. (1976). Green muscardine disease of *Hylobius pales* (Herbst) (Coleoptera: Curculionidae). *Z. Angew. Entomol.* **81**, 413–421.

Schabel, H. G. (1982). Phoretic mites as carriers of entomopathogenic fungi. *J. Invertebr. Pathol.* **39**, 410–412.

Shields, M. S., Lingg, A. J., and Heimsch, R. C. (1981). Identification of a *Penicillium urticae* metabolite which inhibits *Beauveria bassiana*. *J. Invertebr. Pathol.* **38**, 374–377.

Soares, G. G., Marchal, M., and Ferron, P. (1983). Susceptibility of *Otiorhynchus sulcatus* (Coleoptera: Curculionidae) larvae to *Metarhizium anisopliae* and *Metarhizium flavoviride* (Deuteromycotina: Hyphomycetes) at two different temperatures. *Environ. Entomol.* **12**, 1886–1890.

Soper, R. S., Delyzer, A. J., and Smith, L. F. R. (1976). The genus *Massospora* entomopathogenic for cicadas. II. Biology of *Massospora levispora* and its host *Okanagana rimosa*, with notes on *Massospora cicadina* on the periodical cicadas. *Ann. Entomol. Soc. Am.* **69**, 89–95.

Sprenkel, R. K., and Brooks, W. M. (1977). Winter survival of the entomogenous fungus *Nomuraea rileyi* in North Carolina. *J. Invertebr. Pathol.* **29**, 262–266.

Storey, G. K., and Gardner, W. A. (1987). Vertical movement of commercially formulated *Beauveria bassiana* through four Georgia soil types. *Environ. Entomol.* **16**, 178–181.

Sturhan, D., and Schneider, R. (1980). *Hirsutella heteroderae*, ein neuer nematodenparasitärer Pilz. *Phytopathol. Z.* **99**, 105–115.

Veen, K. H., and Ferron, P. (1966). A selective medium for the isolation of *Beauveria tenella* and of *Metarhizium anisopliae*. *J. Invertebr. Pathol.* **8**, 268–269.

Walstad, J. D., Anderson, R. F., and Stambough, W. J. (1970). Effects of environmental conditions on two species of muscardine fungi (*Beauveria bassiana* and *Metarrhizium anisopliae*). *J. Invertebr. Pathol.* **16**, 221–226.

Wartenberg, H., and Freund, K. (1961). Der Konservierungseffekt antibiotischer Mikroorganismen auf Konidien von *Beauveria bassiana* (Bals.) Vuill. *Zentralbl. Bakt. II Abt.* **114**, 718–724.

Wille, H. (1959). Infektionsversuche mit *Rickettsia melolonthae* Krieg und Beiträge zur Histopathologie der "Lorscher Krankheit" der Engerlinge von *Melolontha melolontha* L. *Trans. Int. Conf. Insect Pathol. 1st, 1958, Praha*, pp. 115–119.

Wojciechowska, M., Kmitowa, K., Fedorko, A., and Bajan, C. (1977). Duration of activity of entomopathogenic microorganisms introduced into the soil. *Pol. Ecol. Stud.* **3**, 141–148.

Zimmermann, G. (1980). *Paecilomyces tenuipes* (Peck) Samson, ein seltener insektenpathogener Pilz an Noctuiden. *Anz. Schädlingskd., Pflanzenschutz, Umweltschutz* **53**, 69–72.

Zimmermann, G. (1981). Gewächshausversuche zur Bekämpfung des Gefurchten Dickmaulrüsslers, *Otiorhynchus sulcatus* F., mit dem Pilz *Metarhizium anisopliae* (Metsch.) Sorok. *Nachrichtenbl. Dtsch. Pflanzenschutzdienst* **33**, 103–108.

Zimmermann, G. (1982). Effect of high temperatures and artificial sunlight on the viability of conidia of *Metarhizium anisopliae. J. Invertebr. Pathol.* **40**, 36–40.

Zimmermann, G. (1984). Weitere Versuche mit *Metarhizium anisopliae* (Fungi imperfecti, Moniliales) zur Bekämpfung des Gefurchten Dickmaulrüsslers, *Otiorhynchus sulcatus* F., an Topfpflanzen im Gewächshaus. *Nachrichtenbl. Dtsch. Pflanzenschutzdienst* **36**, 55–59.

Zimmermann, G. (1986). The "*Galleria* bait method" for detection of entomopathogenic fungi in soil. *Z. Angew. Entomol.* **102**, 213–215.

Zimmermann, G., and Bode, E. (1983). Untersuchungen zur Verbreitung des insektenpathogenen Pilzes *Metarhizium anisopliae* (Fungi imperfecti, Moniliales) durch Bodenarthropoden. *Pedobiologia* **25**, 65–71.

Closing Remarks

DR ROY WATLING
President of the British Mycological Society

It is with great pleasure, tinged with sadness, that I accept the task of drawing this Symposium to a close. One can look back with much satisfaction on this association between the Royal Entomological Society of London and the British Mycological Society and it is hard now to appreciate that the flirtation is over. Let us hope it will not be long before it is resumed; I think you will all agree that a great deal has been learnt. It is gratifying, too, that so many foreign guests were able to join us at a symposium which ranged in subject matter from the palatibility of fungi, as far as insects are concerned, to insect control utilizing pathogenic fungi.

It was the late Douglas Hincks who introduced me to the subject of fungivorous insects, especially beetles, and I had many happy times with him collecting fungi and their inhabitants on excursions of the Yorkshire Naturalists' Union. As a taxonomic mycologist, I am now really in the "firing line"! Collecting material for an herbarium can be hazardous indeed unless great care is taken to dry the material properly; if not, mycophagous dipterous larvae can quickly destroy one's collections and waste all one's effort. Small fungal specimens can be dried by immersing them completely in silica gel, in which fungivores and fungi alike are desiccated, but larger specimens must be dried in an air oven; large boletes, particularly, should be sliced to ensure that during the drying process the inner tissue is not totally eaten away leaving only a useless shell. A film drying cabinet with forced ventilation and running at 40–42°C is ideal, although drying over a 100 watt bulb is a useful compromise and is often all that is possible when collecting in foreign parts. Back in the Edinburgh herbarium all material is frozen for 24 hours in a deep freeze at – 18°C to kill concealed fungivores. When sending material from abroad it is always best to pack specimens with paradichlorobenzene or naphthalene crystals to discourage insect activity.

In spite of the care taken in the herbarium, the attractiveness of woody fungi such as polypores for beetles is still sometimes demonstrated, adults emerging even after many years. One such example in Edinburgh was *Ceracis cucculatus* (Coleoptera, Ciidae (H25.52)), apparently "holed up" in a polypore from Kenya. *Trogoderma granarium* (Coleoptera, Dermestidae (H25.21a)) and *Stegobium paniceum* (Coleoptera, Anobiidae (H25.23)), have both been found in fungal material in the herbarium as well as the psocid *Liposceles bostrychophilus* (Liposcelidae (H18)), in fungal material which was supposed to have been frozen. *Octotemnus* sp. (Coleoptera: Ciidae) feeding on *Coriolus hirsutus* (Wulf.: Fr.) Quél. (F5.8), from Papua New Guinea, is one of several insects intercepted during quarantine activities at the Royal Botanic Garden. H. D. Gordon (1938, *Trans. Brit. Mycol. Soc.* **21**, 193–197) described the then rare *Microgramme filum* (Coleoptera, Lathridiidae (H25.47)) from the Edinburgh herbarium in 1938 and periodically it is even today found in old fungi. More recently, material of *Aspergillus versicolor* (Vuill.) Tirab (F6.2) was received for identification, a fungus imperfecti colonizing a pickled squid from which the preservative had evaporated. To add to the mischief, the fungus itself was home for the beetle *Corticaria fulva* (Coleoptera, Lathridiidae). Other enquiries have involved *Dryadaula pactolia* (Tineidae (H32.4)), a moth feeding on a mixture of species of Mycelia Sterilia, the major part previously known as *Rhacodium cellare* Pers.: Fr. (F6.2) (= *Zasmidium*) from a bonded warehouse in Edinburgh (Morrison, B., 1968, *Entomol. Gaz.* **19**, 181–188); honey bees which had "choked" on spores of *Melampsora* (F5.22), a rust fungus; and a member of the genus *Leiodes* (Coleoptera, Leiodidae (H25.4)) feeding on the hypogeous phycomycete *Glomus macrocarpus* Tul. & Tul. (F3.1).

In the first session of the Symposium, J. F. Lawrence considered adaptations in the Coleoptera (H25) for fungivory, particularly adaptations of the mouthparts, illustrating his paper with fascinating scanning electron micrographs. Dipterous fungivory was not covered in detail and it was a great pity that P. J. Chandler was unable to participate, as his is part of the long tradition of studies on mycophagous Diptera (H28) in the British Isles, a tradition which includes the works of P. A. Buxton (1900, *Entomol. Mon. Mag.* **96**, 61–94), F. W. Edwards (1925, *Trans. R. Entomol. Soc. London* **1924**, 505–670) and Chandler (1976, *Proc. R. Irish Acad., Sect. B* **76** (3), 87–110).

Our first session was concluded with a consideration of ecological concepts, admirably covered by I. Hanski, and leading us into the afternoon session dealing with mutualistic relationships between insects and fungi. The first paper by T. G. Wood, was on termite (H9)–*Termitomyces* (F5.10) relationships, a fascinating area of study, especially as the fungus basidiomes are a source of food for human populations in areas where these fungi grow. The newly launched journal of the British Mycological Society, the *Mycologist*, features an article on these exciting agarics (Piearce, G. D., 1986, *Mycologist* **21**, 111–116). J. M. Cherrett successfully replied on behalf of the leaf-cutter ants (H33.9) and their fungi to persuade delegates to the symposium that this relationship is just as interesting and important as that of the termites. R. A. Beaver concluded the day with an account of ambrosia beetles (H25.63), a group causing so much damage to timber in tropical ecosystems almost, we learnt, as soon as the trunks have fallen to the ground; the speaker made us all realize the much greater problems which exist in these areas than we experience in temperate Europe.

Moving into the third session the theme of ambrosia and related beetles was continued firstly with A. A. Berryman posing several questions and offering an overview of the evolution of plant pathogens and insect vectors from his first-hand knowledge of epidemics in north-western United States. Secondly J. Gibbs and Joan Webber narrowed the field still further and brought the audience back nearer home, a sentiment continued by the next speaker, D. Redfern, who refreshingly dragged us all away, for a little while at least, from beetles to consider also the wood wasp–fungus interface.

Finally the arena was open to mycopathogens of insects, and H. C. Evans gave a very illuminating, beautifully illustrated paper on the epigeal and aerial examples, calling on his experiences from collecting in tropical South America. S. Keller and G. Zimmermann jointly introduced the meeting to pathogens of soil insects and described the spread of some of the diseases they cause, and finally J. Eilenberg, substituting at very short notice for a speaker unable to participate, kindly provided an impromptu talk on observations on the Entomophthorales, a group with interesting possibilities for further examples of biological control.

Undoubtedly we all had a stimulating time and on behalf of both Societies I would like to thank the organizers, the speakers and those who kindly mounted posters. I feel strongly that integrated interdisciplinary approaches in biology are essential for the future understanding of our natural heritage; this meeting has confirmed that view. Thank you all for attending.

Appendix: Mycophagy in Insects: a Summary

P. M. HAMMOND AND J. F. LAWRENCE

I. INTRODUCTION

The aims of this appendix are essentially two-fold. The first is to orientate the user of this volume whose knowledge of either fungi or insects is limited, by providing a résumé of how these two groups are classified. The second is to fill a perceived gap in the volume's coverage by providing an easy-to-consult review of those insect–fungus associations in which insects are the exploiters and fungi the resource. To achieve these two aims economically the main substance of the appendix consists of two annotated outline classifications (Tables I and II). Each fungus and insect taxon listed therein is provided with a code number. These code numbers are used extensively in the volume chapters, enabling the reader who wishes to check the systematic position of any fungus or insect mentioned to do so quickly by consulting the tables below.

The fungal classification (Table I) is annotated in simple fashion, each fungus taxon with known insect associates being accompanied by a list of code numbers for the relevant insect taxa. The insect classification (Table II) is more extensively annotated, including for each insect taxon an indication of its "style" of fungal exploitation, examples of particular associations and references to the sources of data, as well as a list, in the form of code numbers, of associated fungal taxa. Taken together, the two outline classifications thus provide, in very condensed form, a systematically arranged summary of known associations in which insects exploit fungi. Knowing that the papers formally presented to the "Insect Mycophagy" section of the Symposium could not cover all relevant insect groups, this summary was conceived originally merely as a means of filling some of the gaps. However, the fact that a comprehensive summary of the topic, one covering all of the Insecta, is currently unavailable elsewhere, has encouraged the production of a summary complete in its coverage, even at the risk of duplicating some information to be found in the volume chapters. Further details concerning the two outline classifications and how they may be consulted are provided below.

II. THE FUNGI

A. Fungi as an Insect Resource

Associations between insects that use fungi as a resource and the fungi exploited are variable in nature. Several authors who have investigated the communities of insects inhabiting the fruiting bodies of the larger fungi (Benick, 1952; Graves, 1960; Höfler, 1960; Klimaszewski and Peck, 1987; etc.) recognize three categories of fungus-associated insects: obligately dependent species ("mycetobionts"), regular but not obligate users of the fungal resource ("mycetophiles"), and occasional or casual users ("mycetoxenes"). In the case of large fungal fruiting bodies the stage of maturation often provides a useful guide to the status of exploiting insects (Scheerpeltz and Höfler, 1948; Hackman and Meinander, 1979; Bruns, 1984; Klimaszewski and Peck, 1987; etc.). Larval stages are of particular importance here. Where larvae exploit "fresh" fruiting bodies the association with the fungus is much more likely to be obligate than for those invading later on, after the onset of decay. Decomposing fruiting bodies are exploited by larvae of a range of generalist insect species, many of them saprophagous, that make equal use of other types of decaying organic matter. Nevertheless, even where the successional stages of a fungus-dwelling insect community are clear-cut, these provide but an imperfect guide to the nature of particular insect–fungus relationships.

The notion of successional stages and the concept of closeness of association with a fungal host (categorization as mycetobionts, etc.) cannot be readily applied to all fungal resources fed on by insects, many of which are small or microscopic. In drawing up the list of fungus-exploiting insects for this chapter (Table II), the emphasis has been first of all on ensuring the inclusion of all associations where available evidence suggests that the insects feed directly and exclusively on fungi. However, a number of insect groups with species in which mycophagy is possibly obligatory, but not necessarily exclusively so, are also included. Such groups and others in which some doubt remains as to whether the species included are truly mycophagous are indicated in Tables I and II by the use of round brackets. Finally, a selection of insect groups with species that are fungus-dwelling but do not feed on living fungal tissue is included in the tables, in square brackets. Whether the species in question are saprophagous, predaceous or parasitic, the chief criterion for inclusion of the taxon to which they belong is prominence in the fungus-inhabiting insect community, particularly during the earlier successional stages.

The food of some lichenivorous insects includes little fungal material as many, especially lepidopterous larvae (Rawlins, 1984), select lichens with a high chlorophyll content or algal-rich areas of lichens on which they feed. However, as the extent of such species' dependence on fungal food is poorly understood, most lichen-associated insect groups are *included* in the tables. The opposite approach has been taken with insects associated with mycorrhizal roots (e.g. some springtails (H1), some pemphigid aphids and coccids (H20, etc.), these generally being omitted. In most cases it is not at all clear that such species depend on the fungi as food. Insects that are associated with the products of activity by yeasts and yeast-like fungi are also difficult to categorize. Most but not all (e.g. Nosodendridae (H25.21)) such associations are excluded from the tables. A similar line has been taken with the many insect groups in which species are associated with dead or dying wood in which rapid changes are taking place because of the presence of rot or other fungi. Although many of these insects depend on fungi—to process the wood so that it forms a suitable food resource—they are mostly omitted from the tables. The exceptions to this are those few cases where there is evidence that fungal tissues form an important part of the diet.

B. Fungi Covered in this Summary

As food and habitats for insects and other invertebrates, fungi have been very unevenly studied. The hymenomycete Basidiomycotina (F5) have been the focus of most attention (Ackerman and Shenefelt, 1973; Graves, 1960; Graves and Graves, 1985; Klimaszewski and Peck, 1987; Matthewman and Pielou, 1971; Minch, 1952; Pielou and Matthewman, 1966; Pielou and Verma, 1968; etc.). Also, many of the available general accounts of fungus-dwelling and fungus-feeding insects (Benick, 1952; Buxton, 1954, 1960; Donisthorpe, 1935; Eisfelder, 1961, 1963, 1970; Hackman and Meinander, 1979; Höfler, 1960; Rehfous, 1955a; Scheerpeltz and Höfler, 1948; Trifourkis, 1977; etc.) make almost exclusive reference to the "macro-fungi" belonging to the Basidiomycotina. In comparison with this major fungus group the number of associated insect taxa (see Table I) recorded for the Ascomycotina (F4), Zygomycotina (F3), etc. is very small. To some extent this may reflect real differences. The fruiting bodies of hymenomycete fungi do, after all, mostly represent large resources that are readily exploitable, if of unpredictable occurrence (Hanski, this volume). Some non-basidiomycete groups of fungi are unlikely to be exploited by insects for obvious reasons. For example, the Spathulosporales (F4.33) are all marine, while the Trichomycetes (F3.2) are all endoparasites of various animals.

The utilization by insects of the small and often scattered fungal resources that abound in various terrestrial situations is largely uncharted territory. This is true of the small or microscopic fungi of leaf litter and other decaying plant material, and also to a large extent of the small saprobic fungi found in association with dead wood, dung, carrion, etc. However, spores, small fruiting bodies and hyphae of many such fungi have been found in the guts of various insect species. The majority of these fungi, mostly saprobic, belong to the Ascomycotina (F4), while others belong to the Zygomycetes (F3.1) or Basidiomycotina (F5). Little is known of insect associations with these fungi, but it is likely that several ascomycete Orders (in particular) for which no or few insect associates are recorded provide food for a wide range of insects (Kimbrough, 1984).

The minute fungi found on the surfaces of living plants, particularly on leaves (so-called phylloplane species) have been reported as food of only a few insects (Broadhead, 1984; Thayer, 1987), but they are probably grazed by many, some of which are likely to specialize in this mode of feeding. Fungi inhabiting fresh water have also been little investigated as invertebrate food but some, especially the oomycete water-moulds (F2.3), are likely to form at least part of the diet of many aquatic insects (Cummins, 1973 and references therein; Lawrence, this volume). Keratin-associated fungi represent a further resource that is likely to provide food for a number of insects. Such fungi are typically Hyphomycetes (F6.2) with teleomorphs in the Gymnoascales (F4.16). In several instances it remains to be established whether supposedly keratin-feeding insects are in fact keratinophagous rather than mycophagous (Robinson, 1986).

In geographical coverage information on exploitation of fungi by insects is also very uneven. Few observations are on record for named species of fungi in tropical regions. Several fungal taxa of largely tropical occurrence, notably of certain lichen-forming groups, may prove to be rich in insect associates when more fully investigated.

C. Fungal Classification

In the absence of any complete fungal classification that reflects the phyletic relations of the majority of included taxa, the classification employed here, the widely used one of Hawksworth *et al.* (1983) has been chosen on purely practical grounds.

The classification provided in Table I is complete down to the level of Class (-etes endings), and also lists all Orders (-ales endings) for the true slime-moulds (Myxomycetes—F1.5–10), Ascomycotina (F4) (for which Hawksworth *et al.* list no Classes), and Basidiomycotina (F5). The division Myxomycota and each of the five subdivisions of the Eumycota are given the code numbers F (for fungus) 1 to 6. A decimal system is employed to denote Classes and/or Orders. Thus the Order Lycoperdales (puffballs), for example, is coded F5.16.

III THE INSECTS

A. Insects Covered in this Summary

In compiling Table II an attempt has been made to cover the two most important insect Orders (in terms of number of included mycophagous species)—Coleoptera and Diptera—as fully as possible. For the remaining Orders, coverage is inevitably uneven. Two groups—Collembola (H1) and Psocoptera (H18)—are particularly inadequately treated (see below). A further inadequacy lies in the general treatment accorded to groups containing mostly detritivorous species (e.g. Thysanura (H5), Blattodea (H8), many Isoptera (H9), Dermaptera (H13), some Orthoptera (H15), Formicoidea (H33), etc). Although the majority of species belonging to such groups are likely to be truly omnivorous or saprophagous, it is likely that some specialize in fungi as food. Unfortunately, adequate conclusions concerning precise diets of these "detritivores" are generally impossible to reach on available evidence, even where gut or faecal contents have been recorded (e.g. Roth and Willis (1960) for Blattodea).

Notes on specific groups follow:

1. Acarina (Mites)

Although one of the major fungus-exploiting arthropod groups, mites, like other non-hexapods, are omitted from mention in Table II. Mycophagy in mites is reviewed by O'Connor (1984), and further information on feeding habits of mites inhabiting litter and soil, including mycophagous species, and their role in decomposition processes is provided by Butcher *et al.* (1971) and Seastedt (1984) (see also references therein).

2. Collembola (H1)

Although clearly deserving of more detailed treatment than that provided in Table II, no attempt has been made here to review springtail–fungus associations below ordinal level. Many species of Collembola, especially those found in leaf litter and humus, are likely to feed at least in part on fungi, but their degree of dependence on fungal material as food is poorly documented. Some Collembola are known to graze mycorrhizal fungi. Fruiting bodies of the larger fungi frequently contain numerous springtails, particularly of the families Entomobryidae, Poduridae and Sminthuridae (Pielou and Matthewman, 1966; Pielou and Verma, 1968; Eisfelder, 1970; etc.). Species of the podurid genus *Hypogastrura* are especially abundant on young hymenomycete (F5) fruiting bodies (Rehfous, 1955a) and may be pests of commercially grown mushrooms (Clift, 1979). The feeding habits of litter and soil dwelling Collembola and their role in decomposition processes are reviewed by Christiansen (1964), Butcher *et al.* (1971) and Seastedt (1984) (see also references therein).

3. Psocoptera (H18)

This Order of insects, like the Collembola, is deserving of more detailed treatment than that provided in Table II, but no attempt has been made here to review associations with fungi under individual psocid families. Arboricolous species of various families are known to graze epiphytic lichens and fungi (New, 1987; etc.), but the extent to which such species specialize on or rely on fungi as food appears to be poorly documented. Detritivorous species also undoubtedly consume fungal material as at least part of their diets, and some litter-inhabiting species may prove to be exclusively mycophagous.

4. Coleoptera (H25)

Many thousands of beetle species are exclusively mycophagous. Many of these feed on small fungal resources, including the spores, small fruiting bodies and hyphae of small fungi, especially Ascomycotina (F4) (Crowson,

1984), as well as the spores of larger fungi (Lawrence, this volume). The Coleoptera are also one of two dominant Orders (the other is Diptera) in the insect communities associated with large fungal fruiting bodies. Although much remains to be discovered, especially concerning the tropical species and the precise food of at least partly fungus-feeding and dead wood associated beetles (Lawrence, this volume), the food of many fungus-associated beetles is moderately well documented. Deductions based on mouthpart structure and gut contents' analyses, as well as direct observations (Lawrence, this volume), are important sources of evidence in this connection.

Several beetle families that *may* prove to have a more than casual association with fungi are omitted from Table II, on the grounds that evidence for obligate mycophagy is presently lacking. Such families fall into two main categories: those associated with the bark and/or sap of dead or dying trees (see Crowson, 1984) where yeasts and other small fungi abound—Boridae, Inopeplidae, Mycteridae, Propalticidae, Pyrochroidae, Pythidae and Synchroidae; and those in which larvae develop in rotten wood where wood-rotting fungi are present—Callirhipidae, Cephaloidae, Cerophytidae, Prostomidae, Scraptiidae, Trictenotomidae, and at least some Aderidae, Elateridae and Oedemeridae.

Much of the information employed in compiling the Coleoptera section of Table II comes from just a few important sources, works by Crowson (1981, 1984, etc.), Fogel and Peck (1975), Lawrence (1982, 1988, etc. and this volume), Lawrence and Newton (1980, 1982) and Newton (1984). Extensive use has also been made of a number of other papers dealing specifically with Coleoptera (Benick, 1952; Donisthorpe, 1935; Eisfelder, 1961, 1963; Höfler, 1960; Klimaszewski and Peck, 1987; Lawrence, 1977; Scheerpeltz and Höfler, 1948), as well as more general works.

5. Diptera (H28)

The larval stages of this insect Order are of major importance as exploiters of the fruiting bodies of the larger fungi (see Hanski, this volume). Rearing records from fruiting bodies, mostly of Basidiomycotina (F5), are the main source of information available concerning mycophagy in the Diptera. However, completeness of coverage is limited by the fact that very few observations of tropical fungus-associated flies are to be found in the literature. Also, observations on actual feeding are scant, more so than for the Coleoptera (H25), so that the precise nature of associations with fungi are often poorly documented.

The sources of detailed information concerning Diptera–fungus associations found to be of greatest assistance in compiling Table II are works by Buxton (1954, 1960, etc.), Chandler (1979), Hackman and Meinander

(1979), Trifourkis (1977) and Ferrar (1987). Considerable use has also been made of papers by Hackman (1976), Rehfous (1955a), Russel-Smith (1979) and Smith (1956), as well as more general works (e.g. Fogel, 1975; Weiss, 1921).

6. Hymenoptera (H33)

Very few Hymenoptera are exclusively mycophagous, and the majority of hymenopterous families receiving mention in Table II contain species that are parasitic on other fungus-dwelling insects. The criteria for inclusion of such parasitic groups in the table are that they contain species that are common constituents of the insect community associated with fungi and/or are specific to an exclusively fungus-dwelling host. Inevitably such criteria involve arbitrary judgements. Families of parasitic Hymenoptera with species that are rather frequently found on or reared from fruiting bodies of large fungi (but not included in Table II) include Cynipidae, Encyrtidae, Mymaridae, Pteromalidae and Scelionidae (see Ferrière, 1955; Pielou and Verma, 1968; etc.). Less frequently associated with fungi, but also possible candidates for inclusion in Table II, are species of Ceraphronidae, Cleonymidae, Perilampidae and Torymidae.

B. Insect Classification

The classification employed in Table II (CSIRO, 1970) covers all of the Hexapoda, i.e. it includes groups such as the Collembola or springtails (H1) that are no longer generally regarded as true insects. The classification is complete to ordinal level and for each Order, whether or not mycophagous associations are recorded, some indication of the size and diversity of the group is provided by listing (in brackets) the number of included suborders, superfamilies and/or families.

Each Order of Hexapoda is given a code number—H (for hexapod) 1 to 33. A decimal system is employed to denote relevant subdivisions, mostly families. Thus the Mycetophilidae (fungus gnats), for example, are coded H28.15. Mycologists using Table II, if unfamiliar with insect classification, should note that -oidea endings denote superfamilies, -idae endings families, -inae endings subfamilies, and -ini endings tribes.

It should be emphasized that the information contained in Table II is in summarized form, and is particularly condensed in the case of those taxa in which mycophagy is common or is the rule. At least one reference (wherever possible) to a published source of information concerning mycophagy is given for each relevant insect taxon. Many additional references, however, may be worthy of consultation in any given case. A number of these, notably those treating the insect communities associated with particular fungus species (e.g. Pielou and Verma, 1968), Order of fungi (e.g. Bruns, 1984), or higher taxa (e.g. Blackwell, 1984), include information concerning a wide range of insect groups (see References below).

Acknowledgements

Several entomologists gave material assistance in the preparation of this appendix by providing advice concerning literature on the feeding biology of various insect groups. In this connection we are grateful to W. R. Dolling (Hemiptera), I. D. Gauld (Hymenoptera), K. M. Harris (Diptera), J. M. Palmer (Thysanoptera), G. S. Robinson (Lepidoptera) and K. G. V. Smith (Diptera).

TABLE I. Outline classification of fungi, following Hawksworth *et al.* (1983). Orders of Ascomycotina marked* include at least some lichen-forming fungi. Associated insect taxa are listed (for code numbers see Table II). Square brackets indicate that the insects occur consistently with the fungi but without mycophagy, while round brackets indicate a similar association and with uncertain mycophagy. Feeding on "ambrosia" fungi is indicated by †.

	Taxon	Insects
F1	MYXOMYCOTA	
1	Prosteliomycetes (1 order)	
2	Ceratiomyxomycetes (1 order)	H1; [H25.8]; H28
3	Dictyosteliomycetes (1 order)	H1
4	Acrasiomycetes (1 order)	
	Myxomycetes (6 orders)	H25.2, 18, 19 (and see under orders)
5	Echinosteliales	
6	Echinosteliopsidales	
7	Liceales	H1; H25.4, 8, 10, 29, 47, [26]; H28.15
8	Trichiales	H1; H25.4, 8, 10, 11, 29, 47; H28.15
9	Stemonitales	H1; H25.4, 8, 10, 11, 29, 46, 47
10	Physarales	H1; H25.4, 8, 10, 11, 29, 46, 47, 50, [9]; (H28.16, 28, 36, 38, 39), [8, 19]
11	Plasmodiophoromycetes (1 order)	
12	Labyrinthulomycetes (1 order)	
	EUMYCOTA	
F2	MASTIGOMYCOTINA	
1	Chytridiomycetes (4 orders)	
2	Hyphochytriomycetes (1 order)	
3	Oomycetes (4 orders)	(H6; H25.11a, ?etc.)

(Continued)

TABLE I. Continued

F3	**ZYGOMYCOTINA**	
1	Zygomycetes (6 orders)	H1; H25.3, 4, 11
2	Trichomycetes (5 orders)	
F4	**ASCOMYCOTINA**	
	[no classes, 37 orders—arranged alphabetically]	H1; H4; H17; H18; H21; H25; H28; H30; H33; (H2; H3; H5; H8; H9; H11; H13; H15; H20)
1	Arthoniales*	
2	Ascosphaerales	
3	Caliciales*	H25.40; H28.39
4	Clavicipitales	H25.40; H28.39
5	Coryneliales	
6	Cyttariales	H25; H28
7	Diaporthales	(H25.61)
8	Diatrypales	H25.62
9	Dothideales*	H18; H21.4; H25.8, 16, 20, 30, 33, 40, 44, 53, 61, 62
10	Elaphomycetales	H25.4
11	Endomycetales	H25.8, 20, 21, †24, 30, †31, 48, †49, †63, (33a, †60; H8; H28.24, 36)
12	Erysiphales	H25.42, 44, (5); H28.10
13	Eurotiales	H25.3, 5, 7, 11, 34, 35, 36, 38, 40, 42, 46, 47, (31, 40, 45, 50)
14	Graphidiales*	H21.4
15	Gyalectales*	
16	Gymnoascales	(H32.4)
17	Helotiales*	H25.8, †63
18	Hypocreales	
19	Laboulbeniales	
20	Lecanidiales*	
21	Lecanorales*	H1; H4; H17; H18; H21.4; H25.57–60, 62, [15]; H28.2, (15); H32.1, 2, 4–12
22	Microascales	
23	Opegraphales*	
24	Ophiostomatales	H25.8, 63
25	Ostropales*	
26	Peltigerales*	H25.57
27	Pertusariales*	
28	Pezizales (= Tuberales)	H25.3, 4, 5, 12, (8, 36); H28.2, 15, 23, 36, 37, (3), [40]
29	Polystigmatales	
30	Pyrenulales*	
31	Rhytismatales	

TABLE I. Continued

32	Sordariales	(H25; etc.)
33	Spathulosporales	
34	Sphaeriales* (= Xylariales)	†H9; H21.4; H25.4, 8, 12, 30, 31, 34, 36, 40, 42, 45, 47–50, 57, 61, 62,†63, (35, 60), [1, 9, 9a]; H28.9, 10, 15, 36, 37, (5, 22, 42), [19]; H32.4, 6, 8; H33.†1, 7, (†9)
35	Taphrinales	H28.27
36	Teloschistales*	
37	Verrucariales*	
F5	BASIDIOMYCOTINA (3 classes)	(H8; H9; H11; H13; H15; H18; H20; and see under orders)
	Hymenomycetes (2 subclasses)	
	Phragmobasidiomycetidae (3 orders)	
1	Tremelalles	H25.5, 8; H28.15, 36
2	Auriculariales	H25.8, 46, 52, 60; H28.10, 15, 37, (16)
3	Septobasidiales	
	Holobasidiomycetidae (9 orders)	
4	Exobasidiales	
5	Brachybasidiales	
6	Dacrymycetales	H28.15
7	Tulasnellales	
8	Aphyllophorales (= Polyporales)	H1; H20.4, (1, 2), [3, 5]; H21.4; H25.3, 4, 4a, 5, 6, 8, 20, 21a–23, 25, 26, 30, 32, 36, 37, 39, 42, 43, 45–47, 50–58, 60, 62, (21b), [4b, 9, 9a, 27]; H28.1, 2, 10, 11, 14, 15, 21, 22, 34, 36, 38, 39, (3, 4, 12, 13, 16, 20, 27, 28, 30–33), [5, 7, 8, 17, 19, 24a, 40, 41]; H32.2–4, 6, 11, 12; H33.†1, [3–8]
9	Cantharellales	H25.6, 8; H28.2, 15, 22, 36, 37, 39, (28)
10	Agaricales	H1; †H9; [H20.3, 5]; H25.6, 8, 10, 30, 45, 46, 50, (12), [1, 7, 9, 9a, 14]; H28.1, 2, 10, 14, 15, 21–23, 34, 36–39, (3, 4, 6, 28, 29, 32), [8, 17, 19, 26, 35, 40]; [H33.6, 8]
11	Boletales	H1; [H20.3, 5]; H25.6, 8, 10, 30, 45, 46, 50, (12), [1, 7, 9, 9a, 14]; H28.1, 2, 10, 14, 15, 21–23, 34, 36–39, (3, 4, 6, 28, 29, 32), [8, 17, 19, 26, 35, 40]; [H33.6, 8]

(Continued)

TABLE I. Continued

12	Russulales	H1; [H20.5]; H25.4, 6, 8, 30, 39, 42, 45, 50, [1, 9, 9a, 46]; H28.1, 2, 9, 14, 15, 22, 34, 36–39, (3, 4), [5, 7, 40]; (H33.2, 4–6)
	Gasteromycetes (9 orders)	
13	Sclerodermatales	H25.36, [8]; H28.2, 9, 10, 15, 22, 37; H32.4
14	Melanogastrales	H25.4
15	Tulostomatales	
16	Lycoperdales	H25.4, 6, 23, 30, 36, [8, 46]; H28.15, 21, 34, 37, (5), [19]; [H33.6]
17	Nidulariales	
18	Phallales	H25.4, 30, [7, 8]; H28.15, 21, 36, 37, 39, (28), [25, 40, 42]
19	Gautieriales	H25.30
20	Hymenogastrales	H25.3, 4, 12
21	Podaxales	
	Uredioniomycetes (1 order)	
22	Uredinales	H21.3; H25.30, 40, 44,(28); H28.10; H32.6, 8
	Ustilaginomycetes (2 orders)	
23	Ustilaginales	H25.38, 40, 62; H28.10
24	Sporidiales	
F6	DEUTEROMYCOTINA (2 classes)	H1; H18; (H2; H3; H5; H8; H9; H11; H13; H15; H20), and see also under classes
1	Coelomycetes (3 orders)	H1; H21.4; H25.8, 30, 34, 47, 48, 50, 63, (57)
2	Hyphomycetes (4 orders)	H1; H18; H21.4; H25.5, 6, 8, 16, 31, 34–36, 38, 40, 42, 44, 46–48, 50, 60–63, (57); H28.27, (9, 10)

Table II. Outline classification of insects (and other hexapod arthropods) with special reference to mycophagy. Classification at ordinal level and above follows CSIRO (1970), that of Coleoptera follows (with minor adjustments) Lawrence (1982), Diptera follows Hennig (1973), and Hymenoptera follows Gauld and Bolton (1988). Square brackets indicate taxa which occur consistently with fungi but in which mycophagy has not been recorded, while round brackets indicate taxa in which mycophagy may occur but is doubtful or facultative. *Abbreviations employed are as follows*: A = adult; F = fungus taxon; f = family; hy = hyphae; L = larva; lich = lichenivorous; mycel = mycelium; myc = mycophagous; ord = order; par = parasitic; phyt = phytophagous; plas = plasmodium feeding; pred = predaceous; sap = saprophagous; sf = subfamily; sp = spores; subord = suborder; supf = superfamily.

		Associated fungal taxa (see Table I)	Feeding habits	Comments	References
	HEXAPODA				
H1	**COLLEMBOLA** (2 subords, 5 fs) (Springtails)	F1.2, 1.3, 1.7, 1.8, 1.9, 1.10; 3.1; 4.21, etc; 5.8, 5.10, 5.11, 5.12, etc; 6.1, 6.2.	Myc(sp/hy/mycel), sap, lich	Many in litter, etc at least facultatively myc; some are common inhabitants of large sporophores	Christiansen, 1964 Butcher *et al.*, 1971 Blackwell, 1984 Pielou and Verma, 1968
H2	(**PROTURA**) (3 fs)	?F4; ?6.	Sap/?myc	In soil, litter, etc; may be partly myc	CSIRO, 1970
H3	(**DIPLURA**) (4 fs)	?F4; ?6.	Sap/?myc	Some (e.g. Campodeidae) may be partly myc	CSIRO, 1970
	INSECTA (33 ords)				
H4	(**ARCHAEOGNATHA**) (2 fs)	F4.21, ?etc	Lich, sap/?myc	Some may be facultatively myc	CSIRO, 1970

(Continued)

Table II. Continued.

		Associated fungal taxa (see Table I)	Feeding habits	Comments	References
H5	(**THYSANURA**) (5 fs)	?F4; ?6.	Sap/?myc	Mostly omnivorous; some may be facultatively myc	CSIRO, 1970
H6	(**EPHEMEROPTERA**) (Mayflies)	?F2.3; ?etc.	Sap/?myc	A do not feed; nymphs (aquatic) that are scavengers may be partly myc	CSIRO, 1970
H7	[**ODONATA**] (3 subords, 24 fs) (Dragonflies and damselflies)	—	(Pred)	(Nymphs aquatic)	—
H8	(**BLATTODEA**) (5 fs) (Cockroaches)	?F4.11, etc.; ?5; ?6.	Sap/?myc	Mainly omnivorous, but diet of many may include fungi	Roth and Willis, 1960
H9	**ISOPTERA** (6 fs) (Termites)	?F4; ?5; ?6.	Varied	Various groups may include some fungi in their diet	— Sands, 1969
		F4.34; 5.10	Myc (ambrosia-fungi in nests)	Most Macrotermitinae	Batra and Batra, 1979 Wood and Thomas, this volume
H10	[**MANTODEA**] (8 fs) (Praying mantids)	—	(Pred)	—	—

H11	**ZORAPTERA** (1 genus)	?F4; ?5; ?6	Myc/?sap/?pred	Many may be at least partly myc	CSIRO, 1970
H12	[**GRYLLOBLATODEA**] (1 f)	—	(Phyt/?sap)	—	—
H13	(**DERMAPTERA**) (3 subords, 7 fs) (Earwigs)	?F4; ?5; ?6	Myc/sap/pred	Mostly scavengers or pred; some likely to be at least facultatively myc	—
H14	[**PLECOPTERA**] (Stoneflies)	—	(?Pred)	(Nymphs aquatic)	—
H15	(**ORTHOPTERA**) (2 subords) (Grasshoppers, crickets, etc)				
	1. ENSIFERA (3 supfs, 9 fs)	?F4; ?5; ?6	Varied: phyt, pred, sap, etc	Omnivorous or scavenging Grylloidea and Gryllacridoidea may be facultatively myc	—
	2. CAELIFERA (3 supfs, 6 fs)	?F4; ?6	Mostly phyt	Some Acridoidea (?etc) include fungal material in their diet	K. Monk, pers. comm.
H16	[**PHASMATODEA**] (2 fs) (Stick-insects, etc)	—	(Phyt)	(Foliage feeders)	—
H17	(**EMBIOPTERA**) (8 fs) (Web-spinners)	?F4.21, ?etc	Phyt, lich/sap	In galleries under bark—? partly myc	Ross, 1970

(*Continued*)

Table II. Continued.

		Associated fungal taxa (see Table I)	Feeding habits	Comments	References
H18	**PSOCOPTERA** (3 subords, 27 fs) (Psocids, booklice)	F4.9, 4.21; ?5; 6.2	Phyt/lich/sap/?myc	Arboricolous spp of various families include ephiphytic fungi and lichen in their diet; others (in litter, etc) may be at least facultatively myc	New, 1987
H19	[**PHTHIRAPTERA**] (3 subords) (Lice)	—	(Ectopar)	(On birds and mammals)	—
H20	**HEMIPTERA** (2 subords) (Bugs)				
	HOMOPTERA (9 supfs, 50 fs)	—	Almost all phyt	—	—
	Fulgoroidea (20 fs)				
	1. (Achilidae)	?F4; ?5; ?6	Nymphs ?myc(hy) (A phyt)	Under bark and in logs (On trees)	O'Brien and Wilson, 1985
	2. (Derbidae)	?F4; ?5; ?6	Nymphs ?myc(hy) (A phyt)	Many under bark (Many on monocots)	O'Brien and Wilson, 1985
	HETEROPTERA (14 supfs, 92 fs)				
	Cimicoidea (8 fs)				

	3. [Miridae]	F5.8, 5.10, 5.11, ?etc	(Pred, phyt)	Various genera found (mostly casually) in large sporophores	Simonet, 1955
	Aradoidea (2 fs)				
	4. Aradidae (incl. Dysodiidae)	F5.8, ?etc	Myc(hy)	?All genera myc, many in arboricolous fungi	Usinger and Matsuda, 1959 Simonet, 1955
	Lygaeoidea (6 fs)				
	5. [Lygaeidae]	F5.8, 5.10, 5.11, 5.12, ?etc	(?All phyt)	Various genera in large fungi, ? all casual visitors	Simonet, 1955
H21	**THYSANOPTERA** (2 subords) (Thrips)				
	TEREBRANTIA (4 fs)		(Mostly phyt)		
	1. (Uzelothripidae)	?F	?Myc	*Uzelothrips*	Ananthakrishnan, 1984
	2. (Merothripidae)	?F	?Myc	*Merothrips*	Ananthakrishnan, 1984
	3. (Thripidae)	F5.22, ?etc	Myc/phyt	A few genera at least partly myc	Ananthakrishnan, 1984
	TUBULIFERA (1 f)				
	4. Phlaeothripidae	F4.9, 4.14, 4.21, 4.34; 5.8; 6.1, 6.2	Myc(sp), Lich(sp), Myc(hy)/?sap	?All Idolothripinae (mainly on F4 and F6); some Phlaeothripinae	Ananthakrishnan, 1984 Ananthakrishnan and Dhileepan, 1984
H22	[**MEGALOPTERA**] (2 fs) (Alderflies)	—	(?All pred)	(Larvae aquatic)	—

(*Continued*)

Table II. Continued.

		Associated fungal taxa (see Table I)	Feeding habits	Comments	References
H23	[**RHAPHIDIOPTERA**] (2 fs) (Snake-flies)	—	(?All pred)	—	—
H24	[**NEUROPTERA**] (5 supfs, 17 fs) (Lacewings)	—	(?All pred)	—	—
H25	**COLEOPTERA** (4 subords) (Beetles)				
	(ARCHOSTEMATA) (3 fs)	?F4; ?5.8	?Myc	Possibly associated with wood-rotting fungi	Weiss and West, 1922 Crowson, 1984 Lawrence, this volume
	ADEPHAGA (8 fs)		(Mostly pred)		
	1. (Carabidae)	F4.34; 5.8, 5.10, 5.12, etc	Pred on various fungivores	Especially arboreal spp in tropical forest (e.g. *Eurycoleus*)	Rehfous, 1955a Erwin and Erwin, 1976
		F5.8	?Myc(hy)	*Mormolyce*	Lieftinck and Wiebes, 1976
	2. Rhysodidae	F1	A + ?L myc(plas)	*Omoglymius* A observed; all live in rotten wood and may be F1 associated	Lawrence, this volume

POLYPHAGA (18 supfs, 138 fs)				
Staphylinoidea (11 fs)				Newton, 1984
3. Ptiliidae	F5.8, 5.20, ?etc	Myc(sp)	*Nossidium* and ?other Nossidiinae	Fogel and Peck, 1975
	F5.8	Myc(sp) in pores	All Nanosellinae	Newton, 1984
	F3.1; 4.13, 4.28; 5.8, 5.10, etc	Myc(sp/hy)/sap	Many Ptiliinae, Acrotrichinae	Newton, 1984 Lawrence, this volume
4. Leiodidae	F1.7, 1.8, 1.9, 1.10	Myc(sp)	Many Anisotomini, Neopelatopini	Lawrence and Newton, 1980
		Myc(plas)	Some *Agathidium*	Wheeler, 1984
	F3.1; 4.10, 4.28; 5.14, 5.18, 5.20	Myc in hypogeal fungi	Many Leiodinae, Catopocerinae and ?all Coloninae	Fogel and Peck, 1975 Newton, 1984
	F4.12, 4.28, 4.34; 5.1, 5.8, 5.10, 5.11, 5.12, 5.16; 6.2	Myc (sp, hy)	Many Leiodinae, Camiarinae, some Cholevinae, including specialists in puffballs, hard brackets, etc	Newton, 1984
	F4.34; 5.10, 5.11, 5.12, etc	(Sap)	Some Cholevinae in old fungi	Klimaszewski and Peck, 1987
4a. (Agyrtidae)	F5.8	Sap/?myc	Some, e.g. *Pelatines* spp, in sporophores	Newton, 1984 Hammond, unpublished
4b. [Scydmaenidae]	F5.8, 5.10, 5.12, 5.16	(Pred on mites)	Spp of various genera	Rehfous, 1955a, etc Hammond, unpublished
5. Micropeplidae	F4.13, 4.28; 5.8, 5.10; 6.2, ?etc	Myc(sp/?hy)	?All myc, some at least are specialists on certain F4	Newton, 1984 P. Nohel, pers. comm.

(Continued)

Table II. Continued.

	Associated fungal taxa (see Table I)	Feeding habits	Comments	References
6. Dasyceridae	F5.8, 5.9, 5.11, 5.12, 5.16; 6.2, ?etc	Myc(sp/hy)	?All myc, mostly on moulds, etc	Newton, 1984 Rehfous, 1955a
7. [Silphidae]	F5.18	?Myc	A *Oecioptoma*, *Necrophila* and *Nicrophorus* attracted to sporophores	Smith, 1956, etc Newton, 1984
	F5.8, 5.10, 5.11, etc	(Sap)	A and ?L of various genera, e.g. *Oecioptoma* and some *Nicrophorus*	Rehfous, 1955a, etc Hammond, unpublished
8. Staphylinidae (incl. Scaphidiinae)	F1.7, 1.8, 1.9, 1.10	Myc(sp)	Many *Baeocera* and *Scaphobaeocera* specialists on slime moulds	Newton, 1984
	F1.8, 1.9, 1.10; 4.11, 4.17, 4.24; 5.1, 5.2, 5.8, 5.9, 5.10, 5.11, 5.12; 6.1, 6.2	Myc(sp/hy)	All Oxyporinae and Gyrophaenina, most Scaphidiinae, some Tachyporinae and Osoriinae, ?Trichophyinae, ?Piestinae, ?etc	Newton, 1984, etc
	F4.9, 4.34; 5.8; 6.2	Myc(sp/hy)	*Neophonus* myc on foliicolous fungi and spores of other fungi present on leaf surfaces	Thayer, 1987

	F1.9, 1.10; 4.28, 4.34; 5.10, 5.11, 5.12, 5.13, 5.16, 5.18	(Sap/?myc)	Various Oxytelinae, Osoriinae and Proteininae, mostly in old fungi	Klimaszewski and Peck, 1987 Hammond, unpublished
	F1.2, 1.9, 1.10; 4.28, 4.34; 5.8, 5.9, 5.10, 5.11, 5.12, 5.13, 5.16, 5.18	(Pred on L of H28, etc)	Various Staphylininae, Aleocharinae, Tachyporinae, etc; some spp constantly fungicolous	Newton, 1984, etc Hammond, unpublished
	F5.8, 5.10, 5.11, 5.12	(L ectopar on pupae of H28)	Some *Aleochara* spp fungicolous	Newton, 1984, etc
Hydrophiloidea (5 fs)				
9. [Hydrophilidae]	F1.10; 4.34; 5.8, 5.10, 5.11, 5.12, etc	(A sap, L pred on L of H28)	Various Sphaeridiinae, e.g. *Cercyon* spp, in fungi	Klimaszewski and Peck, 1987 Rehfous, 1955a, etc
9a. [Histeridae]	F4.34; 5.8, 5.10, 5.11, 5.12	(Pred on L of H28 ?etc)	Some *Hister* spp, etc in fungi, usually decaying	Klimaszewski and Peck, 1987 Rehfous, 1955a, etc
Eucinetoidea (3 fs)				
10. Eucinetidae	F1.7, 1.8, 1.9, 1.10; 5.11, ?etc	Myc(sp)	At least some *Eucinetus* spp and ?other genera are slime mould spore specialists, some *Eucinetus* on Boletales and other fungi	Lawrence and Newton, 1980 Bruns, 1984 Hammond, unpublished
11. Clambidae	F1.8, 1.9, 1.10	Myc(sp),(?plas)	At least some *Clambus*	Lawrence and Newton, 1980 Hammond, unpublished
	F3.1; 4.13, etc	Myc(?sp)	*Acalyptomerus*, *Calyptomerus*, etc	Crowson, 1984

(*Continued*)

Table II. Continued.

	Associated fungal taxa (see Table I)	Feeding habits	Comments	References
11a. (Scirtidae)	F2.3, etc	(L sap/?myc)	L aquatic, scavengers, probably partly myc	Lawrence, 1982
Scarabaeoidea (10 fs)				
12. Geotrupidae	F4.28; 5.20	Myc in hypogeal fungi	Many (?all) Bolboceratinae	Fogel and Peck, 1975
	F5.10, 5.11, etc	?Myc/sap	Brood material for L of some	Hammond, unpublished
13. (Acanthoceridae)	F?4	?Myc	Those living under bark	Crowson, 1984
14. [Scarabaeidae]	F5.8, 5.10, etc	(Sap)	Some *Onthophagus* A (and ?L) in decaying sporophores	Hammond, unpublished
Byrrhoidea (1f)				
15. (Byrrhidae)	F4.21, ?etc	L lich/phyt	Lichens part of diet of various genera	Lawrence, 1982
Dryopoidea (10 fs)				
16. Ptilodactylidae	F4.9; 6.2, ?etc	A myc	Possibly many, on foliage, etc	Lawrence, this volume
Elateroidea (7 fs)				
17. (Throscidae)	F?5	L ?myc	Some *Trixagus* on mycorrhizal roots	Burakowski, 1975
18. (Eucnemidae)	?F1	L ?myc(plas)	L of most genera that occur in rotten wood	Lawrence, this volume
Cantharoidea (10 fs)				
19. (Lycidae)	?F1	L ?myc(plas)	L of many genera that occur in rotten wood and under bark	Lawrence, 1988

Dermestoidea (4 fs)				
20. Derodontidae	F5.8, ?etc	Myc(hy)	*Derodontus* spp, ?etc	Crowson, 1984
	F4.9	Myc	*Nothoderodontus* on sooty moulds on *Nothofagus*	Crowson, 1984; Lawrence, 1988
	F4.11	Myc/sap	*Peltastica* spp on fermenting tree sap	Lawrence, this volume
21. Nosodendridae	F4.11	Myc/sap	?All spp of *Nosodendron* (the only genus) on fermenting tree sap	Lawrence, this volume
21a. Dermestidae	F5.8	Myc(hy)	*Orphilus* spp; (most genera not myc)	Lawrence, 1988; Crowson, 1984
21b. (Jacobsoniidae)	?F4.34, etc	?Myc	*Saphophagus* spp under bark	Crowson, 1984
Bostrichoidea (3 fs)				
22. Endecatomidae	F5.8	Myc(hy)	*Endecatomus* spp	Lawrence, 1988
23. Anobiidae	F5.8	Myc(hy)	Most Dorcatominae in mycelium and hard brackets	Lawrence, 1988; Hammond, unpublished
	F5.16	Myc	*Caenocara* spp puffball specialists	Donisthorpe, 1935; Hammond, unpublished
Lymexyloidea (1 f)				
24. Lymexylidae	F4.11	L myc (ambrosia fungi in wood; sometimes pests of standing timber)	*Hylecoetus* and ?most other genera	Crowson, 1984

(*Continued*)

Table II. Continued.

	Associated fungal taxa (see Table I)	Feeding habits	Comments	References
Cleroidea (7 fs)				
25. Phloiophilidae	F5.8	Myc(hy)	*Phloiophilus* (only genus) on *Phlebia*	Lawrence, 1982
26. Trogossitidae (incl. Peltidae, etc)	F5.8	Myc(hy)	Peltinae, e.g. *Ostoma*, Calitinae, e.g. *Calitys* and ?others	Lawrence, 1988
27. [Cleridae]	F5.8, ?etc	(Pred on various H25)	Many Thaneroclerinae in fungi	Lawrence, 1988
Cucujoidea (24 fs)				
28. (Protocucujidae)	F5.22	A ?myc	*Ericmodes* in galls of *Uromycladium* on *Acacia*	Lawrence, 1982
29. Sphindidae	F1.7, 1.8, 1.9, 1.10	Myc(sp)	All specialist slime mould spore feeders	Lawrence and Newton, 1980
30. Nitidulidae	F4.9	Myc	*Soronia* spp under bark	Lawrence, 1988 Crowson, 1984
	F4.11	Myc/sap	Several genera, e.g. *Cryptarcha*, on fermenting tree sap	
	F5.8, 5.10, 5.11, 5.12, etc	Myc(sp,hy)	Several genera, e.g. *Pallodes*, *Phenolia*, *Cychramus*	Crowson, 1984 Lawrence, 1988
	F5.16	Myc	*Pocadius*, etc puffball specialists	Lawrence, 1988
	F5.18	Myc	*Psilopyga* spp	Lawrence, 1988
	F5.19	Myc in hypogeal fungi	*Thalycra* spp	Fogel and Peck, 1975
	F4.34; 5.22	Myc	*Epuraea*, etc	Lawrence, 1982

31. Rhizophagidae	F4.13; 6.2, ?etc	Myc	Monotominae, e.g. *Monotoma* spp	Lawrence, 1988
	F4.34	Myc	*Bactridium*, *Hesperoaenus*, *?Lenax*, ?etc	Lawrence, 1988 Crowson, 1984
	F4.11, 4.34, ?etc	?Myc	Thioninae may feed on ambrosia fungi of other wood-boring H25	Crowson, 1984 Lawrence, this volume
32. Hobartiidae	F5.8, 5.10	Myc(sp, hy)	*Hobartius* [Australia only]	Lawrence, 1988
33. Phloeostichidae	F4.9	Myc	*Agapytho* on sooty moulds	Lawrence, 1988
33a. (Helotidae)	F4.11	?Myc/sap	All *Helota* (only genus) on fermenting tree sap	Lawrence, 1982
33b. (Cucujidae)	?F4. ?etc	?Myc/pred/?sap	*Pediacus* spp, etc under bark ?myc	Crowson, 1984
34. Laemophloeidae	F4.13, 4.34; 6.2	Myc	*Laemophloeus*, *Placonotus*, etc under bark	Crowson, 1984 Dajoz, 1981
35. Silvanidae	F4.13; 6.2	Myc	Many Silvaninae, etc on moulds	Lawrence, 1988
	?F4.34, ?etc	?Myc	Various Uleiotinae, Silvaninae, etc under bark	Crowson, 1984
36. Cryptophagidae	F4.13; 6.2	Myc	Many Cryptophaginae, etc on moulds	Lawrence, this volume
	F4.28, 4.34; 5.8, 5.10, 5.16, ?etc	Myc(sp, ?hy)	Some *Cryptophagus* specialists on particular sp of F4, and puffballs	Hingley, 1971 Lawrence, this volume

(*Continued*)

Table II. Continued.

	Associated fungal taxa (see Table I)	Feeding habits	Comments	References
37. Lamingtoniidae	F5.8	Myc(hy)	*Lamingtonium* (one species only)	J. F. Lawrence, pers. comm.
38. Languriidae	F4.13; 6.2	Myc	*Cryptophilus* spp, etc on moulds	Lawrence, this volume
	F5.23	Myc(sp)	*Leucohimatium* sp on smuts	Lawrence, 1988
39. Erotylidae	F5.8	Myc(hy)	*Dacne*, *Microsternus*, etc	Lawrence, 1988
	F5.10, 5.12	Myc(sp)	Many Triplacini, etc	Hammond, unpublished
	F5	Myc	Almost all Erotylidae exclusive to F5 basidocarps	
40. Phalacridae	F4.4, 4.13; 6.2	Myc	*Acylomus* and *?Stilbus* spp graze moulds and feed on ergots	Steiner, 1984
	F4.34	Myc(sp/hy)	*Litochropus* sp on *Daldinia*	Steiner, 1984
	F4.9	Myc	*Cyclaxyra* sp on sooty moulds	Lawrence, this volume
	F5.22	Myc(sp)	Some *Phalacrus* and *Phalacropsis* on various rusts	Steiner, 1984
	F5.23	Myc(sp)	Some *Phalacrus* on smuts	Steiner, 1984
41. (Cerylonidae)	?F4; ?6	?Myc/sap	Spp of most genera, including *Cerylon* etc under bark, may be myc	Crowson, 1984 Lawrence, 1988

42. Corylophidae	F4.13; 5.8, 5.10, 5.12; 6.2, ?etc	Myc(sp)	*Sericoderus*, *Orthoperus* and probably many other genera, primarily mould feeders	Lawrence, this volume
	F4.12	Myc	*Corylophodes* spp	Lawrence, this volume
	F4.34	Myc	*Arthrolips*, *Molamba*, ?etc	Crowson, 1984
	F5.8	Myc(sp)	*Holopsis* spp on *Ganoderma*	Lawrence, this volume
43. Discolomidae	F5.8	Myc(hy)	*Aphanocephalus* spp. ?etc; [most genera under bark]	Lawrence, 1988
44. Coccinellidae	F4.12	Myc	Psylloborini, e.g. *Psyllobora* on mildews	Hodek, 1973
	F6.2	Myc/phyt	*Tytthaspis* spp on moulds	Ricci *et al.*, 1983
	F4.9; 5.22	Myc/phyt/pred	*Rhyzobius* sp on rusts, etc	Ricci, 1986
45. Sphaerosomatidae	F4.34; 5.8, 5.10, 5.11, 5.12	Myc	*Sphaerosoma* spp	Lawrence, 1988
46. Endomychidae	F4.13, ?etc	Myc	*Mycetaea* and ?other genera	Hammond, unpublished
	F5.2	Myc	*Endomychus* sp and ?others	Lawrence, 1988
	F5.8	Myc(hy)	*Eumorphus*, *Symbiotes*, etc	Lawrence, 1988
	F5.10	Myc(sp)	Various genera	Lawrence, 1988
	F5.16	Myc	*Lycoperdina* spp puffball specialists	Lawrence, this volume

(*Continued*)

Table II. Continued.

	Associated fungal taxa (see Table I)	Feeding habits	Comments	References
47. Lathridiidae	F1.7, 1.8, 1.9, 1.10	Myc(sp)	Some *Enicmus* spp	Lawrence and Newton, 1980
	F4.13; 6.2, etc	Myc(sp)	Spp of many genera, e.g. *Lathridius*, Corticariinae [the latter also on pollen], etc	Lawrence, 1988 Hammond, unpublished
	F4.34; 6.2	Myc(sp)	Some *Enicmus*, etc on *Nummularia, Xylaria, Cryptostroma*, etc	Lawrence, 1977 Dajoz, 1981 Russell, 1979 Hammond, unpublished
	F5.8	Myc(sp)	Some *Corticaria* spp	Lawrence, 1988
48. Biphyllidae	F4.11	Myc	*Anchorius*, etc	Lawrence, 1988
	F4.34; 6.2	Myc(sp)	*Biphyllus, Diplocoelus*, etc on *Daldinia, Nummularia, Xylaria, Cryptostroma*, etc, often as specialists	Lawrence, 1988 Dajoz, 1981 Hingley, 1971 Ratti, 1978 Hammond, unpublished
49. Bothrideridae	F4.34	Myc	*Xylariophilus*, ?etc	Lawrence, 1988
	F4.11, 4.34, etc	Myc	Teredini feed on ambrosia fungi of other wood-boring H25	Lawrence, this volume

Tenebrionoidea (28 fs)				
50. Mycetophagidae	F1.10; 5.8, 5.10, 5.11, 5.12	Myc(sp, hy)	*Mycetophagus*, *Triphyllus*, etc, especially on soft polypores	Lawrence, 1988 Hammond, unpublished
	F4.34; 6.2	Myc	*Litargus*, etc on *Daldinia, Nummularia, Cryptostroma*, etc	Hingley, 1971 Dajoz, 1981 Hammond, unpublished
51. Tetratomidae	F5.8	Myc(hy)	*Tetratoma* spp in soft polypores, and some in small epiphytic fruiting bodies	Lawrence, 1988 Hammond, unpublished
52. Ciidae	F5.8	Myc(hy)	Spp of most genera, many in hard brackets	Lawrence, 1973
	F5.2	Myc	*Orthocis*	Lawrence, 1973
53. Melandryidae (= Serropalpidae)	F4.9	L myc	Some Orchesiini [some also in rotten wood]	Lawrence, 1988
	F5.8	L myc(hy)	Most Hallomeninae, Eustrophinae and Orchesiini, many in soft brackets	Lawrence, 1988 Hammond, unpublished
54. Mordellidae	F5.8	L myc(hy)	*Curtimorda* etc; [most L in plant stems or rotten wood]	Lawrence, 1988
55. Archeocrypticidae	F5.8	L myc(hy)	*Enneboeopsis*, ?etc in soft brackets; [other genera not myc]	Lawrence, 1988

(Continued)

Table II. Continued.

	Associated fungal taxa (see Table I)	Feeding habits	Comments	References
56. Pterogeniidae	F5.8	Myc(hy)	All spp in woody brackets	Lawrence, 1988 Hammond, unpublished
57. Colydiidae	F4.34; 6.2, ?etc	Myc	*Bitoma, Cicones, Synchita*, etc on *Xylaria, Nummularia, Cryptostroma*, etc	Crowson, 1984 Dajoz, 1981 Hammond, unpublished
	F4.26, ?etc	Lich	*Orthocerus* sp	Crowson, 1984
	F5.8	Myc(hy)	Various genera	Lawrence, 1988
	F4.34; 6.1, 6.2, ?etc	?Myc/sap	Colydiinae in ambrosia beetle burrows	Lawrence, 1982
58. Zopheridae	F4.21, ?etc	Lich	*Latometus*, etc	Lawrence, 1988
	F5.8	Myc(hy)	Ulodinae and A Zopherinae	Lawrence, 1988
59. Chalcodryidae	F4.21, ?etc	Lich	*Chalcodrya* sp	Lawrence, 1988
60. Tenebrionidae	F4.21, ?etc	Lich	*Amarygmus, Titaena*, etc	Lawrence, 1988
	F5.2	Myc	*Platydema* spp	Lawrence, 1988
	F5.8	Myc(hy)	Toxicini, Bolitophagini, Diaperini, etc	Lawrence, 1988
	F4.11, 4.34, ?etc	?Myc	Some *Corticeus* in burrows of H63 may feed on ambrosia fungi	Lawrence, this volume

	61. Salpingidae	?F4.7, 4.34; 6.2	Myc	*Rhinosimus* spp, etc on lignicolous ascomycetes	Crowson, 1984
		F4.9	Myc	*Rabocerus*, ?etc	Crowson, 1984
	Curculionoidea (10 fs)				
	62. Anthribidae	F4.8, 4.9, 4.34, ?etc	Myc(sp, ?hy)	Many Anthribinae, e.g. *Platyrhinus*, on lignicolous ascomycetes	Crowson, 1984 Hingley, 1971
		F4.21, ?etc	L lich	*Lichenobius* spp, etc	Crowson, 1984
		F2.3; 5.22, 5.23	Myc	*Araecerus* spp on rusts, *Brachytarsoides* on smuts, etc	Crowson, 1984 Lawrence, 1988
		F5.8	Myc(hy)	Eupariini in hard brackets	Matthewman and Pielou, 1971
	63. Curculionidae	F4.11, 4.17, 4.24, 4.34; 5; 6.1, 6.2	L myc (ambrosia fungi in wood; sometimes pests of standing timber)	All Platypodinae, Xyleborini and some spp of other scolytine tribes	Beaver, this volume
			(Phyt)	Most genera, but other wood-associated spp may be partially myc	Crowson, 1984
H26	[**STREPSIPTERA**] (5 fs)	—	(L par of H20, H33, etc)	—	—
H27	[**MECOPTERA**] (2 subords, 7 fs)	—	(Pred, sap)	(Some larvae aquatic)	—

(Continued)

Table II. Continued.

	Associated fungal taxa (see Table I)	Feeding habits	Comments	References
H28 **DIPTERA** (2 subords) (Two-winged flies)				
NEMATOCERA (4 infraords)				
TIPULOMORPHA (2 supfs)				
1. Trichoceridae	F5.8, 5.10, 5.11, 5.12, ?etc	L myc, sap	A few *Trichocera*	Chandler, 1979 Hackman and Meinander, 1979
2. Limoniidae	F4.28; 5.8, 5.9, 5.10, 5.11, 5.12, 5.13, ?etc	L myc	Some *Limonia, Ula,* etc	Chandler, 1979 Hackman and Meinander, 1979
	F4.21	L lich	*Geranomyia* in marine lichens	Stubbs, 1979
PSYCHODOMORPHA (4 supfs, 6 fs)				
3. (Psychodidae)	F4.28; 5.8, 5.10, 5.11, 5.12, ?etc	L ?myc, sap	Many *Psychoda* spp in decaying fungi, *P. lobata* also in fresh fungi	Chandler, 1979 Hackman and Meinander, 1979
CULICIMORPHA (2 supfs, 7 fs)				
Chironomoidea (4 fs)				
4. (Ceratopogonidae)	F5.8, 5.10, 5.11, 5.12	L ?myc/sap	Some *Culicoides*	Chandler, 1979
	F5.8, 5.10, 5.11, ?etc	L sap	Some *Forcipomyia*	Hackman and Meinander, 1979

5. (Chironomidae)	F4.34; 5.16	L ?myc	*Bryophaenocladius*	Chandler, 1979
	F5.8, 5.10, 5.12, ?etc	L sap	Various *Smittia*	
BIBIONOMORPHA (4 sections, 17 fs)				
6. Perissommatidae	F5.11	L ?myc	*Perissomma* (1 Australian sp) in *Boletus*	K.G.V. Smith, pers. comm.
7. [Anisopodidae]	F5.8, 5.10, 5.11, 5.12	L sap	*Sylvicola* in decaying fungi	Chandler, 1979 Bruns, 1984
8. (Scatopsidae)	F1.10; 5.8, 5.10, 5.11	L sap	*Coboldia*, etc	Chandler, 1979
9. Lestremiidae	F4.34; 5.10, 5.12, 5.13, ?etc; ?6.2	L myc	All/almost all probably myc, mostly fleshy fungi; *Brittenia* and *Heteropeza* spp also in *Daldinia*	Hennig, 1973 Chandler, 1979 K.M. Harris, pers. comm.
	F5.10	Pests of cultivated mushrooms	Some *Heteropeza, Lestremia* and *Mycophila*	Clift, 1979, etc
10. Cecidomyiidae	F4.34; 5.2, 5.8, 5.10, 5.11, 5.13, 5.22, ?etc	L myc, ?sap	Some Cecidomyiinae and possibly all Porricondylinae; some *Mycocecis* gall-making	Hennig, 1973 Chandler, 1979 K.M. Harris, pers. comm.
	F4.12; ?6.2, ?etc	L myc	*Mycodiplosis*	K.G.V. Smith, pers. comm.
	F4.34; 5.8, 5.10, 5.11	L pred on mites & insects	*Lestodiplosis*	Hingley, 1971
11. Ditomyiidae	F5.8	L myc	*Ditomyia* and ?others	Chandler, 1979
12. (Diadocidiidae)	F5.8	L ?myc	*Diadocidia* [usually in rotten wood]	Chandler, 1979

(*Continued*)

Table II. Continued.

	Associated fungal taxa (see Table I)	Feeding habits	Comments	References
13. (Keroplatidae)	F5.8	L ?myc/pred	*Keroplatus* mostly pred but at least facultatively myc	Chandler, 1979 K.G.V. Smith, pers. comm.
14. Bolitophilidae	F5.8, 5.10, 5.11, 5.12, ?etc	L myc	All probably myc, mostly in soft fruiting bodies	Chandler, 1979 Hackman and Meinander, 1979
15. Mycetophilidae	F1.7, 1.8; ?4.21, 4.28, 4.34; 5.1, 5.2, 5.6, 5.8, 5.9, 5.10, 5.11, 5.12, 5.13, 5.16, 5.18	L myc (sp/hy/plas), sap	Most probably myc, some apparently sap (mostly in F5); *Mycetophila* sp gall-making in F5.12	Chandler, 1979 Hackman and Meinander, 1979 Russel-Smith, 1979 Trifourkis, 1977
16. (Sciaridae)	F1.10; 5.2, 5.8, 5.10	L ?myc, sap	Some genera with fungicolous spp, these mostly sap in decaying fungi	Chandler, 1979 Hackman and Meinander, 1979
	F5.10	Pests of cultivated mushrooms	Some *Lycoriella*	Clift, 1979, etc
BRACHYCERA (3 infraords)				
HOMOEODACTYLA (4 supfs. 12 fs)				
17. [Stratiomyidae]	F5.10, 5.11	L ?pred	*Pctecticus* sp	Bruns, 1984
	F5.8	(A ?pred)	*Sargus* spp visit sporophores	Chandler, 1979

ASILOMORPHA (2 sections. 7 fs)				
18. [Scenopinidae]	F5.8	L pred on L of H25 and H32	*Scenopinus*	Chandler, 1979
19. [Empididae]	F1.10; 4.34; 5.8, 5.10, 5.11, 5.16	L pred	A few, e.g. some *Tachypeza*, fungicolous	Chandler, 1979
20. (Dolichopodidae)	F5.8	L ?myc	A few, e.g. *Systenus*, reared from lignicolous fungi	Chandler, 1979
CYCLORRHAPHA (3 sections)				
ASCHIZA (2 supfs, 6 fs)				
21. Platypezidae	F5.8, 5.10, 5.11, 5.16, 5.18	L myc(hy)	At least some Opetiinae and all Platypezinae s.l. probably myc; *Agathomyia* sp gall-making in *Ganoderma*	Chandler, 1979 Ferrar, 1987
22. Phoridae	F4.34; 5.8, 5.9, 5.10, 5.11, 5.12, 5.13, ?etc	L myc, sap	Most fungicolous spp sap; some, e.g. *Megaselia*, myc in fresh fungi; some, e.g. *Thaumatoxena*, may be myc in fungus gardens of H9	Chandler, 1979 Bruns, 1984 Ferrar, 1987
	F5.10	Pests of cultivated mushrooms	Some *Megaselia*	Clift, 1979, etc
23. Syrphidae	F4.28; 5.10, 5.11	L myc/(sap)	Some *Cheilosia*, mostly in fleshy fungi	Chandler, 1979 Hackman and Meinander, 1979
		(L pred, sap)	(Most genera not fungicolous)	Ferrar, 1987

(*Continued*)

Table II. Continued.

	Associated fungal taxa (see Table I)	Feeding habits	Comments	References
SCHIZOPHORA (2 subsections)				
ACALYPTRATAE (11 supfs, 65 fs)				
24. (Platystomatidae)	F5.10	L sap	Spp of a few genera, e.g. *Platystoma*, possibly facultative-myc	Chandler, 1979 Ferrar, 1987
	?F4.11	L ?myc	L of sap flow spp may feed on yeasts	
24a. [Otitidae]	F5.8	L ?sap (L sap, phyt)	*Pseudonephritis* sp in fungi (Most genera not fungicolous)	Weiss and West, 1922 Ferrar, 1987
24b. [Sciomyzidae]	F5.10	L ?sap	*Eurotocus* sp [Helo-sciomyzinae] in fungi	Ferrar, 1987
		(L sap, ?pred)	(Most genera not fun-gicolous)	
25. [Dryomyzidae]	F5.18	L sap	*Dryomyza* sp in decaying *Phallus*	Chandler, 1979
26. [Sepsidae]	F1; ?5.11	L sap (L sap)	*Nemopoda* sp in fungi (L of most genera in dung, etc)	Chandler, 1979 Ferrar, 1987
27. Lauxaniidae	?F4; ?6.2	A myc(hy/sp)	Spp of many genera graze phylloplane fungi on leaf sur-faces	Broadhead, 1984

	F5.8	L ?myc	*Lyciella* sp in *Phlebia*	Chandler, 1979
	F4.35	L myc	*Sapromyza* sp reared from witches' broom on black spruce	Ferrar, 1987
28. (Lonchaeidae)	F1.10; 5.8, 5.9, 5.10, 5.11, 5.18	L ?myc/sap	Some *Lonchaea*	Chandler, 1979 Bruns, 1984
		(L sap, phyt, pred)	(Most genera not fungicolous)	Ferrar, 1987
29. (Piophilidae)	F5.10, 5.11, ?etc	L ?myc/sap	Some *Amphipogon*, *Mycetaulus* in fungi	Chandler, 1979 Ferrar, 1987
		(L sap)	(Most genera in carrion, etc)	
30. (Asteiidae)	F5.8, 5.10	L ?myc	Some *Leiomyza* in lignicolous fungi	Chandler, 1979 Ferrar, 1987
		(L ?sap)	(Most genera associated with wood)	
31. (Acartophthalmidae)	F5.8	L ?myc/sap	A *Acartophthalmus* visit decaying fungi; L possibly fungicolous	Chandler, 1979 Ferrar, 1987
32. (Anthomyzidae)	F5.8, 5.10		Visited by A *Anthomyza* sp	Chandler, 1979
	F5.11	L ?myc	*Anthomyza* sp reared	
33. (Odiniidae)	F5.8	L ?myc/sap/pred	A of some *Odinia* spp visit lignicolous fungi [generally associated with beetle borings], L of some may live therein	Chandler, 1979 Ferrar, 1987

(*Continued*)

Table II. Continued.

	Associated fungal taxa (see Table I)	Feeding habits	Comments	References
34. Chloropidae	F5.8, 5.10, 5.11, 5.12, 5.16	L myc	Spp of a few genera, e.g. *Botanobia, Gaurax, Tricimba*, reared from fungi	Chandler, 1979 Bruns, 1984 Pielou and Verma, 1968 Ferrar, 1987
		(L phyt, sap, pred)	(Most genera not fungicolous)	
35. (Carnidae)	F5.11	L ?sap	*Meoneura* sp reared from *Leccinum*	Hackman and Meinander, 1979
36. Drosophilidae	F1.10; 4.28, 4.34; 5.1, 5.8, 5.9, 5.10, 5.11, 5.12, 5.18	L and A myc, sap	Some, e.g. *Leucophenga* and *Drosophila* spp, myc in fresh fungi, other fungicolous spp sap	Chandler, 1979 Lacy, 1984 Ferrar, 1987
			Amiota sp myc in *Daldinia*	Hingley, 1971
	F5.10	L myc	*Drosophila* sp gall-making in *Psathyrella*	K.G.V. Smith, pers. comm.
	?F4.11	L and A ?myc/sap	Sap flow spp, e.g. *Amiota* and *Chymomyza*, consume yeasts	Begon, 1982
37. Heleomyzidae	F4.34; 5.2, 5.9, 5.10, 5.11, 5.12, 5.13, 5.16, 5.18, ?etc	L myc	Possibly all Suilliinae myc in fresh fungi	Chandler, 1979 Ferrar, 1987
	F4.28	L myc	L of *Suillia* spp may ruin truffles	Janvier, 1963
		(L sap)	(Most Heleomyzinae)	

38. Sphaeroceridae	F1.10; 5.8, 5.10, 5.11, 5.12	L myc, sap	Spp of some genera, e.g. *Copromyza*, *Leptocera*, mostly in decaying fungi; some, e.g. some *Limosina*, in fresh	Chandler, 1979 Buxton, 1954 Hackman and Meinander, 1979
		(L sap)	(Most genera not fungicolous)	
CALYPTRATAE (4 supfs, 17 fs)				
39. Anthomyiidae	F1.10; 5.8, 5.9, 5.10, 5.11, 5.12, 5.18	L myc, sap	Several genera, e.g. *Pegomya*, with fungicolous spp	Chandler, 1979 Hackman and Meinander, 1979
	F4.4	L pred	*Mycophaga* spp	Bruns, 1984
		L and A myc-(sp/hy/etc)	*Phorbia* and *Pegohylemyia* spp	Chandler, 1979 Kohlmeyer and Kohlmeyer, 1974
		(L phyt)	(Most genera not fungicolous)	
40. [Muscidae]	F4.28; 5.8, 5.10, 5.11, 5.12, 5.18	L sap	Some Fanniinae, e.g. *Fannia* and *Piezura* spp, reared from fungi, mostly decaying	Chandler, 1979 Ferrar, 1987 Skidmore, 1985
		L pred/sap	Some Muscinae, e.g. *Muscina* and *Mydaea* spp fungus associated, mostly pred, some facultatively sap	

(Continued)

Table II. Continued.

	Associated fungal taxa (see Table I)	Feeding habits	Comments	References
41. [Tachinidae]	F5.8	L endopar on L of H32.4, ?etc	Spp of a few genera, e.g. *Actia, Elfia, Elodia*, in fungi	Chandler, 1979
		(L endopar on L of H32, H25, etc)	(Most genera not fungicolous)	Ferrar, 1987
42. Calliphoridae	?F4.34; 5.10	L sap, ? myc	Several *Hemigymnochaeta* spp reared from fungi, and from termite (H9) fungus gardens	Ferrar, 1987
	F5.18	A ?sap	A *Calliphora*, etc visit *Phallus*	Smith, 1956 Scheerpeltz and Höfler, 1948
		(L sap)	(Most genera in carrion, etc)	
H29 [**SIPHONAPTERA**] (2 superfs, 17 fs) (Fleas)	—	(Ectopar)	—	—
H30 [**TRICHOPTERA**] (24 fs) (Caddis-flies)	—	(L phyt, sap/pred)	L. aquatic; A take only nectar, water, etc	—
H31 [**ZEUGLOPTERA**] (1 f)	—	(L phyt/sap)	—	Rawlins, 1984

H32	[**LEPIDOPTERA**] (3 subords) (Moths, butterflies)				
	DITRYSIA (18 supfs, 101 fs)				
	Tortricoidea (2 fs)				
	1. (Tortricidae)	F4.21, ?etc	L lich	Mostly phyt or detritivorous; a few at least facultatively lich	Rawlins, 1984 Powell, 1980
	Tineoidea (7 fs)				
	2. (Psychidae)	F4.21, etc F5.8	L lich L myc(hy)	Several genera obligate; may feed occasionally on fleshy polypores	Rawlins, 1984 Simonet, 1955 Rehfous, 1955a
	3. Arrhenophanidae	F5.8	L myc(hy)	*Arrhenophanes* and ?others	Costa Lima, 1945
	4. Tineidae	F4.34; 5.8, 5.13, ?etc	L myc(hy)	Obligately myc spp in various subfs, especially Scardiinae and Nemapogoninae	Zagulajev, 1972 Rawlins, 1984 Midtgaard, 1985 Robinson, 1986
		F4.16?	L myc(hy)	Some birds' nests Tineinae	G.S. Robinson, pers. comm.
		F4.21, ?etc	L lich	In several subfs, especially Meessiinae	Rawlins, 1984
	Gelechioidea (18 fs)				
	5. (Cosmopterygidae)	F4.21, etc	L lich/sap L myc/sap	Some facultatively myc and lich (e.g. *Hyposmocoma*)	Rawlins, 1984

(*Continued*)

Table II. Continued.

	Associated fungal taxa (see Table I)	Feeding habits	Comments	References
6. Oecophoridae	F4.34; 5.8	L myc(hy)	Several genera, mostly Oecophorinae	Lawrence and Powell, 1969
	F5.22	L myc	*Stathmopoda* spp in rust galls on *Acacia*	Rawlins, 1984
	F4.21, ?etc	L lich	Several genera of Xyloryctinae	Rawlins, 1984
7. Blastobasidae	F4.21, ?etc	L lich	Symmocinae	Rawlins, 1984
Pyraloidea (5 fs)				
8. Pyralidae	F4.34	L myc	*Myelois* sp specific to *Daldinia; Apomyelois* on *Hypoxylon*	Hingley, 1971 Powell, 1967 Rawlins, 1984
	F5.22	L ?myc	*Dioryctria* on blister rusts	Rawlins, 1984
	F4.21, ?etc	L lich	Several genera obligate	
Papilionoidea (5 fs)	F5.10, etc	A (myc)	Some visit sporophores to feed on exudates	Rawlins, 1984
9. Lycaenidae	F4.21, ?etc	L lich	A few obligate (e.g. *Epitola*)	Rawlins, 1984
Geometroidea				
10. Geometridae	F4.21, ?etc	L lich	A few possibly obligate (e.g. *Dichromodes*)	Rawlins, 1984

	Noctuoidea (7 fs)				
	11. Noctuidae	F5.8, ?etc	L myc(hy)	Some obligate in various subfs (e.g. Hypenodinae and Ophiderinae)	Rawlins, 1984
	12. Arctiidae	F4.21, ?etc	L lich	Many in several subfs	Rawlins, 1984
		F4.21, ?etc	L lich	Many in Lithosiinae	Rawlins, 1984
		F5.8	L ?myc(hy)	Lich spp may occasionally feed on polypores, etc	Simonet, 1955 Rehfous, 1955b
H33	**HYMENOPTERA** (2 subords) (Bees, ants, wasps)				
	SYMPHYTA (6 superfs, 14 fs)				
	Syricoidea (3 fs)				
	1. Syricidae	F4.34; 5.8	L myc (ambrosia fungi in wood; sometimes pests of standing timber)	*Sirex* and *Urocerus* on *Amylostereum* (F5.8), *Xiphydria* on *Cerrena* (F4.34)	Gilbertson, 1984
	APOCRITA PARASITICA (9 superfs, 48 fs)				
	Cynipoidea (6 fs)				
	2. [Eucoilidae]	F5.10, 5.12, ?etc	L endopar on L of H28.40	Some, e.g. *Eucoila* spp, in fungi	Ferrière, 1955 Gauld and Bolton, 1988

(*Continued*)

Table II. Continued.

	Associated fungal taxa (see Table I)	Feeding habits	Comments	References
Chalcidoidea (21 fs)				
3. [Eulophidae]	F5.8, ?etc	L par on L of H25.52, etc	Some, e.g. *Astichus*, in fungi	Ferrière, 1955 Bouçek and Askew, 1968
Proctotrupoidea (11 fs)				
4. [Proctotrupidae]	F5.8, 5.10, 5.12, etc	L par on L of H25, H28.15, etc	Some, e.g. *Cryptoserphus*, etc	Ferrière, 1955 Gauld and Bolton, 1988
5. [Diapriidae] (incl. Belytidae)	F5.8, 5.10, 5.12, etc	L par on L of H28.15, H28.16, etc	Belytinae mostly in fungi; also some Diapriinae	Ferriére, 1955 Gauld and Bolton, 1988
Ichneumonoidea (2 fs)				
6. [Ichneumonidae]	F5.8, 5.10, 5.11, 5.12, 5.16, etc	L par on L of H25, H28.15, 28.26	Many Orthocentrinae and Oxytorinae in fungi; Rhissini associated with H33.1 burrows	Ferrière, 1955 Gauld and Bolton, 1988 Pielou and Verma, 1968
7. [Braconidae]	F4.34; 5.8, 5.10, etc	L par on L of H25.53, H28.22, 28.48, etc	Many Alysiinae, some Blacinae, Euphorinae, etc in fungi	Ferriére, 1955 Wharton, 1984 Gauld and Bolton, 1988 Pielou and Verma, 1968 Matthewman and Pielou, 1971

ACULEATA (3 supfs, 21 fs)				
Chrisidoidea (7 fs)				
8. [Bethylidae]	F5.8, 5.11, etc	L ectopar on L of H25.52, H32.3, etc	Various, e.g. *Goniozus*, *Plastanoxus*, etc in fungi	Evans, 1978 M. C. Day pers. comm. Pielou and Verma, 1968
Vespoidea (12 fs)				
9. Formicidae (Ants)	?F4.34, 5.10; 6.2	Myc (ambrosia-fungi in nests)	Attinae, e.g. *Atta*, *Acromyrmex*, etc, only	Weber, 1979 Cherrett *et al.*, this volume

REFERENCES

Ackerman, J. K., and Shenefelt, R. D. (1973). Organisms, especially insects, associated with wood rotting higher fungi (Basidiomycetes) in Wisconsin forests. *Trans. Wis. Acad. Sci., Arts and Lett.* **61**, 185–206.

Ananthakrishnan, T. N. (1984). "Biology of Thrips". Indara Publishing House, Oak Park, Michigan.

Ananthakrishnan, T. N., and Dhileepan, K. (1984). Thrips–fungus association with special reference to the sporophagous *Bactrothrips idolomorphus* (Karny). (Tubulifera: Thysanoptera.) *Proc. Indian Acad. Sci., Anim. Sci.* **93**, 243–249.

Batra, L. R., and Batra, S. W. T. (1979). Termite–fungus mutualism. *In* "Insect–Fungus Symbiosis, Nutrition, Mutualism, and Commensalism" (L. R. Batra, ed.), pp. 117–163. Allanheld, Osman, Montclair.

Begon, M. (1982). Yeasts and *Drosophila. In.* "The Genetics and Biology of *Drosophila*, Vol. 3b" (M. Ashburner, H. L. Carson and J. N. Thompson, eds), pp. 345–384. Academic Press, London.

Benick, L. (1952). Pilzkäfer and Käferpilze: Ökologishe und statistische Untersuchungen. *Acta Zool. Fenn.* **70**, 1–250.

Blackwell, M. (1984). Myxomycetes and their arthropod associates. *In* "Fungus–Insect Relationships, Perspectives in Ecology and Evolution" (Q. D. Wheeler and M. Blackwell, eds), pp. 67–90. Columbia Univ. Press, New York.

Bouçek, Z., and Askew, R. R. (1968). "Index of Entomophagous Insects 3. Palearctic Eulophidae (excl. Tetrastichinae)". Paris.

Broadhead, E. C. (1984). Adaptations for fungal grazing in lauxaniid flies. *J. Nat. Hist.* **18**, 639–649.

Bruns, T. D. (1984). Insect mycophagy in the Boletales: fungivore diversity and the mushroom habitat. *In* "Fungus–Insect Relationships, Perspectives in Ecology and Evolution (Q. D. Wheeler and M. Blackwell, eds), pp. 91–129. Columbia Univ. Press, New York.

Burakowski, B. (1975). Development, distribution and habits of *Trixagus dermestoides* (L.), with notes on the Throscidae and Lissomidae (Coleoptera, Elateroidea). *Ann. Zool. Warsz.* **32**, 375–405.

Butcher, J. W., Snider, R., and Snider, R. J. (1971). Bioecology of edaphic Collembola and Acarina. *Annu. Rev. Entomol.* **16**, 249–288.

Buxton, F. A. (1954). British Diptera associated with fungi. Part II. Diptera bred from Myxomycetes. *Proc. R. Entomol. Soc. London, Ser. A* **29**, 163–171.

Buxton, P. A. (1960). British Diptera associated with fungi. III. Flies of all families reared from about 150 species of fungi. *Entomol. Mon. Mag.* **96**, 61–94.

Chandler, P. (1979). Fungi. *In* "A Dipterist's Handbook" (A. Stubbs and P. Chandler, eds), pp. 199–211. Amateur Entomologists Society, Hanworth, Middlesex, UK.

Christiansen, K. (1964). Bionomics of Collembola. *Annu. Rev. Entomol.* **9**, 147–148.

Clift, A. D. (1979). The identity, economic importance and control of insect pests of mushrooms in New South Wales, Australia. *Mushroom Sci.* **10**, 367–383.

Costa Lima, A. da (1945). Insetos do Brasil, 5. Lepidópteros 1. *Série Didát. Escola Nac. Agr.* **7**, 1–379.

Crowson, R. A. (1981). "The Biology of the Coleoptera." Academic Press, New York.

Crowson, R. A. (1984). The associations of Coleoptera with Ascomycetes. *In* "Fungus–Insect Relationships, Perspectives in Ecology and Evolution" (Q. D. Wheeler and M. Blackwell, eds), pp. 256–285. Columbia Univ. Press, New York.

CSIRO (1970). "The Insects of Australia. A handbook for students and research workers." Melbourne University Press, Carlton, Victoria.

Cummins, K. W. (1973). Trophic relations of aquatic insects. *Annu. Rev. Entomol.* **18**, 183–206.

Dajoz, R. (1981). Note sur les Coléoptères d'un champignon Ascomycète de Tunisie. *L'Entomologiste* **37**, 203–211.

Donisthorpe, H. (1935). The British fungicolous Coleoptera. *Entomol. Mon. Mag.* **71**, 21–31.

Eisfelder, I. (1961). Käferpilze und Pilzkäfer. *Z. Pilzkd.* **27**, 44–54.

Eisfelder, I. (1963). Käfer als Pilz bewohner. *Z. Pilzkd.* **29**, 77–97.

Eisfelder, I. (1970). Apterygoten (Urinsekten) in und an Pilze. *Z. Pilzkd.* **36**, 171–184.

Erwin, T., and Erwin, L. J. M. (1976). Relations of predaceous beetles to tropical forest wood decay. Part II. The natural history of Neotropical *Eurycoleus macularis* Chevrolat (Carabidae: Lebiini) and its implications in the evolution of ectoparasitoidism. *Biotropica* **8**, 215–224.

Evans, H. E. (1978). The Bethylidae of America north of Mexico. *Mem. Am. Entomol. Inst.* **27**, 1–332.

Ferrar, P. (1987). "A Guide to the Breeding Habits and Immature Stages of Diptera Cyclorrhapha" (Part 1: text). Entomonograph 8, 1–478. E. J. Brill/Scandinavian Science Press, Leiden/Copenhagen.

Ferrière, C. (1955). Note sur les Hymenoptères des champignons. *Mitt. Schweiz. Entomol. Ges.* **28**, 106–108.

Fogel, R. (1975). "Insect Mycophagy: A Preliminary Bibliography." U.S. Dep. Agric. For. Serv., Gen. Tech. Rept. RNW-36.

Fogel, R., and Peck, S. B. (1975). Ecological studies of hypogeous fungi. Part I. Coleoptera associated with sporocarps. *Mycologia* **67**, 741–747.

Gauld, I., and Bolton, B. (1988). "An Introduction to the Hymenoptera". British Museum (Natural History) and Oxford University Press, London and Oxford.

Gilbertson, R. L. (1984). Relationships between insects and wood-rotting Basidiomycetes. *In* "Fungus–Insect Relationships, Perspectives in Ecology and Evolution" (Q. D. Wheeler and M. Blackwell, eds), pp. 130–165. Columbia Univ. Press, New York.

Graves, R. C. (1960). Ecological observations on the insects and other inhabitants of woody shelf fungi (Basidiomycetes: Polyporaceae) in the Chicago area. *Ann. Entomol. Soc. Am.* **53**, 61–78.

Graves, R. C., and Graves, A. C. F. (1985). Diptera associated with shelf fungi and certain other micro-habitats in the highlands area of western North Carolina. *Entomol. News* **96**, 87–92.

Hackman, W. (1976). The biology of anthomyiid flies feeding as larvae on fungi (Diptera). *Notul. Entomol.* **56**, 129–134.

Hackman, W., and Meinander, M. (1979). Diptera feeding as larvae on macro-fungi in Finland. *Ann. Zool. Fenn.* **16**, 50–83.

Hawksworth, D. L., Sutton, B. C., and Ainsworth, G. C. (1983). "Ainsworth & Bisby's Dictionary of the Fungi". Seventh Edition. Commonwealth Mycological Institute, Kew.

Hennig, W. (1973). Diptera (Zweiflügler). *Hand. Zool. Berl.* **4**(2) (31), 1–337.

Hingley, M. R. (1971). The ascomycete fungus *Daldinia concentrica*, as a habitat for animals. *J. Anim. Ecol.* **40**, 17–32.

Hodek, I. (1973). "Biology of Coccinellidae." W. Junk, The Hague.

Höfler, K. (1960). Pilzkäfer und Käferpilze. *Verh. Zool. Bot. Ges. Wien.* **1**, 74–83.

Janvier, H. (1963). La mouche de la truffe (*Helomyza tuberiperda* Rondani). *Bull. Soc. Entomol. Fr.* **68**, 140–147.

Kimbrough, J. W. (1984). Life cycles and natural history of Ascomycetes. *In* "Fungus–Insect Relationships, Perspectives in Ecology and Evolution" (Q. D. Wheeler and M. Blackwell, eds), pp. 184–210. Columbia Univ. Press, New York.

Klimaszewski, J. and Peck, S. B. (1987). Succession and phenology of beetle faunas

(Coleoptera) in the fungus *Polyporellus squamosus* (Huds.:Fr.) Karst. (Polyporaceae) in Silesia, Poland. *Can. J. Zool.* **65**, 542–550.

Kohlmeyer, J., and Kohlmeyer, E. (1974). Distribution of *Epichloe typhina* (Ascomycetes) and its parasitic fly. *Mycologia* **66**, 77–86.

Lacy, R. C. (1984). Ecological and genetic responses to mycophagy in Drosophilidae (Diptera). *In* "Fungus–Insect Relationships, Perspectives in Ecology and Evolution" (Q. D. Wheeler and M. Blackwell, eds), pp. 286–301. Columbia Univ. Press, New York.

Lawrence, J. F. (1973). Host preference in ciid beetles (Coleoptera: Ciidae) inhabiting the fruiting bodies of Basidiomycetes in North America. *Bull. Mus. Comp. Zool.* **145**, 163–212.

Lawrence, J. F. (1977). Coleoptera associated with an *Hypoxylon* species (Ascomycetes: Xylariaceae) on oak. *Coleopt. Bull.* **31**, 309–312.

Lawrence, J. F. (1982). Coleoptera. *In* "Synopsis and Classification of Living Organisms" (S. P. Parker, ed.), pp. 482–553. McGraw-Hill, New York.

Lawrence, J. F. (1988). Coleoptera. *In* "Immature Insects", Vol. 2 (F. W. Stehr, ed.), Kendall-Hunt, Dubuque, Iowa.

Lawrence, J. F., and Newton, A. F. (1980). Coleoptera associated with the fruiting bodies of slime molds (Myxomycetes). *Coleopt. Bull.* **34**, 129–143.

Lawrence, J. F., and Newton, A. F. (1982). Evolution and classification of beetles. *Annu. Rev. Ecol. Syst.* **13**, 261–290.

Lawrence, J. F., and Powell, J. (1969). Host relationships in North American fungus-feeding moths (Oecophoridae, Oinophilidae, Tineidae). *Bull. Mus. Comp. Zool.* **138**, 29–51.

Lieftinck, M. A., and Wiebes, J. T. (1968). Notes on the genus *Mormolyce* Hagenbach (Coleoptera, Carabidae). *Bijdr. Dierkunde* **38**, 59–68.

Matthewman, W. G., and Pielou, D. P. (1971). Arthropods inhabiting the sporophores of *Fomes fomentarius* (Polyporaceae) in Gatineau Park, Quebec, *Can. Entomol.* **103**, 775–847.

Midtgaard, F. (1985). Fungivorous moths (Lep., Tineidae & Oecophoridae). *Fauna* (Blinden) **38**, 50–52.

Minch, F. L. (1952). Insect inhabitants of *Polyporus betulinus*. *J. N.Y. Entomol. Soc.* **60**, 31–35.

New, T. R. (1987). Biology of the Psocoptera. *Orient. Insects* **21**, 1–109.

Newton, A. F. (1984). Mycophagy in Staphylinoidea (Coleoptera). *In* "Fungus–Insect Relationships. Perspectives in Ecology and Evolution (Q. D. Wheeler and M. Blackwell, eds), pp. 302–353. Columbia Univ. Press, New York.

O'Brien, L. B., and Wilson, S. W. (1985). Planthopper systematics and external morphology. *In* "The Leafhoppers and Planthoppers" (L. R. Nault and J. G. Rodriguez, eds), pp. 61–102. John Wiley, New York.

O'Connor, B. M. (1984). Acarine-fungal relationships: the evolution of symbiotic associations. *In* "Fungus–Insect Relationships, Perspectives in Ecology and evolution" (Q. D. Wheeler and M. Blackwell, eds), pp. 354–381. Columbia Univ. Press, New York.

Pielou, D. P., and Matthewman, W. G. (1966). The fauna of *Fomes fomentarius* (Linnaeus ex Fries) Kickx growing on dead birch in Gatineau Park, Quebec. *Can. Entomol.* **98**, 1308–1312.

Pielou, D. P., and Verma, A. N. (1968). The arthropod fauna associated with the birch bracket fungus, *Polyporus betulinus* in eastern Canada. *Can. Entomol.* **100**, 1179–1199.

Powell, J. A. (1967). *Apomyelois bistriatella*: A moth which feeds in an ascomycete fungus (Lepidoptera: Pyralidae). *J. N.Y. Entomol. Soc.* **75**, 190–194.

Powell, J. A. (1980). Evolution of larval food preferences in microlepidoptera. *Annu. Rev. Entomol.* **25**, 133–159.

Ratti, E. (1978). Findings of Coleoptera on carpophores of *Gyromitra esculenta* (Ascomycetes) in the Dolomites, with record of a Ptiliidae new to Italy. *Soc. Venez. Sci. Nat. Lav.* **3**, 46–48.

Rawlins, J. E. (1984). Mycophagy in Lepidoptera. *In* "Fungus–Insect Relationships, Perspectives in Ecology and Evolution" (Q. D. Wheeler and M. Blackwell, eds), pp. 382–423. Columbia Univ. Press, New York.

Rehfous, M. (1955a). Contribution a l'étude des insectes des champignons. *Mitt. Schweiz. Entomol. Ges.* **28**, 1–106.

Rehfous, M. (1955b). Note sur les Lépidoptères des Champignons. *Mitt. Schweiz. Entomol. Ges.* **28**, 109–110.

Ricci, C. (1986). Seasonal food preferences and behaviour of *Rhyzobius litura*. *In* "Ecology of Aphidophaga" (I. Hodek, ed.), pp. 119–123. Academia and W. Junk, Prague and Dordrecht.

Ricci, C., Fiori, G., and Colazza, S. (1983). Regime alimentare dell'adulto di *Tytthaspis sedecimpunctata* (L.) (Coleoptera Coccinellidae) in ambiente a antropica primaria: prato polifita. *Atti XIII Congr. Naz. It. Entomol., Sestriere-Torino 1983*, 691–698.

Robinson, G. S. (1986). Fungus moths. A review of the Scardiinae. *Bull. Br. Mus. Nat. Hist. Entomol.* **52**, 37–181.

Ross, E. S. (1970). Biosystematics of the Embioptera. *Annu. Rev. Entomol.* **15**, 157–172.

Roth, L. M., and Willis, E. R. (1960). The biotic associations of cockroaches. *Smithson. Misc. Collect.* **141**, 1–470.

Russell, L. K. (1979). Beetles associated with slime molds (Mycetozoa) in Oregon and California (Coleoptera: Leiodidae, Sphindidae, Lathridiidae). *Pan-Pacif. Entomol.* **55**, 1–9.

Russel-Smith, A. (1979). A study of fungus flies (Diptera: Mycetophilidae) in beech woodland. *Ecol. Entomol.* **4**, 355–364.

Sands, W. A. (1969). The association of termites and fungi. *In* "Biology of Termites" (K. Krishna and F. M. Weesner, eds), pp. 495–524. Academic Press, New York.

Scheerpeltz, O., and Höfler, K. (1948). "Käfer und Pilze". Verlag fur Jugend und Volk, Vienna.

Seastedt, T. R. (1984). The role of microarthropods in decomposition and mineralization processes. *Annu. Rev. Entomol.* **29**, 25–46.

Simonet, J. (1955). Note relative aux Hémiptères capturés sur les Champignons. *Mitt. Schweiz. Entomol. Ges.* **28**, 111–114.

Skidmore, P. (1985). The biology of the Muscidae of the world. *Series Entomol.* **29**, 1–550.

Smith, K. G. V. (1956). On the Diptera associated with the stinkhorn (*Phallus impudicus* Pers.) with notes on other insects and invertebrates found on this fungus. *Proc. R. Entomol. Soc. London, Ser. A* **31**, 49–55.

Steiner, W. E. (1984). A review of the biology of phalacrid beetles (Coleoptera). *In* "Fungus–Insect Relationships, Perspectives in Ecology and Evolution" (Q. D. Wheeler and M. Blackwell, eds), pp. 424–445. Columbia Univ. Press, New York.

Stubbs, A. (1979). Mosses, lichens and liverworts. *In* "A Dipterist's Handbook" (A. Stubbs and P. Chandler, eds), p. 212. Amateur Entomologists Society, Hanworth, Middlesex, U.K.

Thayer, M. (1987). Biology and phylogenetic relationships of *Neophonus bruchi*, an anomalous south Andean staphylinid (Coleoptera). *Syst. Entomol.* **12**, 389–404.

Trifourkis, S. (1977). The bionomics and taxonomy of the larval Mycetophilidae and other fungicolous Diptera I: 1–393, II: 394–792. N.E.L.P. Faculty of Science. Ph.D. Thesis, University of London.

Usinger, R. L., and Matsuda, R. (1959). "Classification of the Aradidae (Hemiptera-Heteroptera)." British Museum (Natural History), London.

Weber, N. A. (1979). Fungus-culturing ants. *In* "Insect–Fungus Symbiosis, Nutrition, Mutualism, and Commensalism" (L. R. Batra, ed.), pp. 77–116. Allanheld, Osman, Montclair.

Weiss, H. B. (1921). A bibliography of fungus insects and their hosts. *Entomol. News* **32**, 45–47.

Weiss, H. B., and West, E. (1922). Notes on fungus insects. *Can. Entomol.* **54**, 198–199.

Wharton, R. A. (1984). Biology of the Alysiini (Hymenoptera: Braconidae) parasitoids of cyclorrhaphous Diptera. *Tech. Monogr. Texas Agric. Exp. Stn* **11**, 1–39.

Wheeler, Q. D. (1984). Evolution of slime mold feeding in leiodid beetles. *In* "Fungus–Insect Relationships, Perspectives in Ecology and Evolution" (Q. D. Wheeler and M. Blackwell, eds), pp. 446–478. Columbia Univ. Press, New York.

Zagulajev, A. K. (1972). Food relations and evolution of feeding types in Tineidae (Lepidoptera). *Zool. Zh.* **41**, 1507–1516.

Index

Numbers in italics refer to figures

B

C

D

E

H

I

J

K

L

M

P

Q

R

S

T

U

V

W

X

Y

Z